A HISTORY OF
THE WARFARE OF SCIENCE
WITH THEOLOGY
IN CHRISTENDOM

A HISTORY OF THE WARFARE OF SCIENCE WITH THEOLOGY

IN CHRISTENDOM

BY

ANDREW DICKSON WHITE

IN TWO VOLUMES

VOL. I

GLOUCESTER, MASS.

PETER SMITH

1978

This edition is an unabridged and unaltered republication of the first edition that appeared in 1896.

Reprinted, 1978, by
Peter Smith Publisher, Inc.

Library of Congress Catalog Card Number: 60-2524

Manufactured in the United States of America

ISBN 0-8446-3170-1

To the Memory of

EZRA CORNELL

I DEDICATE THIS BOOK.

Thoughts that great hearts once broke for, we
  Breathe cheaply in the common air.—LOWELL.

Discipulus est prioris posterior dies.—PUBLIUS SYRUS.

Truth is the daughter of Time.—BACON.

The Truth shall make you free.—ST. JOHN, viii, 32.

INTRODUCTION.

My book is ready for the printer, and as I begin this preface my eye lights upon the crowd of Russian peasants at work on the Neva under my windows. With pick and shovel they are letting the rays of the April sun into the great ice barrier which binds together the modern quays and the old granite fortress where lie the bones of the Romanoff Czars.

This barrier is already weakened; it is widely decayed, in many places thin, and everywhere treacherous; but it is, as a whole, so broad, so crystallized about old boulders, so imbedded in shallows, so wedged into crannies on either shore, that it is a great danger. The waters from thousands of swollen streamlets above are pressing behind it; wreckage and refuse are piling up against it; every one knows that it must yield. But there is danger that it may resist the pressure too long and break suddenly, wrenching even the granite quays from their foundations, bringing desolation to a vast population, and leaving, after the subsidence of the flood, a widespread residue of slime, a fertile breeding-bed for the germs of disease.

But the patient *mujiks* are doing the right thing. The barrier, exposed more and more to the warmth of spring by the scores of channels they are making, will break away gradually, and the river will flow on beneficent and beautiful.

My work in this book is like that of the Russian *mujik* on the Neva. I simply try to aid in letting the light of historical truth into that decaying mass of outworn thought which attaches the modern world to mediæval conceptions of Christianity, and which still lingers among us—a most serious barrier to religion and morals, and a menace to the whole normal evolution of society.

For behind this barrier also the flood is rapidly rising —the flood of increased knowledge and new thought; and this barrier also, though honeycombed and in many places thin, creates a danger—danger of a sudden breaking away, distressing and calamitous, sweeping before it not only outworn creeds and noxious dogmas, but cherished principles and ideals, and even wrenching out most precious religious and moral foundations of the whole social and political fabric.

My hope is to aid—even if it be but a little—in the gradual and healthful dissolving away of this mass of unreason, that the stream of "religion pure and undefiled" may flow on broad and clear, a blessing to humanity.

And now a few words regarding the evolution of this book.

It is something over a quarter of a century since I labored with Ezra Cornell in founding the university which bears his honored name.

Our purpose was to establish in the State of New York an institution for advanced instruction and research, in which science, pure and applied, should have an equal place with literature; in which the study of literature, ancient and modern, should be emancipated as much as possible from pedantry; and which should be free from various useless trammels and vicious methods which at that period hampered many, if not most, of the American universities and colleges.

We had especially determined that the institution should be under the control of no political party and of no single religious sect, and with Mr. Cornell's approval I embodied stringent provisions to this effect in the charter.

It had certainly never entered into the mind of either of us that in all this we were doing anything irreligious or unchristian. Mr. Cornell was reared a member of the Society of Friends; he had from his fortune liberally aided every form of Christian effort which he found going on about him, and among the permanent trustees of the public library which he had already founded, he had named all the clergymen of the town—Catholic and Protestant. As for myself, I had been bred a churchman, had recently been elected a trustee of one church college, and a professor in another; those nearest and dearest to me were devoutly religious; and, if I may be allowed to speak of a matter so personal to myself, my most cherished friendships were among deeply religious men and women, and my greatest sources of enjoyment were ecclesiastical architecture, religious music, and the more devout forms of poetry. So far from wishing to injure Christianity, we both hoped to promote it; but we did not confound religion with sectarianism, and we saw in the sectarian character of American colleges and universities, as a whole, a reason for the poverty of the advanced instruction then given in so many of them.

It required no great acuteness to see that a system of control which, in selecting a Professor of Mathematics or Language or Rhetoric or Physics or Chemistry, asked first and above all to what sect or even to what wing or branch of a sect he belonged, could hardly do much to advance the moral, religious, or intellectual development of mankind.

The reasons for the new foundation seemed to us, then, so cogent that we expected the co-operation of all good citizens, and anticipated no opposition from any source.

As I look back across the intervening years, I know not whether to be more astonished or amused at our simplicity.

Opposition began at once. In the State Legislature it confronted us at every turn, and it was soon in full blaze throughout the State—from the good Protestant bishop who proclaimed that all professors should be in holy orders, since to the Church alone was given the command, "Go, teach all nations," to the zealous priest who published a charge that Goldwin Smith—a profoundly Christian scholar—had come to Cornell in order to inculcate the "infidelity of the *Westminster Review*"; and from the eminent divine who went from city to city denouncing the "atheistic and pantheistic tendencies" of the proposed education, to the perfervid minister who informed a denominational synod that Agassiz, the last great opponent of Darwin, and a devout theist, was "preaching Darwinism and atheism" in the new institution.

As the struggle deepened, as hostile resolutions were introduced into various ecclesiastical bodies, as honored clergymen solemnly warned their flocks first against the "atheism," then against the "infidelity," and finally against the "indifferentism" of the university, as devoted pastors endeavoured to dissuade young men from matriculation, I took the defensive, and, in answer to various attacks from pulpits and religious newspapers, attempted to allay the fears of the public. "Sweet reasonableness" was fully tried. There was established and endowed in the university perhaps the most effective Christian pulpit, and one of the most vigorous branches of the Christian Association, then in the United States; but all this did nothing to ward off the attack. The clause in the charter of the university forbidding it to give predominance to the doctrines of any sect, and above all the fact that much prominence was given to instruction in various branches of science, seemed to prevent all compromise, and it soon became clear that to stand on the defensive only made matters worse. Then it was that there was borne in upon me a sense of the real difficulty—the antagonism between the theological and scientific view of the universe and of education in relation to it; therefore it was that, having been invited to deliver a lecture in the great hall of the Cooper Institute at New York, I took as my subject *The Battlefields of Science*, maintaining this thesis which follows:

In all modern history, interference with science in the supposed interest of religion, no matter how conscientious such interference may have been, has resulted in the direst evils both to religion and to science, and invariably; and, on the other hand, all untrammelled scientific investigation, no matter how dangerous to religion some of its stages may have seemed for the time to be, has invariably resulted in the highest good both of religion and of science.

The lecture was next day published in the *New York Tribune* at the request of Horace Greeley, its editor, who was also one of the Cornell University trustees. As a result of this widespread publication and of sundry attacks which it elicited, I was asked to maintain my thesis before various university associations and literary clubs; and I shall always remember with gratitude that among those who stood by me and presented me on the lecture platform with words of approval and cheer was my revered instructor, the Rev. Dr. Theodore Dwight Woolsey, at that time President of Yale College.

My lecture grew—first into a couple of magazine articles, and then into a little book called *The Warfare of Science*, for which, when republished in England, Prof. John Tyndall wrote a preface.

Sundry translations of this little book were published, but the most curious thing in its history is the fact that a very friendly introduction to the Swedish translation was written by a Lutheran bishop.

Meanwhile Prof. John W. Draper published his book on *The Conflict between Science and Religion*, a work of great ability, which, as I then thought, ended the matter, so far as my giving it further attention was concerned.

But two things led me to keep on developing my own work in this field: First, I had become deeply interested in it, and could not refrain from directing my observation and study to it; secondly, much as I admired Draper's treatment of the questions involved, his point of view and mode of looking at history were different from mine.

He regarded the struggle as one between Science and Religion. I believed then, and am convinced now, that it was a struggle between Science and Dogmatic Theology.

More and more I saw that it was the conflict between two epochs in the evolution of human thought—the theological and the scientific.

So I kept on, and from time to time published *New Chapters in the Warfare of Science* as magazine articles in *The Popular Science Monthly*. This was done under many difficulties. For twenty years, as President of Cornell University and Professor of History in that institution, I was immersed in the work of its early development. Besides this, I could not hold myself entirely aloof from public affairs, and was three times sent by the Government of the United States to do public duty abroad: first as a commissioner to Santo Domingo, in 1870; afterward as minister to Germany, in 1879; finally, as minister to Russia, in 1892; and was also called upon by the State of New York to do considerable labor in connection with international exhibitions at Philadelphia and at Paris. I was also obliged from time to time to throw off by travel the effects of overwork.

The variety of residence and occupation arising from these causes may perhaps explain some peculiarities in this book which might otherwise puzzle my reader.

While these journeyings have enabled me to collect materials over a very wide range—in the New World, from Quebec to Santo Domingo and from Boston to Mexico, San Francisco, and Seattle, and in the Old World from Trondhjem to Cairo and from St. Petersburg to Palermo—they have often obliged me to write under circumstances not very favorable: sometimes on an Atlantic steamer, sometimes on a Nile boat, and not only in my own library at Cornell, but in those of Berlin, Helsingfors, Munich, Florence, and the British Museum. This fact will explain to the benevolent reader not only the citation of different editions of the same authority in different chapters, but some iterations which in the steady quiet of my own library would not have been made.

It has been my constant endeavour to write for the general reader, avoiding scholastic and technical terms as much as possible and stating the truth simply as it presents itself to me.

That errors of omission and commission will be found here and there is probable—nay, certain; but the substance of the book will, I believe, be found fully true. I am encouraged in this belief by the fact that, of the three bitter attacks which this work in its earlier form has already encountered, one was purely declamatory, objurgatory, and hortatory, and the others based upon ignorance of facts easily pointed out.

And here I must express my thanks to those who have aided me. First and above all to my former student and dear friend, Prof. George Lincoln Burr, of Cornell University, to whose contributions, suggestions, criticisms, and cautions I am most deeply indebted; also to my friends U.

G. Weatherly, formerly Travelling Fellow of Cornell, and now Assistant Professor in the University of Indiana,—Prof. and Mrs. Earl Barnes and Prof. William H. Hudson, of Stanford University,—and Prof. E. P. Evans, formerly of the University of Michigan, but now of Munich, for extensive aid in researches upon the lines I have indicated to them, but which I could never have prosecuted without their co-operation. In libraries at home and abroad they have all worked for me most effectively, and I am deeply grateful to them.

This book is presented as a sort of *Festschrift*—a tribute to Cornell University as it enters the second quarter-century of its existence, and probably my last tribute.

The ideas for which so bitter a struggle was made at its foundation have triumphed. Its faculty, numbering over one hundred and fifty; its students, numbering but little short of two thousand; its noble buildings and equipment; the munificent gifts, now amounting to millions of dollars, which it has received from public-spirited men and women; the evidences of public confidence on all sides; and, above all, the adoption of its cardinal principles and main features by various institutions of learning in other States, show this abundantly. But there has been a triumph far greater and wider. Everywhere among the leading modern nations the same general tendency is seen. During the quarter-century just past the control of public instruction, not only in America but in the leading nations of Europe, has passed more and more from the clergy to the laity. Not only are the presidents of the larger universities in the United States, with but one or two exceptions, laymen, but the same thing is seen in the old European strongholds of metaphysical theology. At my first visit to Oxford and Cambridge, forty years ago, they were entirely under ecclesiastical control. Now, all this is changed. An eminent member of the present British Government has recently said, "A candidate for high university position is handicapped by holy orders." I refer to this with not the slightest feeling of hostility toward the clergy, for I have none; among them are many of my dearest friends; no one honours their proper work more than I; but the above fact is simply noted as proving the continuance of that evolution which I have endeavoured to describe in this series of monographs—an evolution, indeed, in which the warfare of Theology against Science has been one of the most active and powerful agents. My belief is that in the field left to them—their proper field—the clergy will more and more, as they cease to struggle against scientific methods and conclusions, do work even nobler and more beautiful than anything they have heretofore done. And this is saying much. My conviction is that Science, though it has evidently conquered Dogmatic Theology based on biblical texts and ancient modes of thought, will go hand in hand with Religion; and that, although theological control will continue to diminish, Religion, as seen in the recognition of "a Power in the universe, not ourselves, which makes for righteousness," and in the love of God and of our neighbor, will steadily grow stronger and stronger, not only in the American institutions of learning but in the world at large. Thus may the declaration of Micah as to the requirements of Jehovah, the definition by St. James of "pure religion and undefiled," and, above all, the precepts and ideals of the blessed Founder of Christianity himself, be brought to bear more and more effectively on mankind.

I close this preface some days after its first lines were written. The sun of spring has done its work on the Neva; the great river flows tranquilly on, a blessing and a joy; the *mujiks* are forgotten.

A. D. W.

LEGATION OF THE UNITED STATES, ST. PETERSBURG,
April 14, 1894.

P. S.—Owing to a wish to give more thorough revision to some parts of my work, it has been withheld from the press until the present date.

A. D. W.

CORNELL UNIVERSITY, ITHACA, N. Y.,
August 15, 1895.

CONTENTS OF THE FIRST VOLUME.

CHAPTER I.

FROM CREATION TO EVOLUTION.

CHAPTER II.

GEOGRAPHY.

CHAPTER III.

ASTRONOMY.

CHAPTER IV.

FROM "SIGNS AND WONDERS" TO LAW IN THE HEAVENS.

CHAPTER V.

FROM GENESIS TO GEOLOGY.

CHAPTER VI.

THE ANTIQUITY OF MAN, EGYPTOLOGY, AND ASSYRIOLOGY.

CHAPTER VII.

THE ANTIQUITY OF MAN AND PREHISTORIC ARCHÆOLOGY.

CHAPTER VIII.

THE "FALL OF MAN" AND ANTHROPOLOGY.

CHAPTER IX.

THE "FALL OF MAN" AND ETHNOLOGY.

CHAPTER X.

THE "FALL OF MAN" AND HISTORY.

CHAPTER XI.

FROM "THE PRINCE OF THE POWER OF THE AIR" TO METEOROLOGY.

CHAPTER XII.

FROM MAGIC TO CHEMISTRY AND PHYSICS.

A HISTORY OF
THE WARFARE OF SCIENCE
WITH THEOLOGY
IN CHRISTENDOM

THE WARFARE OF SCIENCE
WITH THEOLOGY.

CHAPTER I.

FROM CREATION TO EVOLUTION.

I. THE VISIBLE UNIVERSE.

AMONG those masses of cathedral sculpture which preserve so much of mediæval theology, one frequently recurring group is noteworthy for its presentment of a time-honoured doctrine regarding the origin of the universe.

The Almighty, in human form, sits benignly, making the sun, moon, and stars, and hanging them from the solid firmament which supports the "heaven above" and overarches the "earth beneath."

The furrows of thought on the Creator's brow show that in this work he is obliged to contrive; the knotted muscles upon his arms show that he is obliged to toil; naturally, then, the sculptors and painters of the mediæval and early modern period frequently represented him as the writers whose conceptions they embodied had done—as, on the seventh day, weary after thought and toil, enjoying well-earned repose and the plaudits of the hosts of heaven.

In these thought-fossils of the cathedrals, and in other revelations of the same idea through sculpture, painting, glass-staining, mosaic work, and engraving, during the Middle Ages and the two centuries following, culminated a belief which had been developed through thousands of years, and which has determined the world's thought until our own time.

Its beginnings lie far back in human history; we find

them among the early records of nearly all the great civilizations, and they hold a most prominent place in the various sacred books of the world. In nearly all of them is revealed the conception of a Creator of whom man is an imperfect image, and who literally and directly created the visible universe with his hands and fingers.

Among these theories, of especial interest to us are those which controlled theological thought in Chaldea. The Assyrian inscriptions which have been recently recovered and given to the English-speaking peoples by Layard, George Smith, Sayce, and others, show that in the ancient religions of Chaldea and Babylonia there was elaborated a narrative of the creation which, in its most important features, must have been the source of that in our own sacred books. It has now become perfectly clear that from the same sources which inspired the accounts of the creation of the universe among the Chaldeo-Babylonian, the Assyrian, the Phœnician, and other ancient civilizations came the ideas which hold so prominent a place in the sacred books of the Hebrews. In the two accounts imperfectly fused together in Genesis, and also in the account of which we have indications in the book of Job and in the Proverbs, there is presented, often with the greatest sublimity, the same early conception of the Creator and of the creation—the conception, so natural in the childhood of civilization, of a Creator who is an enlarged human being working literally with his own hands, and of a creation which is "the work of his fingers." To supplement this view there was developed the belief in this Creator as one who, having

> . . . "from his ample palm
> Launched forth the rolling planets into space,"

sits on high, enthroned "upon the circle of the heavens," perpetually controlling and directing them.

From this idea of creation was evolved in time a somewhat nobler view. Ancient thinkers, and especially, as is now found, in Egypt, suggested that the main agency in creation was not the hands and fingers of the Creator, but his *voice*. Hence was mingled with the earlier, cruder belief regarding the origin of the earth and heavenly bodies by the Almighty the more impressive idea that "he spake and they were made"—that they were brought into existence by his *word.**

Among the early fathers of the Church this general view of creation became fundamental; they impressed upon Christendom more and more strongly the belief that the universe was created in a perfectly literal sense by the hands or voice of God. Here and there sundry theologians of larger mind attempted to give a more spiritual view regarding some parts of the creative work, and of these were St. Gregory of Nyssa and St. Augustine. Ready as they were to accept the literal text of Scripture, they revolted against the conception of an actual creation of the universe by the hands and fingers of a Supreme Being, and in this they were followed by Bede and a few others; but the more material conceptions prevailed, and we find these taking shape not only in the sculptures and mosaics and stained glass of cathedrals, and in the illuminations of missals and psalters, but later, at the close of the Middle Ages, in the pictured Bibles and in general literature.

Into the Anglo-Saxon mind this ancient material conception of the creation was riveted by two poets whose works

* Among the many mediæval representations of the creation of the universe, I especially recall from personal observation those sculptured above the portals of the cathedrals of Freiburg and Upsala, the paintings on the walls of the Campo Santo at Pisa, and, most striking of all, the mosaics of the Cathedral of Monreale and those in the Cappella Palatina at Palermo. Among peculiarities showing the simplicity of the earlier conception the representation of the repose of the Almighty on the seventh day is very striking. He is shown as seated in almost the exact attitude of the "Weary Mercury" of classic sculpture—bent, and with a very marked expression of fatigue upon his countenance and in the whole disposition of his body.

The Monreale mosaics are pictured in the great work of Gravina, and the Pisa frescoes in Didron's *Iconographie*, Paris, 1843, p. 598. For an exact statement of the resemblances which have settled the question among the most eminent scholars in favour of the derivation of the Hebrew cosmogony from that of Assyria, see Jensen, *Die Kosmologie der Babylonier*, Strassburg, 1890, pp. 304, 306; also Franz Lukas, *Die Grundbegriffe in den Kosmographien der alten Völker*, Leipsic, 1893, pp. 35–46; also George Smith's *Chaldean Genesis*, especially the German translation with additions by Delitzsch, Leipsic, 1876, and Schrader, *Die Keilinschriften und das Alte Testament*, Giessen, 1883, pp. 1–54, etc. See also Renan, *Histoire du peuple d'Israel*, vol. i, chap. i, *L'antique influence babylonienne*. For Egyptian views regarding creation, and especially for the transition from the idea of creation by the hands and fingers of the Creator to creation by his *voice* and his "word," see Maspero and Sayce, *The Dawn of Civilization*, pp. 145–146.

appealed especially to the deeper religious feelings. In the seventh century Cædmon paraphrased the account given in Genesis, bringing out this material conception in the most literal form; and a thousand years later Milton developed out of the various statements in the Old Testament, mingled with a theology regarding "the creative Word" which had been drawn from the New, his description of the creation by the second person in the Trinity, than which nothing could be more literal and material:

> "He took the golden compasses, prepared
> In God's eternal store, to circumscribe
> This universe and all created things.
> One foot he centred, and the other turned
> Round through the vast profundity obscure,
> And said, 'Thus far extend, thus far thy bounds:
> This be thy just circumference, O world!'" *

So much for the orthodox view of the *manner* of creation.

The next point developed in this theologic evolution had reference to the *matter* of which the universe was made, and it was decided by an overwhelming majority that no material substance existed before the creation of the material universe—that "God created everything out of nothing." Some venturesome thinkers, basing their reasoning upon the first verses of Genesis, hinted at a different view—namely, that the mass, "without form and void," existed before the universe; but this doctrine was soon swept out of sight. The vast majority of the fathers were explicit on this point. Tertullian especially was very severe against those who took any other view than that generally accepted as orthodox: he declared that, if there had been any pre-existing matter out of which the world was formed, Scripture would have mentioned it; that by not mentioning it God has given us a clear proof that there was no such thing; and, after a manner not unknown in other theological controversies, he threatens Hermogenes, who takes the opposite view, with

---

* For Gregory of Nyssa, Augustine, and the general subject of the development of an evolution theory among the Greeks, see the excellent work by Dr. Osborn, *From the Greeks to Darwin*, pp. 33 and following; for Cædmon, see any edition—I have used Bouterwek's, Gütersloh, 1854; for Milton, see *Paradise Lost*, book vii, lines 225–231.

"the woe which impends on all who add to or take away from the written word."

St. Augustine, who showed signs of a belief in a pre-existence of matter, made his peace with the prevailing belief by the simple reasoning that, "although the world has been made of some material, that very same material must have been made out of nothing."

In the wake of these great men the universal Church steadily followed. The Fourth Lateran Council declared that God created everything out of nothing; and at the present hour the vast majority of the faithful—whether Catholic or Protestant—are taught the same doctrine; on this point the syllabus of Pius IX and the Westminster Catechism fully agree.*

Having thus disposed of the manner and matter of creation, the next subject taken up by theologians was the *time* required for the great work.

Here came a difficulty. The first of the two accounts given in Genesis extended the creative operation through six days, each of an evening and a morning, with much explicit detail regarding the progress made in each. But the second account spoke of "*the day*" in which "the Lord God made the earth and the heavens." The explicitness of the first account and its naturalness to the minds of the great mass of early theologians gave it at first a decided advantage; but Jewish thinkers, like Philo, and Christian thinkers, like Origen, forming higher conceptions of the Creator and his work, were not content with this, and by them was launched upon the troubled sea of Christian theology the idea that the creation was instantaneous, this idea being strengthened not only by the second of the Genesis legends, but by the great text, "He spake, and it was done; he commanded, and it stood fast"—or, as it appears in the Vulgate and in most translations, "He spake, and they were made; he commanded, and they were created."

---

* For Tertullian, see *Tertullian against Hermogenes*, chaps. xx and xxii; for St. Augustine regarding "creation from nothing," see the *De Genesi contra Manichæos*, lib. i, cap. vi; for St. Ambrose, see the *Hexameron*, lib. i, cap. iv; for the decree of the Fourth Lateran Council, and the view received in the Church to-day, see the article *Creation* in Addis and Arnold's *Catholic Dictionary*.

As a result, it began to be held that the safe and proper course was to believe literally *both* statements; that in some mysterious manner God created the universe in six days, and yet brought it all into existence in a moment. In spite of the outcries of sundry great theologians, like Ephrem Syrus, that the universe was created in exactly six days of twenty-four hours each, this compromise was promoted by St. Athanasius and St. Basil in the East, and by St. Augustine and St. Hilary in the West.

Serious difficulties were found in reconciling these two views, which to the natural mind seem absolutely contradictory; but by ingenious manipulation of texts, by dexterous play upon phrases, and by the abundant use of metaphysics to dissolve away facts, a reconciliation was effected, and men came at least to believe that they believed in a creation of the universe instantaneous and at the same time extended through six days.*

Some of the efforts to reconcile these two accounts were so fruitful as to deserve especial record. The fathers, Eastern and Western, developed out of the double account in Genesis, and the indications in the Psalms, the Proverbs, and the book of Job, a vast mass of sacred science bearing upon this point. As regards the whole work of creation, stress was laid upon certain occult powers in numerals. Philo Judæus, while believing in an instantaneous creation, had also declared that the world was created in six days because "of all numbers six is the most productive"; he had explained the creation of the heavenly bodies on the fourth day by "the harmony of the number four"; of the animals on the fifth day by the five senses; of man on the sixth day by the same virtues in the number six which had caused it to be set as a limit to the creative work; and, greatest of all, the rest on the seventh day by the vast mass of mysterious virtues in the number seven.

St. Jerome held that the reason why God did not pronounce the work of the second day "good" is to be found

* For Origen, see his *Contra Celsum*, cap. xxxvi, xxxvii; also his *De Principibus*, cap. v; for St. Augustine, see his *De Genesi contra Manichæos* and *De Genesi ad Litteram*, *passim;* for Athanasius, see his *Discourses against the Arians*, ii, 48, 49.

in the fact that there is something essentially evil in the number two, and this was echoed centuries afterward, afar off in Britain, by Bede.

St. Augustine brought this view to bear upon the Church in the following statement: "There are three classes of numbers—the more than perfect, the perfect, and the less than perfect, according as the sum of them is greater than, equal to, or less than the original number. Six is the first perfect number: wherefore we must not say that six is a perfect number because God finished all his works in six days, but that God finished all his works in six days because six is a perfect number."

Reasoning of this sort echoed along through the mediæval Church until a year after the discovery of America, when the *Nuremberg Chronicle* re-echoed it as follows: "The creation of things is explained by the number six, the parts of which, one, two, and three, assume the form of a triangle."

This view of the creation of the universe as instantaneous and also as in six days, each made up of an evening and a morning, became virtually universal. Peter Lombard and Hugo of St. Victor, authorities of vast weight, gave it their sanction in the twelfth century, and impressed it for ages upon the mind of the Church.

Both these lines of speculation—as to the creation of everything out of nothing, and the reconciling of the instantaneous creation of the universe with its creation in six days—were still further developed by other great thinkers of the Middle Ages.

St. Hilary of Poictiers reconciled the two conceptions as follows: "For, although according to Moses there is an appearance of regular order in the fixing of the firmament, the laying bare of the dry land, the gathering together of the waters, the formation of the heavenly bodies, and the arising of living things from land and water, yet the creation of the heavens, earth, and other elements is seen to be the work of a single moment."

St. Thomas Aquinas drew from St. Augustine a subtle distinction which for ages eased the difficulties in the case: he taught in effect that God created the substance of things in a moment, but gave to the work of separating, shaping, and adorning this creation, six days.*

The early reformers accepted and developed the same view, and Luther especially showed himself equal to the occasion. With his usual boldness he declared, first, that Moses "spoke properly and plainly, and neither allegorically nor figuratively," and that therefore "the world with all creatures was created in six days." And he then goes on to show how, by a great miracle, the whole creation was also instantaneous.

Melanchthon also insisted that the universe was created out of nothing and in a mysterious way, both in an instant and in six days, citing the text: "He spake, and they were made."

Calvin opposed the idea of an instantaneous creation, and laid especial stress on the creation in six days: having called attention to the fact that the biblical chronology shows the world to be not quite six thousand years old and that it is now near its end, he says that "creation was extended through six days that it might not be tedious for us to occupy the whole of life in the consideration of it."

Peter Martyr clinched the matter by declaring: "So important is it to comprehend the work of creation that we see the creed of the Church take this as its starting point. Were this article taken away there would be no original sin, the promise of Christ would become void, and all the vital force of our religion would be destroyed." The Westminster divines in drawing up their Confession of Faith

* For Philo Judæus, see his *Creation of the World*, chap. iii; for St. Augustine on the powers of numbers in creation, see his *De Genesi ad Litteram*, iv, chap. ii; for Peter Lombard, see the *Sententiæ*, lib. ii, dist. xv, 5; and for Hugo of St. Victor, see *De Sacramentis*, lib. i, pars i; also, *Annotat. Elucidat. in Pentateuchum*, cap. v, vi, vii; for St. Hilary, see *De Trinitate*, lib. xii; for St. Thomas Aquinas, see his *Summa Theologica*, quest. lxxxiv, arts. i and ii; the passage in the *Nuremberg Chronicle*, 1493, is in fol. iii; for Bossuet, see his *Discours sur l'Histoire Universelle;* for the sacredness of the number seven among the Babylonians, see especially Schrader, *Die Keilinschriften und das Alte Testament*, pp. 21, 22; also George Smith *et al.;* for general ideas on the occult powers of various numbers, especially the number seven, and the influence of these ideas on theology and science, see my chapter on astronomy. As to mediæval ideas on the same subject, see Detzel, *Christliche Ikonographie*, Freiburg, 1894, pp. 44 and following.

specially laid it down as necessary to believe that all things visible and invisible were created not only out of nothing but in exactly six days.

Nor were the Roman divines less strenuous than the Protestant reformers regarding the necessity of holding closely to the so-called Mosaic account of creation. As late as the middle of the eighteenth century, when Buffon attempted to state simple geological truths, the theological faculty of the Sorbonne forced him to make and to publish a most ignominious recantation which ended with these words: "I abandon everything in my book respecting the formation of the earth, and generally all which may be contrary to the narrative of Moses."

Theologians, having thus settled the manner of the creation, the matter used in it, and the time required for it, now exerted themselves to fix its *date*.

The long series of efforts by the greatest minds in the Church, from Eusebius to Archbishop Usher, to settle this point are presented in another chapter. Suffice it here that the general conclusion arrived at by an overwhelming majority of the most competent students of the biblical accounts was that the date of creation was, in round numbers, four thousand years before our era; and in the seventeenth century, in his great work, Dr. John Lightfoot, Vice-Chancellor of the University of Cambridge, and one of the most eminent Hebrew scholars of his time, declared, as the result of his most profound and exhaustive study of the Scriptures, that "heaven and earth, centre and circumference, were created all together, in the same instant, and clouds full of water," and that "this work took place and man was created by the Trinity on October 23, 4004 B. C., at nine o'clock in the morning."

Here was, indeed, a triumph of Lactantius's method, the result of hundreds of years of biblical study and theological thought since Bede in the eighth century, and Vincent of Beauvais in the thirteenth, had declared that creation must have taken place in the spring. Yet, alas! within two centuries after Lightfoot's great biblical demonstration as to the exact hour of creation, it was discovered that at that hour an exceedingly cultivated people, enjoying all the fruits of a highly developed civilization, had long been swarming in the great cities of Egypt, and that other nations hardly less advanced had at that time reached a high development in Asia.*

But, strange as it may seem, even after theologians had thus settled the manner of creation, the matter employed in it, the time required for it, and the exact date of it, there remained virtually unsettled the first and greatest question of all; and this was nothing less than the question, WHO actually created the universe?

Various theories more or less nebulous, but all centred in texts of Scripture, had swept through the mind of the Church. By some theologians it was held virtually that the actual creative agent was the third person of the Trinity, who, in the opening words of our sublime creation poem, "moved upon the face of the waters." By others it was held that the actual Creator was the second person of the Trinity, in behalf of whose agency many texts were cited from the New Testament. Others held that the actual Creator was the first person, and this view was embodied in the two great formulas known as the Apostles' and Nicene Creeds, which explicitly assigned the work to "God the Father Almighty, Maker of heaven and earth." Others, finding a deep meaning in the words "Let *us* make," ascribed in Genesis to the Creator, held that the entire Trinity directly created all things; and still others, by curious metaphysical processes, seemed to arrive at the idea that peculiar combinations of two persons of the Trinity achieved the creation.

In all this there would seem to be considerable courage

* For Luther, see his *Commentary on Genesis*, 1545, introduction, and his comments on chap. i, verse 12; the quotations from Luther's commentary are taken mainly from the translation by Henry Cole, D. D., Edinburgh, 1858; for Melanchthon, see *Loci Theologici*, in Melanchthon, *Opera*, ed. Bretschneider, vol. xxi, pp. 269, 270, also pp. 637, 638—in quoting the text (Ps. xxiii, 9) I have used, as does Melanchthon himself, the form of the Vulgate; for the citations from Calvin, see his *Commentary on Genesis* (*Opera omnia*, Amsterdam, 1671, tom. i, cap. ii, p. 8); also in the *Institutes*, Allen's translation, London, 1838, vol. i, chap. xv, pp. 126, 127; for Peter Martyr, see his *Commentary on Genesis*, cited by Zöckler, vol. i, p. 690; for the articles in the Westminster Confession of Faith, see chap. iv; for Buffon's recantation, see Lyell, *Principles of Geology*, chap. iii, p. 57. For Lightfoot's declaration, see his works, edited by Pitman, London, 1822.

in view of the fearful condemnations launched in the Athanasian Creed against all who should "confound the persons" or "divide the substance of the Trinity."

These various stages in the evolution of scholastic theology were also embodied in sacred art, and especially in cathedral sculpture, in glass-staining, in mosaic working, and in missal painting.

The creative Being is thus represented sometimes as the third person of the Trinity, in the form of a dove brooding over chaos; sometimes as the second person, and therefore a youth; sometimes as the first person, and therefore fatherly and venerable; sometimes as the first and second persons, one being venerable and the other youthful; and sometimes as three persons, one venerable and one youthful, both wearing papal crowns, and each holding in his lips a tip of the wing of the dove, which thus seems to proceed from both and to be suspended between them.

Nor was this the most complete development of the mediæval idea. The Creator was sometimes represented with a single body, but with three faces, thus showing that Christian belief had in some pious minds gone through substantially the same cycle which an earlier form of belief had made ages before in India, when the Supreme Being was represented with one body but with the three faces of Brahma, Vishnu, and Siva.

But at the beginning of the modern period the older view in its primitive Jewish form was impressed upon Christians by the most mighty genius in art the world has known; for in 1512, after four years of Titanic labour, Michael Angelo uncovered his frescoes within the vault of the Sistine Chapel.

They had been executed by the command and under the sanction of the ruling Pope, Julius II, to represent the conception of Christian theology then dominant, and they remain to-day in all their majesty to show the highest point ever attained by the older thought upon the origin of the visible universe.

In the midst of the expanse of heaven the Almighty Father—the first person of the Trinity—in human form, august and venerable, attended by angels and upborne by mighty winds, sweeps over the abyss, and, moving through successive compartments of the great vault, accomplishes the work of the creative days. With a simple gesture he divides the light from the darkness, rears on high the solid firmament, gathers together beneath it the seas, or summons into existence the sun, moon, and planets, and sets them circling about the earth.

In this sublime work culminated the thought of thousands of years; the strongest minds accepted it or pretended to accept it, and nearly two centuries later this conception, in accordance with the first of the two accounts given in Genesis, was especially enforced by Bossuet, and received a new lease of life in the Church, both Catholic and Protestant.*

But to these discussions was added yet another, which, beginning in the early days of the Church, was handed down the ages until it had died out among the theologians of our own time.

In the first of the biblical accounts light is created and the distinction between day and night thereby made on the first day, while the sun and moon are not created until the fourth day. Masses of profound theological and pseudo-scientific reasoning have been developed to account for this —masses so great that for ages they have obscured the simple fact that the original text is a precious revelation to us of one of the most ancient of recorded beliefs—the belief that light and darkness are entities independent of the heavenly bodies, and that the sun, moon, and stars exist not merely to increase light but to "divide the day from the night, to be for signs and for seasons, and for days and for years," and "to rule the day and the night."

---

* For strange representations of the Creator and of the creation by one, two, or three persons of the Trinity, see Didron, *Iconographie Chrétienne*, pp. 35, 178, 224, 483, 567–580, and elsewhere; also Detzel as already cited. The most naïve of all survivals of the mediæval idea of creation which the present writer has ever seen was exhibited in 1894 on the banner of one of the guilds at the celebration of the four-hundredth anniversary of the founding of the Munich Cathedral. Jesus of Nazareth, as a beautiful boy and with a nimbus encircling his head, was shown turning and shaping the globe on a lathe, which he keeps in motion with his foot. The emblems of the Passion are about him, God the Father looking approvingly upon him from a cloud, and the dove hovering between the two. The date upon the banner was 1727.

Of this belief we find survivals among the early fathers, and especially in St. Ambrose. In his work on creation he tells us: "We must remember that the light of day is one thing and the light of the sun, moon, and stars another—the sun by his rays appearing to add lustre to the daylight. For before sunrise the day dawns, but is not in full refulgence, for the sun adds still further to its splendour." This idea became one of the "treasures of sacred knowledge committed to the Church," and was faithfully received by the Middle Ages. The mediæval mysteries and miracle plays give curious evidences of this: In a performance of the creation, when God separates light from darkness, the stage direction is, "Now a painted cloth is to be exhibited, one half black and the other half white." It was also given more permanent form. In the mosaics of San Marco at Venice, in the frescoes of the Baptistery at Florence and of the Church of St. Francis at Assisi, and in the altar carving at Salerno, we find a striking realization of it—the Creator placing in the heavens two disks or living figures of equal size, each suitably coloured or inscribed to show that one represents light and the other darkness. This conception was without doubt that of the person or persons who compiled from the Chaldean and other earlier statements the accounts of the creation in the first of our sacred books.*

Thus, down to a period almost within living memory, it was held, virtually "always, everywhere, and by all," that the universe, as we now see it, was created literally and

* For scriptural indications of the independent existence of light and darkness, compare with the first verses of the first chapter of Genesis such passages as Job xxxviii, 19, 24; for the general prevalence of this early view, see Lukas, *Kosmogonie*, pp. 31, 33, 41, 74, and *passim*; for the view of St. Ambrose regarding the creation of light and of the sun, see his *Hexameron*, lib. 4, cap. iii; for an excellent general statement, see Huxley, *Mr. Gladstone and Genesis*, in the *Nineteenth Century*, 1886, reprinted in his *Essays on Controverted Questions*, London, 1892, note, pp. 126 *et seq.*; for the acceptance in the miracle plays of the scriptural idea of light and darkness as independent creations, see Wright, *Essays on Archæological Subjects*, vol. ii, p. 178; for an account, with illustrations, of the mosaics, etc., representing this idea, see Tikkanen, *Die Genesis-mosaiken von San Marco*, Helsingfors, 1889, pp. 14 and 16 of text and Plates I and II. Very naïvely the Salerno carver, not wishing to colour the ivory which he wrought, has inscribed on one disk the word "LUX" and on the other "NOX." See also Didron, *Iconographie*, p. 482.

directly by the voice or hands of the Almighty, or by both —out of nothing—in an instant or in six days, or in both—about four thousand years before the Christian era—and for the convenience of the dwellers upon the earth, which was at the base and foundation of the whole structure.

But there had been implanted along through the ages germs of another growth in human thinking, some of them even as early as the Babylonian period. In the Assyrian inscriptions we find recorded the Chaldeo-Babylonian idea of *an evolution* of the universe out of the primeval flood or "great deep," and of the animal creation out of the earth and sea. This idea, recast, partially at least, into monotheistic form, passed naturally into the sacred books of the neighbours and pupils of the Chaldeans—the Hebrews; but its growth in Christendom afterward was checked, as we shall hereafter find, by the more powerful influence of other inherited statements which appealed more intelligibly to the mind of the Church.

Striking, also, was the effect of this idea as rewrought by the early Ionian philosophers, to whom it was probably transmitted from the Chaldeans through the Phœnicians. In the minds of Ionians like Anaximander and Anaximenes it was most clearly developed: the first of these conceiving of the visible universe as the result of processes of evolution, and the latter pressing further the same mode of reasoning, and dwelling on agencies in cosmic development recognised in modern science.

This general idea of evolution in Nature thus took strong hold upon Greek thought and was developed in many ways, some ingenious, some perverse. Plato, indeed, withstood it; but Aristotle sometimes developed it in a manner which reminds us of modern views.

Among the Romans Lucretius caught much from it, extending the evolutionary process virtually to all things.

In the early Church, as we have seen, the idea of a creation direct, material, and by means like those used by man, was all-powerful for the exclusion of conceptions based on evolution. From the more simple and crude of the views of creation given in the Babylonian legends, and thence incorporated into Genesis, rose the stream of orthodox thought on the subject, which grew into a flood and swept on through the Middle Ages and into modern times. Yet here and there in the midst of this flood were high grounds of thought held by strong men. Scotus Erigena and Duns Scotus, among the schoolmen, bewildered though they were, had caught some rays of this ancient light, and passed on to their successors, in modified form, doctrines of an evolutionary process in the universe.

In the latter half of the sixteenth century these evolutionary theories seemed to take more definite form in the mind of Giordano Bruno, who evidently divined the fundamental idea of what is now known as the "nebular hypothesis"; but with his murder by the Inquisition at Rome this idea seemed utterly to disappear—dissipated by the flames which in 1600 consumed his body on the Campo dei Fiori.

Yet within the two centuries divided by Bruno's death the world was led into a new realm of thought in which an evolution theory of the visible universe was sure to be rapidly developed. For there came, one after the other, five of the greatest men our race has produced—Copernicus, Kepler, Galileo, Descartes, and Newton—and when their work was done the old theological conception of the universe was gone. "The spacious firmament on high"—"the crystalline spheres"—the Almighty enthroned upon "the circle of the heavens," and with his own hands, or with angels as his agents, keeping sun, moon, and planets in motion for the benefit of the earth, opening and closing the "windows of heaven," letting down upon the earth the "waters above the firmament," "setting his bow in the cloud," hanging out "signs and wonders," hurling comets, "casting forth lightnings" to scare the wicked, and "shaking the earth" in his wrath: all this had disappeared.

These five men had given a new divine revelation to the world; and through the last, Newton, had come a vast new conception, destined to be fatal to the old theory of creation, for he had shown throughout the universe, in place of almighty caprice, all-pervading law. The bitter opposition of theology to the first four of these men is well known; but the fact is not so widely known that Newton, in spite of his deeply religious spirit, was also strongly opposed. It was vigorously urged against him that by his statement of the law of gravitation he "took from God that direct action on his works so constantly ascribed to him in Scripture and transferred it to material mechanism," and that he "substituted gravitation for Providence." But, more than this, these men gave a new basis for the theory of evolution as distinguished from the theory of creation.

Especially worthy of note is it that the great work of Descartes, erroneous as many of its deductions were, and, in view of the lack of physical knowledge in his time, must be, had done much to weaken the old conception. His theory of a universe brought out of all-pervading matter, wrought into orderly arrangement by movements in accordance with physical laws—though it was but a provisional hypothesis—had done much to draw men's minds from the old theological view of creation; it was an example of intellectual honesty arriving at errors, but thereby aiding the advent of truths. Crippled though Descartes was by his almost morbid fear of the Church, this part of his work was no small factor in bringing in that attitude of mind which led to a reception of the thoughts of more unfettered thinkers.

Thirty years later came, in England, an effort of a different sort, but with a similar result. In 1678 Ralph Cudworth published his *Intellectual System of the Universe.* To this day he remains, in breadth of scholarship, in strength of thought, in tolerance, and in honesty, one of the greatest glories of the English Church, and his work was worthy of him. He purposed to build a fortress which should protect Christianity against all dangerous theories of the universe, ancient or modern. The foundations of the structure were laid with old thoughts thrown often into new and striking forms; but, as the superstructure arose more and more into view, while genius marked every part of it, features appeared which gave the rigidly orthodox serious misgivings. From the old theories of direct personal action on the universe by the Almighty he broke utterly. He dwelt on the action of law, rejected the continuous exercise of miraculous intervention, pointed out the fact that in the natural world there are "errors" and "bungles," and argued vigorously in favour of the origin and maintenance of the universe as a slow and gradual development of Nature in obedience to an inward principle. The Balaks of seventeenth-century orthodoxy might well condemn this honest Balaam.

Toward the end of the next century a still more profound genius, Immanuel Kant, presented the nebular theory, giving it, in the light of Newton's great utterances, a consistency which it never before had; and about the same time Laplace gave it yet greater strength by mathematical reasonings of wonderful power and extent, thus implanting firmly in modern thought the idea that our own solar system and others—suns, planets, satellites, and their various movements, distances, and magnitudes—necessarily result from the obedience of nebulous masses to natural laws.

Throughout the theological world there was an outcry at once against "atheism," and war raged fiercely. Herschel and others pointed out many nebulous patches apparently gaseous. They showed by physical and mathematical demonstrations that the hypothesis accounted for the great body of facts, and, despite clamour, were gaining ground, when the improved telescopes resolved some of the patches of nebulous matter into multitudes of stars. The opponents of the nebular hypothesis were overjoyed; they now sang pæans to astronomy, because, as they said, it had proved the truth of Scripture. They had jumped to the conclusion that all nebulæ must be alike; that, if *some* are made up of systems of stars, *all* must be so made up; that none can be masses of attenuated gaseous matter, because some are not.

Science halted for a time. The accepted doctrine became this: that the only reason why all the nebulæ are not resolved into distinct stars is that our telescopes are not sufficiently powerful. But in time came the discovery of the spectroscope and spectrum analysis, and thence Fraunhofer's discovery that the spectrum of an ignited gaseous body is non-continuous, with interrupting lines; and Draper's discovery that the spectrum of an ignited solid is continuous, with no interrupting lines. And now the spectroscope was turned upon the nebulæ, and many of them were found to be gaseous. Here, then, was ground for the inference that in these nebulous masses at different stages of condensation—some apparently mere patches of mist, some with luminous centres—we have the process of development actually going on, and observations like those of Lord Rosse and Arrest gave yet further confirmation to this view. Then came the great contribution of the nineteenth century to physics, aiding to explain important parts of the vast process by the mechanical theory of heat.

Again the nebular hypothesis came forth stronger than ever, and about 1850 the beautiful experiment of Plateau on the rotation of a fluid globe came in apparently to illustrate if not to confirm it. Even so determined a defender of orthodoxy as Mr. Gladstone at last acknowledged some form of a nebular hypothesis as probably true.

Here, too, was exhibited that form of surrendering theological views to science under the claim that science concurs with theology, which we have seen in so many other fields; and, as typical, an example may be given, which, however restricted in its scope, throws light on the process by which such surrenders are obtained. A few years since one of the most noted professors of chemistry in the city of New York, under the auspices of one of its most fashionable churches, gave a lecture which, as was claimed in the public prints and in placards posted in the streets, was to show that science supports the theory of creation given in the sacred books ascribed to Moses. A large audience assembled, and a brilliant series of elementary experiments with oxygen, hydrogen, and carbonic acid was concluded by the Plateau demonstration. It was beautifully made. As the coloured globule of oil, representing the earth, was revolved in a transparent medium of equal density, as it became flattened at the poles, as rings then broke forth from it and revolved about it, and, finally, as some of these rings broke into satellites, which for a moment continued to circle about the central mass, the audience, as well they might, rose and burst into rapturous applause.

Thereupon a well-to-do citizen arose and moved the thanks of the audience to the eminent professor for "this perfect demonstration of the exact and literal conformity of the statements given in Holy Scripture with the latest re sults of science." The motion was carried unanimously and with applause, and the audience dispersed, feeling that a great service had been rendered to orthodoxy. *Sancta simplicitas!*

What this incident exhibited on a small scale has been seen elsewhere with more distinguished actors and on a broader stage. Scores of theologians, chief among whom of late. in zeal if not in knowledge, has been Mr. Gladstone, have endeavoured to "reconcile" the two accounts in Genesis with each other and with the truths regarding the origin of the universe gained by astronomy, geology, geography, physics, and chemistry. The result has been recently stated by an eminent theologian, the Hulsean Professor of Divinity at the University of Cambridge. He declares, "No attempt at reconciling Genesis with the exacting requirements of modern sciences has ever been known to succeed without entailing a degree of special pleading or forced interpretation to which, in such a question, we should be wise to have no recourse."*

The revelations of another group of sciences, though sometimes bitterly opposed and sometimes "reconciled" by

---

* For an interesting reference to the outcry against Newton, see McCosh, *The Religious Aspect of Evolution*, New York, 1890, pp. 103, 104; for germs of an evolutionary view among the Babylonians, see George Smith, *Chaldean Account of Genesis*, New York, 1876, pp. 74, 75; for a germ of the same thought in Lucretius, see his *De Natura Rerum*, lib. v, pp. 187–194, 447–454; for Bruno's conjecture (in 1591), see Jevons, *Principles of Science*, London, 1874, vol. ii, p. 299; for Kant's statement, see his *Naturgeschichte des Himmels*; for his part in the nebular hypothesis, see Lange, *Geschichte des Materialismus*, vol. i, p. 266; for value of Plateau's beautiful experiment, very cautiously estimated, see Jevons, vol. ii, p. 36; also Elisée Reclus, *The Earth*, translated by Woodward, vol. i, pp. 14–18, for an estimate still more careful; for a general account of discoveries of the nature of nebulæ by spectroscope, see Draper, *Conflict between Religion and Science*; for a careful discussion regarding the spectra of solid, liquid, and gaseous bodies, see Schellen, *Spectrum Analysis*, pp. 100 *et seq.*; for a very thorough discussion of the bearings of discoveries made by spectrum analysis upon the nebular hypothesis, ibid., pp. 532–537; for a presentation of the difficulties yet unsolved, see an article by Plummer in the *London Popular Science Review* for January, 1875; for an excellent short summary of recent observations and thought on this subject, see T. Sterry Hunt, *Address at the Priestley Centennial*, pp. 7, 8; for an interesting modification of this hypothesis, see Proctor's writings; for a still more recent view see Lockyer's two articles on *The Sun's Place in Nature*, in *Nature* for February 14 and 25, 1895.

theologians, have finally set the whole question at rest. First, there have come the biblical critics—earnest Christian scholars, working for the sake of truth—and these have revealed beyond the shadow of a reasonable doubt the existence of at least two distinct accounts of creation in our book of Genesis, which can sometimes be forced to agree, but which are generally absolutely at variance with each other. These scholars have further shown the two accounts to be not the cunningly devised fables of priestcraft, but evidently fragments of earlier legends, myths, and theologies, accepted in good faith and brought together for the noblest of purposes by those who put in order the first of our sacred books.

Next have come the archæologists and philologists, the devoted students of ancient monuments and records; of these are such as Rawlinson, George Smith, Sayce, Oppert, Jensen, Schrader, Delitzsch, and a phalanx of similarly devoted scholars, who have deciphered a multitude of ancient texts, especially the inscriptions found in the great library of Assurbanipal at Nineveh, and have discovered therein an account of the origin of the world identical in its most important features with the later accounts in our own book of Genesis.

These men have had the courage to point out these facts and to connect them with the truth that these Chaldean and Babylonian myths, legends, and theories were far earlier than those of the Hebrews, which so strikingly resemble them, and which we have in our sacred books; and they have also shown us how natural it was that the Jewish accounts of the creation should have been obtained at that remote period when the earliest Hebrews were among the Chaldeans, and how the great Hebrew poetic accounts of creation were drawn either from the sacred traditions of these earlier peoples or from antecedent sources common to various ancient nations.

In a summary which for profound thought and fearless integrity does honour not only to himself but to the great position which he holds, the Rev. Dr. Driver, Professor of Hebrew and Canon of Christ Church at Oxford, has recently stated the case fully and fairly. Having pointed out the fact that the Hebrews were one people out of many who thought upon the origin of the universe, he says that they "framed theories to account for the beginnings of the earth and man"; that "they either did this for themselves or borrowed those of their neighbours"; that "of the theories current in Assyria and Phœnicia fragments have been preserved, and these exhibit points of resemblance with the biblical narrative sufficient to warrant the inference that both are derived from the same cycle of tradition."

After giving some extracts from the Chaldean creation tablets he says: "In the light of these facts it is difficult to resist the conclusion that the biblical narrative is drawn from the same source as these other records. The biblical historians, it is plain, derived their materials from the best human sources available. . . . The materials which with other nations were combined into the crudest physical theories or associated with a grotesque polytheism were vivified and transformed by the inspired genius of the Hebrew historians, and adapted to become the vehicle of profound religious truth."

Not less honourable to the sister university and to himself is the statement recently made by the Rev. Dr. Ryle, Hulsean Professor of Divinity at Cambridge. He says that to suppose that a Christian "must either renounce his confidence in the achievements of scientific research or abandon his faith in Scripture is a monstrous perversion of Christian freedom." He declares: "The old position is no longer tenable; a new position has to be taken up at once, prayerfully chosen, and hopefully held." He then goes on to compare the Hebrew story of creation with the earlier stories developed among kindred peoples, and especially with the pre-existing Assyro-Babylonian cosmogony, and shows that they are from the same source. He points out that any attempt to explain particular features of the story into harmony with the modern scientific ideas necessitates "a non-natural" interpretation; but he says that, if we adopt a natural interpretation, "we shall consider that the Hebrew description of the visible universe is unscientific as judged by modern standards, and that it shares the limitations of the imperfect knowledge of the age at which it was committed to writing." Regarding the account in Genesis of man's physical origin, he says that it "is expressed in the simple terms of prehistoric legend, of unscientific pictorial description."

In these statements and in a multitude of others made by eminent Christian investigators in other countries is indicated what the victory is which has now been fully won over the older theology.

Thus, from the Assyrian researches as well as from other sources, it has come to be acknowledged by the most eminent scholars at the leading seats of Christian learning that the accounts of creation with which for nearly two thousand years all scientific discoveries have had to be "reconciled"—the accounts which blocked the way of Copernicus, and Galileo, and Newton, and Laplace—were simply transcribed or evolved from a mass of myths and legends largely derived by the Hebrews from their ancient relations with Chaldea, rewrought in a monotheistic sense, imperfectly welded together, and then thrown into poetic forms in the sacred books which we have inherited.

On one hand, then, we have the various groups of men devoted to the physical sciences all converging toward the proofs that the universe, as we at present know it, is the result of an evolutionary process—that is, of the gradual working of physical laws upon an early condition of matter; on the other hand, we have other great groups of men devoted to historical, philological, and archæological science whose researches all converge toward the conclusion that our sacred accounts of creation were the result of an evolution from an early chaos of rude opinion.

The great body of theologians who have so long resisted the conclusions of the men of science have claimed to be fighting especially for "the truth of Scripture," and their final answer to the simple conclusions of science regarding the evolution of the material universe has been the cry, "The Bible is true." And they are right—though in a sense nobler than they have dreamed. Science, while conquering them, has found in our Scriptures a far nobler truth than that literal historical exactness for which theologians have so long and so vainly contended. More and more as we consider the results of the long struggle in this field we are brought to the conclusion that the inestimable value of the great sacred books of the world is found in their revelation of the steady striving of our race after higher conceptions, beliefs, and aspirations, both in morals and religion. Unfolding and exhibiting this long-continued effort, each of the great sacred books of the world is precious, and all, in the highest sense, are true. Not one of them, indeed, conforms to the measure of what mankind has now reached in historical and scientific truth; to make a claim to such conformity is folly, for it simply exposes those who make it and the books for which it is made to loss of their just influence.

That to which the great sacred books of the world conform, and our own most of all, is the evolution of the highest conceptions, beliefs, and aspirations of our race from its childhood through the great turning-points in its history. Herein lies the truth of all bibles, and especially of our own. Of vast value they indeed often are as a record of historical outward fact; recent researches in the East are constantly increasing this value; but it is not for this that we prize them most: they are eminently precious, not as a record of outward fact, but as a mirror of the evolving heart, mind, and soul of man. They are true because they have been developed in accordance with the laws governing the evolution of truth in human history, and because in poem, chronicle, code, legend, myth, apologue, or parable they reflect this development of what is best in the onward march of humanity. To say that they are not true is as if one should say that a flower or a tree or a planet is not true; to scoff at them is to scoff at the law of the universe. In welding together into noble form, whether in the book of Genesis, or in the Psalms, or in the book of Job, or elsewhere, the great conceptions of men acting under earlier inspiration, whether in Egypt, or Chaldea, or India, or Persia, the compilers of our sacred books have given to humanity a possession ever becoming more and more precious; and modern science, in substituting a new heaven and a new earth for the old—the reign of law for the reign of caprice, and the idea of evolution for that of creation—has added and is steadily adding a new revelation divinely inspired.

In the light of these two evolutions, then—one of the visible universe, the other of a sacred creation-legend—science and theology, if the master minds in both are wise, may at last be reconciled. A great step in this reconciliation was recently seen at the main centre of theological thought among English-speaking people, when, in the collection of essays entitled *Lux Mundi*, emanating from the college established in these latter days as a fortress of orthodoxy at Oxford, the legendary character of the creation accounts in our sacred books was acknowledged, and when the Archbishop of Canterbury asked, "May not the Holy Spirit at times have made use of myth and legend?" *

II. THEOLOGICAL TEACHINGS REGARDING THE ANIMALS AND MAN.

In one of the windows of the cathedral at Ulm a mediæval glass-stainer has represented the Almighty as busily engaged in creating the animals, and there has just left the divine hands an elephant fully accoutred, with armour, harness, and housings, ready for war. Similar representations appear in illuminated manuscripts and even in early printed books, and, as the culmination of the whole, the Almighty is shown as fashioning the first man from a hillock of clay and extracting from his side, with evident effort, the first woman.

This view of the general process of creation had come from far, appearing under varying forms in various ancient cosmogonies. In the Egyptian temples at Philæ and Den-

* For the first citations above made, see *The Cosmogony of Genesis*, by the Rev. S. R. Driver, D. D., Canon of Christ Church and Regius Professor of Hebrew at Oxford, in *The Expositor* for January, 1886; for the second series of citations, see *The Early Narratives of Genesis*, by Herbert Edward Ryle, Hulsean Professor of Divinity at Cambridge, London, 1892. For evidence that even the stiffest of Scotch Presbyterians have now come to discard the old literal biblical narrative of creation and to regard the declaration of the Westminster Confession thereon as a "disproved theory of creation," see Principal John Tulloch, in *Contemporary Review*, March, 1877, on *Religious Thought in Scotland*—especially page 550.

derah may still be seen representations of the Nile gods modelling lumps of clay into men, and a similar work is ascribed in the Assyrian tablets to the gods of Babylonia. Passing into our own sacred books, these ideas became the starting point of a vast new development of theology.*

The fathers of the Church generally received each of the two conflicting creation legends in Genesis literally, and then, having done their best to reconcile them with each other and to mould them together, made them the final test of thought upon the universe and all things therein. At the beginning of the fourth century Lactantius struck the key-note of this mode of subordinating all other things in the study of creation to the literal text of Scripture, and he enforces his view of the creation of man by a bit of philology, saying the final being created "is called man because he is made from the ground—*homo ex humo*."

In the second half of the same century this view as to the literal acceptance of the sacred text was reasserted by St. Ambrose, who, in his work on the creation, declared that "Moses opened his mouth and poured forth what God had said to him." But a greater than either of them fastened this idea into the Christian theologies. St. Augustine, preparing his *Commentary on the Book of Genesis*, laid down in one famous sentence the law which has lasted in the Church until our own time: "Nothing is to be accepted save on the authority of Scripture, since greater is that authority than all the powers of the human mind." The vigour of the sentence in its original Latin carried it ringing down the centuries: "*Major est Scripturæ auctoritas quam omnis humani ingenii capacitas.*"

Through the mediæval period, in spite of a revolt led by no other than St. Augustine himself, and followed by a

---

* For representations of Egyptian gods creating men out of lumps of clay, see Maspero and Sayce, *The Dawn of History*, p. 156; for the Chaldean legends of the creation of men and animals, see ibid., p. 543; also George Smith, *Chaldean Account of Genesis*, Sayce's edition, pp. 36, 72, and 93; also for similar legends in other ancient nations, Lenormant, *Origines de l'Histoire*, pp. 17 *et seq.*; for mediæval representations of the creation of man and woman, see Didron, *Iconographie*, pp. 35, 178, 224, 537.

series of influential churchmen, contending, as we shall hereafter see, for a modification of the accepted view of creation, this phrase held the minds of men firmly. The great Dominican encyclopædist, Vincent of Beauvais, in his *Mirror of Nature*, while mixing ideas brought from Aristotle with a theory drawn from the Bible, stood firmly by the first of the accounts given in Genesis, and assigned the special virtue of the number six as a reason why all things were created in six days; and in the later Middle Ages that eminent authority, Cardinal d'Ailly, accepted everything regarding creation in the sacred books literally. Only a faint dissent is seen in Gregory Reisch, another authority of this later period, who, while giving, in his book on the beginning of things, a full length woodcut showing the Almighty in the act of extracting Eve from Adam's side, with all the rest of new-formed Nature in the background, leans in his writings, like St. Augustine, toward a belief in the pre-existence of matter.

At the Reformation the vast authority of Luther was thrown in favour of the literal acceptance of Scripture as the main source of natural science. The allegorical and mystical interpretations of earlier theologians he utterly rejected. "Why," he asks, "should Moses use allegory when he is not speaking of allegorical creatures or of an allegorical world, but of real creatures and of a visible world, which can be seen, felt, and grasped? Moses calls things by their right names, as we ought to do. . . . I hold that the animals took their being at once upon the word of God, as did also the fishes in the sea."

Not less explicit in his adherence to the literal account of creation given in Genesis was Calvin. He warns those who, by taking another view than his own, "basely insult the Creator, to expect a judge who will annihilate them." He insists that all species of animals were created in six days, each made up of an evening and a morning, and that no new species has ever appeared since. He dwells on the production of birds from the water as resting upon certain warrant of Scripture, but adds, "If the question is to be argued on physical grounds, we know that water is more akin to air than the earth is." As to difficulties in the scriptural account of creation, he tells us that God "wished by these to give proofs of his power which should fill us with astonishment."

The controlling minds in the Roman Church steadfastly held this view. In the seventeenth century Bossuet threw his vast authority in its favour, and in his *Discourse on Universal History*, which has remained the foundation not only of theological but of general historical teaching in France down to the present republic, we find him calling attention to what he regards as the culminating act of creation, and asserting that, literally, for the creation of man earth was used, and "the finger of God applied to corruptible matter."

The Protestant world held this idea no less persistently. In the seventeenth century Dr. John Lightfoot, Vice-Chancellor of the University of Cambridge, the great rabbinical scholar of his time, attempted to reconcile the two main legends in Genesis by saying that of the "clean sort of beasts there were seven of every kind created, three couples for breeding and the odd one for Adam's sacrifice on his fall, which God foresaw"; and that of unclean beasts only one couple was created.

So literal was this whole conception of the work of creation that in these days it can scarcely be imagined. The Almighty was represented in theological literature, in the pictured Bibles, and in works of art generally, as a sort of enlarged and venerable Nuremberg toymaker. At times the accounts in Genesis were illustrated with even more literal exactness; thus, in connection with a well-known passage in the sacred text, the Creator was shown as a tailor, seated, needle in hand, diligently sewing together skins of beasts into coats for Adam and Eve. Such representations presented no difficulties to the docile minds of the Middle Ages and the Reformation period; and in the same spirit, when the discovery of fossils began to provoke thought, these were declared to be "models of his works approved or rejected by the great Artificer," "outlines of future creations," "sports of Nature," or "objects placed in the strata to bring to naught human curiosity"; and this kind of explanation lingered on until in our own time an eminent naturalist, in his anxiety to save the literal account in Genesis, has urged that Jehovah tilted and twisted the strata, scattered the fossils through them, scratched the glacial furrows upon them, spread over them the marks of erosion by water, and set Niagara pouring—all in an instant—thus mystifying the world "for some inscrutable purpose, but for his own glory." *

The next important development of theological reasoning had regard to the *divisions* of the animal kingdom.

Naturally, one of the first divisions which struck the inquiring mind was that between useful and noxious creatures, and the question therefore occurred, How could a good God create tigers and serpents, thorns and thistles? The answer was found in theological considerations upon *sin.* To man's first disobedience all woes were due. Great men for eighteen hundred years developed the theory that before Adam's disobedience there was no death, and therefore neither ferocity nor venom.

Some typical utterances in the evolution of this doctrine are worthy of a passing glance. St. Augustine expressly confirmed and emphasized the view that the vegetable as well as the animal kingdom was cursed on account of man's sin. Two hundred years later this utterance had been echoed on from father to father of the Church until it was caught by Bede; he declared that before man's fall animals were harmless, but were made poisonous or hurtful by Adam's sin, and he said, "Thus fierce and poisonous animals were created for terrifying man (because God foresaw that

* For the citation from Lactantius, see *Divin. Instit.*, lib. ii, cap. xi, in Migne, tome vi, pp. 311, 312; for St. Augustine's great phrase, see the *De Genes. ad litt.*, ii, 5; for St. Ambrose, see lib. i, cap. ii; for Vincent of Beauvais, see the *Speculum Naturale*, lib. i, cap. ii, and lib. ii, cap. xv and xxx; also Bourgeat, *Études sur Vincent de Beauvais*, Paris, 1856, especially chaps. vii, xii, and xvi; for Cardinal d'Ailly, see the *Imago Mundi*, and for Reisch, see the various editions of the *Margarita Philosophica*; for Luther's statements, see Luther's *Schriften*, ed. Walch, Halle, 1740, *Commentary on Genesis*, vol. i; for Calvin's view of the creation of the animals, including the immutability of species, see the *Comm. in Gen.*, tome i of his *Opera omnia*, Amst., 1671, cap. i, v, xx, p. 5, also cap. ii, v, ii, p. 8, and elsewhere; for Bossuet, see his *Discours sur l'Histoire universelle* (in his *Œuvres*, tome v, Paris, 1846); for Lightfoot, see his works, edited by Pitman, London, 1822; for Bede, see the *Hexæmeron*, lib. i, in Migne, tome xci, p. 21; for Mr. Gosse's modern defence of the literal view, see his *Omphalos*, London, 1857, *passim*.

he would sin), in order that he might be made aware of the final punishment of hell."

In the twelfth century this view was incorporated by Peter Lombard into his great theological work, the *Sentences*, which became a text-book of theology through the middle ages. He affirmed that "no created things would have been hurtful to man had he not sinned; they became hurtful for the sake of terrifying and punishing vice or of proving and perfecting virtue; they were created harmless, and on account of sin became hurtful."

This theological theory regarding animals was brought out in the eighteenth century with great force by John Wesley. He declared that before Adam's sin "none of these attempted to devour or in any wise hurt one another"; "the spider was as harmless as the fly, and did not lie in wait for blood." Not only Wesley, but the eminent Dr. Adam Clarke and Dr. Richard Watson, whose ideas had the very greatest weight among the English Dissenters, and even among leading thinkers in the Established Church, held firmly to this theory; so that not until, in our own time, geology revealed the remains of vast multitudes of carnivorous creatures, many of them with half-digested remains of other animals in their stomachs, all extinct long ages before the appearance of man upon earth, was a victory won by science over theology in this field.

A curious development of this doctrine was seen in the belief drawn by sundry old commentators from the condemnation of the serpent in Genesis—a belief, indeed, perfectly natural, since it was evidently that of the original writers of the account preserved in the first of our sacred books. This belief was that, until the tempting serpent was cursed by the Almighty, all serpents stood erect, walked, and talked.

This belief was handed down the ages as part of "the sacred deposit of the faith" until Watson, the most prolific writer of the evangelical reform in the eighteenth century and the standard theologian of the evangelical party, declared: "We have no reason at all to believe that the animal had a serpentine form in any mode or degree until its transformation; that he was then degraded to a reptile to go upon his belly imports, on the contrary, an entire loss and alteration of the original form." Here, again, was a ripe result of the theologic method diligently pursued by the strongest thinkers in the Church during nearly two thousand years; but this "sacred deposit" also faded away when the geologists found abundant remains of fossil serpents dating from periods long before the appearance of man.

Troublesome questions also arose among theologians regarding animals classed as "superfluous." St. Augustine was especially exercised thereby. He says: "I confess I am ignorant why mice and frogs were created, or flies and worms. . . . All creatures are either useful, hurtful, or superfluous to us. . . . As for the hurtful creatures, we are either punished, or disciplined, or terrified by them, so that we may not cherish and love this life." As to the "superfluous animals," he says, "Although they are not necessary for our service, yet the whole design of the universe is thereby completed and finished." Luther, who followed St. Augustine in so many other matters, declined to follow him fully in this. To him a fly was not merely superfluous, it was noxious—sent by the devil to vex him when reading.

Another subject which gave rise to much searching of Scripture and long trains of theological reasoning was the difference between the creation of man and that of other living beings.

Great stress was laid by theologians, from St. Basil and St. Augustine to St. Thomas Aquinas and Bossuet, and from Luther to Wesley, on the radical distinction indicated in Genesis, God having created man "in his own image." What this statement meant was seen in the light of the later biblical statement that "Adam begat Seth in his own likeness, after his image."

In view of this and of well-known texts incorporated from older creation legends into the Hebrew sacred books it came to be widely held that, while man was directly moulded and fashioned separately by the Creator's hand, the animals generally were evoked in numbers from the earth and sea by the Creator's voice.

A question now arose naturally as to the *distinctions of species* among animals. The vast majority of theologians agreed in representing all animals as created "in the beginning," and named by Adam, preserved in the ark, and continued ever afterward under exactly the same species. This belief ripened into a dogma. Like so many other dogmas in the Church, Catholic and Protestant, its real origins are to be found rather in pagan philosophy than in the Christian Scriptures; it came far more from Plato and Aristotle than from Moses and St. Paul. But this was not considered: more and more it became necessary to believe that each and every difference of species was impressed by the Creator "in the beginning," and that no change had taken place or could have taken place since.

Some difficulties arose here and there as zoölogy progressed and revealed ever-increasing numbers of species; but through the Middle Ages, and indeed long after the Reformation, these difficulties were easily surmounted by making the ark of Noah larger and larger, and especially by holding that there had been a human error in regard to its measurement.*

But naturally there was developed among both ecclesiastics and laymen a human desire to go beyond these special points in the history of animated beings—a desire to know what the creation really *is*.

Current legends, stories, and travellers' observations, poor as they were, tended powerfully to stimulate curiosity in this field.

Three centuries before the Christian era Aristotle had made the first really great attempt to satisfy this curiosity, and had begun a development of studies in natural history which remains one of the leading achievements in the story of our race.

---

* For St. Augustine, see *De Genesi* and *De Trinitate*, *passim*; for Bede, see *Hexæmeron*, lib. i, in Migne, tome xci, pp. 21, 36–38, 42; and *De Sex Dierum Creatione*, in Migne, tome xciii, p. 215; for Peter Lombard on "noxious animals," see his *Sententiæ*, lib. ii, dist. xv, 3, Migne, tome cxcii, p. 682; for Wesley, Clarke, and Watson, see quotations from them and notes thereto in my chapter on *Geology*; for St. Augustine on "superfluous animals," see the *De Genesi*, lib. i, cap. xvi, 26; on Luther's view of flies, see the *Table Talk* and his famous utterance, "*Odio muscas quia sunt imagines diaboli et hæreticorum*"; for the agency of Aristotle and Plato in fastening the belief in the fixity of species into Christian theology, see Sachs, *Geschichte der Botanik*, München, 1875, p. 107 and note, also p. 113.

But the feeling which we have already seen so strong in the early Church—that all study of Nature was futile in view of the approaching end of the world—indicated so clearly in the New Testament and voiced so powerfully by Lactantius and St. Augustine—held back this current of thought for many centuries. Still, the better tendency in humanity continued to assert itself. There was, indeed, an influence coming from the Hebrew Scriptures themselves which wrought powerfully to this end; for, in spite of all that Lactantius or St. Augustine might say as to the futility of any study of Nature, the grand utterances in the Psalms regarding the beauties and wonders of creation, in all the glow of the truest poetry, ennobled the study even among those whom logic drew away from it.

But, as a matter of course, in the early Church and throughout the Middle Ages all such studies were cast in a theologic mould. Without some purpose of biblical illustration or spiritual edification they were considered futile; too much prying into the secrets of Nature was very generally held to be dangerous both to body and soul; only for showing forth God's glory and his purposes in the creation were such studies praiseworthy. The great work of Aristotle was under eclipse. The early Christian thinkers gave little attention to it, and that little was devoted to transforming it into something absolutely opposed to his whole spirit and method; in place of it they developed the *Physiologus* and the *Bestiaries*, mingling scriptural statements, legends of the saints, and fanciful inventions with pious intent and childlike simplicity. In place of research came authority—the authority of the Scriptures as interpreted by the *Physiologus* and the *Bestiaries*—and these remained the principal source of thought on animated Nature for over a thousand years.

Occasionally, indeed, fear was shown among the rulers in the Church, even at such poor prying into the creation as this, and in the fifth century a synod under Pope Gelasius administered a rebuke to the *Physiologus*; but the interest in Nature was too strong: the great work on *Creation* by St. Basil had drawn from the *Physiologus* precious illustrations of Holy Writ, and the strongest of the early popes, Gregory the Great, virtually sanctioned it.

Thus was developed a sacred science of creation and of the divine purpose in Nature, which went on developing from the fourth century to the nineteenth—from St. Basil to St. Isidore of Seville, from Isidore to Vincent of Beauvais, and from Vincent to Archdeacon Paley and the Bridgewater Treatises.

Like all else in the Middle Ages, this sacred science was developed purely by theological methods. Neglecting the wonders which the dissection of the commonest animals would have afforded them, these naturalists attempted to throw light into Nature by ingenious use of scriptural texts, by research among the lives of the saints, and by the plentiful application of metaphysics. Hence even such strong men as St. Isidore of Seville treasured up accounts of the unicorn and dragons mentioned in the Scriptures and of the phœnix and basilisk in profane writings. Hence such contributions to knowledge as that the basilisk kills serpents by his breath and men by his glance, that the lion when pursued effaces his tracks with the end of his tail, that the pelican nourishes her young with her own blood, that serpents lay aside their venom before drinking, that the salamander quenches fire, that the hyena can talk with shepherds, that certain birds are born of the fruit of a certain tree when it happens to fall into the water, with other masses of science equally valuable.

As to the method of bringing science to bear on Scripture, the *Physiologus* gives an example, illustrating the passage in the book of Job which speaks of the old lion perishing for lack of prey. Out of the attempt to explain an unusual Hebrew word in the text there came a curious development of error, until we find fully evolved an account of the "ant-lion," which, it gives us to understand, was the lion mentioned by Job, and it says: "As to the ant-lion, his father hath the shape of a lion, his mother that of an ant; the father liveth upon flesh and the mother upon herbs; these bring forth the ant-lion, a compound of both and in part like to either; for his fore part is like that of a lion and his hind part like that of an ant. Being thus composed, he is neither able to eat flesh like his father nor herbs like his mother, and so he perisheth."

In the middle of the thirteenth century we have a triumph of this theological method in the great work of the English Franciscan Bartholomew on *The Properties of Things*. The theological method as applied to science consists largely in accepting tradition and in spinning arguments to fit it. In this field Bartholomew was a master. Having begun with the intent mainly to explain the allusions in Scripture to natural objects, he soon rises logically into a survey of all Nature. Discussing the "cockatrice" of Scripture, he tells us: "He drieth and burneth leaves with his touch, and he is of so great venom and perilous that he slayeth and wasteth him that nigheth him without tarrying; and yet the weasel overcometh him, for the biting of the weasel is death to the cockatrice. Nevertheless the biting of the cockatrice is death to the weasel if the weasel eat not rue before. And though the cockatrice be venomous without remedy while he is alive, yet he looseth all the malice when he is burnt to ashes. His ashes be accounted profitable in working of alchemy, and namely in turning and changing of metals."

Bartholomew also enlightens us on the animals of Egypt, and says, "If the crocodile findeth a man by the water's brim he slayeth him, and then he weepeth over him and swalloweth him."

Naturally this good Franciscan naturalist devotes much thought to the "dragons" mentioned in Scripture. He says: "The dragon is most greatest of all serpents, and oft he is drawn out of his den and riseth up into the air, and the air is moved by him, and also the sea swelleth against his venom, and he hath a crest, and reareth his tongue, and hath teeth like a saw, and hath strength, and not only in teeth but in tail, and grieveth with biting and with stinging. Whom he findeth he slayeth. Oft four or five of them fasten their tails together and rear up their heads, and sail over the sea to get good meat. Between elephants and dragons is everlasting fighting; for the dragon with his tail spanneth the elephant, and the elephant with his nose throweth down the dragon. . . . The cause why the dragon desireth his blood is the coldness thereof, by the which the dragon desireth to cool himself. Jerome saith that the dragon is a full thirsty beast, insomuch that he openeth his mouth against the wind to quench the burning of his thirst in that wise. Therefore, when he seeth ships in great wind he flieth against the sail to take the cold wind, and overthroweth the ship."

These ideas of Friar Bartholomew spread far and struck deep into the popular mind. His book was translated into the principal languages of Europe, and was one of those most generally read during the Ages of Faith. It maintained its position nearly three hundred years; even after the invention of printing it held its own, and in the fifteenth century there were issued no less than ten editions of it in Latin, four in French, and various versions of it in Dutch, Spanish, and English. Preachers found it especially useful in illustrating the ways of God to man. It was only when the great voyages of discovery substituted ascertained fact for theological reasoning in this province that its authority was broken.

The same sort of science flourished in the *Bestiaries*, which were used everywhere, and especially in the pulpits, for the edification of the faithful. In all of these, as in that compiled early in the thirteenth century by an ecclesiastic, William of Normandy, we have this lesson, borrowed from the *Physiologus*: "The lioness giveth birth to cubs which remain three days without life. Then cometh the lion, breatheth upon them, and bringeth them to life. . . . Thus it is that Jesus Christ during three days was deprived of life, but God the Father raised him gloriously."

Pious use was constantly made of this science, especially by monkish preachers. The phœnix rising from his ashes proves the doctrine of the resurrection; the structure and mischief of monkeys proves the existence of demons; the fact that certain monkeys have no tails proves that Satan has been shorn of his glory; the weasel, which "constantly changes its place, is a type of the man estranged from the word of God, who findeth no rest."

The moral treatises of the time often took the form of works on natural history, in order the more fully to exploit these religious teachings of Nature. Thus from the book *On Bees*, of the Dominican Thomas of Cantimpré, we learn that "wasps persecute bees and make war on them out of natural hatred"; and these, he tells us, typify the demons who dwell in the air and with lightning and tempest assail and vex mankind—whereupon he fills a long chapter with anecdotes of such demonic warfare on mortals. In like manner his fellow-Dominican, the inquisitor Nider, in his book *The Ant Hill*, teaches us that the ants in Ethiopia, which are said to have horns and to grow so large as to look like dogs, are emblems of atrocious heretics, like Wyclif and the Hussites, who bark and bite against the truth; while the ants of India, which dig up gold out of the sand with their feet and hoard it, though they make no use of it, symbolize the fruitless toil with which the heretics dig out the gold of Holy Scripture and hoard it in their books to no purpose.

This pious spirit not only pervaded science; it bloomed out in art, and especially in the cathedrals. In the gargoyles overhanging the walls, in the grotesques clambering about the towers or perched upon pinnacles, in the dragons prowling under archways or lurking in bosses of foliage, in the apocalyptic beasts carved upon the stalls of the choir, stained into the windows, wrought into the tapestries, illuminated in the letters and borders of psalters and missals, these marvels of creation suggested everywhere morals from the Physiologus, the Bestiaries, and the Exempla.*

* For the *Physiologus*, *Bestiaries*, etc., see Berger de Xivrey, *Traditions Tératologiques*; also Hippeau's edition of the *Bestiaire de Guillaume de Normandie*, Caen, 1852, and such mediæval books of Exempla as the *Lumen Naturæ;* also Hoefer, *Histoire de la Zoologie*; also Rambaud, *Histoire de la Civilisation Française*, Paris, 1885, vol. i, pp. 368, 369; also Cardinal Pitra, preface to the *Spicilegium Solismense*, Paris, 1885. *passim*; also Carus, *Geschichte der Zoologie;* and, for an admirable summary, the article *Physiologus* in the *Encyclopædia Britannica*. In the illuminated manuscripts in the Library of Cornell University are some very striking examples of grotesques. For admirably illustrated articles on the *Bestiaries*, see Cahier and Martin, *Mélanges d'Archéologie*, Paris, 1851, 1852, and 1856, vol. ii of the first series, pp. 85–232, and second series, volume on *Curiosités Mystérieuses*, pp. 106–164; also J. R. Allen, *Early Christian Symbolism in Great Britain and Ireland* (London, 1887), lecture vi; for an exhaustive discussion of the subject, see *Das Thierbuch des normannischen Dichters Guillaume le Clerc*, herausgegeben von Reinisch, Leipsic, 1890; and, for an Italian example, Goldstaub und Wendriner, *Ein Tosco-Venezianischer Bestiarius*, Halle, 1892, where is given, on pp. 369–371, a very pious but very comical tradition regarding the beaver, hardly mentionable to ears polite. For Friar Bartholomew, see (besides his book itself) *Medieval Lore*, edited by Robert Steele, London, 1893, pp. 118–138.

Here and there among men who were free from church control we have work of a better sort. In the twelfth and thirteenth centuries Abd Allatif made observations upon the natural history of Egypt which showed a truly scientific spirit, and the Emperor Frederick II attempted to promote a more fruitful study of Nature; but one of these men was abhorred as a Mussulman and the other as an infidel. Far more in accordance with the spirit of the time was the ecclesiastic Giraldus Cambrensis, whose book on the topography of Ireland bestows much attention upon the animals of the island, and rarely fails to make each contribute an appropriate moral. For example, he says that in Ireland "eagles live for so many ages that they seem to contend with eternity itself; so also the saints, having put off the old man and put on the new, obtain the blessed fruit of everlasting life." Again, he tells us: "Eagles often fly so high that their wings are scorched by the sun; so those who in the Holy Scriptures strive to unravel the deep and hidden secrets of the heavenly mysteries, beyond what is allowed, fall below, as if the wings of the presumptuous imaginations on which they are borne were scorched."

In one of the great men of the following century appeared a gleam of healthful criticism: Albert the Great, in his work on the animals, dissents from the widespread belief that certain birds spring from trees and are nourished by the sap, and also from the theory that some are generated in the sea from decaying wood.

But it required many generations for such scepticism to produce much effect, and we find among the illustrations in an edition of Mandeville published just before the Reformation not only careful accounts but pictured representations both of birds and of beasts produced in the fruit of trees.*

This general employment of natural science for pious purposes went on after the Reformation. Luther frequently made this use of it, and his example controlled his followers.

* For Giraldus Cambrensis, see the edition in the Bohn Library, London, 1863, p. 30; for Abd Allatif and Frederick II, see Hoefer, as above; for Albertus Magnus, see the *De Animalibus*, lib. xxiii; for the illustrations in Mandeville, see the Strasburg edition, 1484; for the history of the myth of the tree which produces birds, see Max Müller's *Lectures on the Science of Language*, second series, lect. xii.

In 1612, Wolfgang Franz, Professor of Theology at Luther's university, gave to the world his sacred history of animals, which went through many editions. It contained a very ingenious classification, describing "natural dragons," which have three rows of teeth to each jaw, and he piously adds, "the principal dragon is the Devil."

Near the end of the same century, Father Kircher, the great Jesuit professor at Rome, holds back the sceptical current, insists upon the orthodox view, and represents among the animals entering the ark sirens and griffins.

Yet even among theologians we note here and there a sceptical spirit in natural science. Early in the same seventeenth century Eugène Roger published his *Travels in Palestine.* As regards the utterances of Scripture he is soundly orthodox: he prefaces his work with a map showing, among other important points referred to in biblical history, the place where Samson slew a thousand Philistines with the jawbone of an ass, the cavern which Adam and Eve inhabited after their expulsion from paradise, the spot where Balaam's ass spoke, the place where Jacob wrestled with the angel, the steep place down which the swine possessed of devils plunged into the sea, the position of the salt statue which was once Lot's wife, the place at sea where Jonah was swallowed by the whale, and "the exact spot where St. Peter caught one hundred and fifty-three fishes."

As to natural history, he describes and discusses with great theological acuteness the basilisk. He tells us that the animal is about a foot and a half long, is shaped like a crocodile, and kills people with a single glance. The one which he saw was dead, fortunately for him, since in the time of Pope Leo IV—as he tells us—one appeared in Rome and killed many people by merely looking at them; but the Pope destroyed it with his prayers and the sign of the cross. He informs us that Providence has wisely and mercifully protected man by requiring the monster to cry aloud two or three times whenever it leaves its den, and that the divine wisdom in creation is also shown by the fact that the monster is obliged to look its victim in the eye, and at a certain fixed distance, before its glance can penetrate the victim's brain and so pass to his heart. He also gives a reason for supposing that the same divine mercy has provided that the crowing of a cock will kill the basilisk.

Yet even in this good and credulous missionary we see the influence of Bacon and the dawn of experimental science; for, having been told many stories regarding the salamander, he secured one, placed it alive upon the burning coals, and reports to us that the legends concerning its power to live in the fire are untrue. He also tried experiments with the chameleon, and found that the stories told of it were to be received with much allowance: while, then, he locks up his judgment whenever he discusses the letter of Scripture, he uses his mind in other things much after the modern method.

In the second half of the same century Hottinger, in his *Theological Examination of the History of Creation*, breaks from the belief in the phœnix; but his scepticism is carefully kept within the limits imposed by Scripture. He avows his doubts, first, "because God created the animals in couples, while the phœnix is represented as a single, unmated creature"; secondly, "because Noah, when he entered the ark, brought the animals in by sevens, while there were never so many individuals of the phœnix species"; thirdly, because "no man is known who dares assert that he has ever seen this bird"; fourthly, because "those who assert there is a phœnix differ among themselves."

In view of these attacks on the salamander and the phœnix, we are not surprised to find, before the end of the century, scepticism regarding the basilisk: the eminent Prof. Kirchmaier, at the University of Wittenberg, treats phœnix and basilisk alike as old wives' fables. As to the phœnix, he denies its existence, not only because Noah took no such bird into the ark, but also because, as he pithily remarks, "birds come from eggs, not from ashes." But the unicorn he can not resign, nor will he even concede that the unicorn is a rhinoceros; he appeals to Job and to Marco Polo to prove that this animal, as usually conceived, really exists, and says, "Who would not fear to deny the existence of the unicorn, since Holy Scripture names him with distinct praises?" As to the other great animals mentioned in Scripture, he is so rationalistic as to admit that behemoth was an elephant and leviathan a whale.

But these germs of a fruitful scepticism grew, and we soon find Dannhauer going a step further and declaring his disbelief even in the unicorn, insisting that it was a rhinoceros—only that and nothing more. Still, the main current continued strongly theological. In 1712 Samuel Bochart published his great work upon the animals of Holy Scripture. As showing its spirit we may take the titles of the chapters on the horse:

"Chapter VI. Of the Hebrew Name of the Horse."

"Chapter VII. Of the Colours of the Six Horses in Zechariah."

"Chapter VIII. Of the Horses in Job."

"Chapter IX. Of Solomon's Horses, and of the Texts wherein the Writers praise the Excellence of Horses."

"Chapter X. Of the Consecrated Horses of the Sun."

Among the other titles of chapters are such as: Of Balaam's Ass; Of the Thousand Philistines slain by Samson with the Jawbone of an Ass; Of the Golden Calves of Aaron and Jeroboam; Of the Bleating, Milk, Wool, External and Internal Parts of Sheep mentioned in Scripture; Of Notable Things told regarding Lions in Scripture; Of Noah's Dove and of the Dove which appeared at Christ's Baptism. Mixed up in the book, with the principal mass drawn from Scripture, were many facts and reasonings taken from investigations by naturalists; but all were permeated by the theological spirit.*

The inquiry into Nature having thus been pursued nearly two thousand years theologically, we find by the middle of the sixteenth century some promising beginnings of a different method—the method of inquiry into Nature scientifically—the method which seeks not plausibilities but facts. At

* For Franz and Kircher, see Perrier, *La Philosophie Zoologique avant Darwin*, Paris, 1884, p. 29; for Roger, see his *La Terre Saincte*, Paris, 1664, pp. 89–92, 130, 218, etc.; for Hottinger, see his *Historiæ Creationis Examen theologico-philologicum*, Heidelberg, 1659, lib. vi, quæst. lxxxiii; for Kirchmaier, see his *Disputationes Zoologicæ* (published collectively after his death), Jena, 1736; for Dannhauer, see his *Disputationes Theologicæ*, Leipsic, 1707, p. 14; for Bochart, see his *Hierozoikon, sive De Animalibus Sacræ Scripturæ*, Leyden, 1712.

that time Edward Wotton led the way in England and Conrad Gesner on the Continent, by observations widely extended, carefully noted, and thoughtfully classified.

This better method of interrogating Nature soon led to the formation of societies for the same purpose. In 1560 was founded an Academy for the Study of Nature at Naples, but theologians, becoming alarmed, suppressed it, and for nearly one hundred years there was no new combined effort of that sort, until in 1645 began the meetings in London of what was afterward the Royal Society. Then came the Academy of Sciences in France, and the Accademia del Cimento in Italy; others followed in all parts of the world, and a great new movement was begun.

Theologians soon saw a danger in this movement. In Italy, Prince Leopold de' Medici, a protector of the Florentine Academy, was bribed with a cardinal's hat to neglect it, and from the days of Urban VIII to Pius IX a similar spirit was there shown. In France, there were frequent ecclesiastical interferences, of which Buffon's humiliation for stating a simple scientific truth was a noted example. In England, Protestantism was at first hardly more favourable toward the Royal Society, and the great Dr. South denounced it in his sermons as irreligious.

Fortunately, one thing prevented an open breach between theology and science: while new investigators had mainly given up the mediæval method so dear to the Church, they had very generally retained the conception of direct creation and of design throughout creation—a design having as its main purpose the profit, instruction, enjoyment, and amusement of man.

On this the naturally opposing tendencies of theology and science were compromised. Science, while somewhat freed from its old limitations, became the handmaid of theology in illustrating the doctrine of creative design, and always with apparent deference to the Chaldean and other ancient myths and legends embodied in the Hebrew sacred books.

About the middle of the seventeenth century came a great victory of the scientific over the theologic method. At that time Francesco Redi published the results of his inquiries into the doctrine of spontaneous generation. For ages a widely accepted doctrine had been that water, filth, and carrion had received power from the Creator to generate worms, insects, and a multitude of the smaller animals; and this doctrine had been especially welcomed by St. Augustine and many of the fathers, since it relieved the Almighty of making, Adam of naming, and Noah of living in the ark with these innumerable despised species. But to this fallacy Redi put an end. By researches which could not be gainsaid, he showed that every one of these animals came from an egg; each, therefore, must be the lineal descendant of an animal created, named, and preserved from "the beginning."

Similar work went on in England, but under more distinctly theological limitations. In the same seventeenth century a very famous and popular English book was published by the naturalist John Ray, a fellow of the Royal Society, who produced a number of works on plants, fishes, and birds; but the most widely read of all was entitled *The Wisdom of God manifested in the Works of Creation.* Between the years 1691 and 1827 it passed through nearly twenty editions.

Ray argued the goodness and wisdom of God from the adaptation of the animals not only to man's uses but to their own lives and surroundings.

In the first years of the eighteenth century Dr. Nehemiah Grew, of the Royal Society, published his *Cosmologia Sacra* to refute anti-scriptural opinions by producing evidences of creative design. Discussing "the ends of Providence," he says, "A crane, which is scurvy meat, lays but two eggs in the year, but a pheasant and partridge, both excellent meat, lay and hatch fifteen or twenty." He points to the fact that "those of value which lay few at a time sit the oftener, as the woodcock and the dove." He breaks decidedly from the doctrine that noxious things in Nature are caused by sin, and shows that they, too, are useful; that, "if nettles sting, it is to secure an excellent medicine for children and cattle"; that, "if the bramble hurts man, it makes all the better hedge"; and that, "if it chances to prick the owner, it tears the thief." "Weasels, kites, and other hurtful animals induce us to watchfulness; thistles and moles, to good husbandry; lice oblige us to cleanliness in our bodies, spiders in our houses, and the moth in our clothes." This very optimistic view, triumphing over the theological theory of noxious animals and plants as effects of sin, which prevailed with so much force from St. Augustine to Wesley, was developed into nobler form during the century by various thinkers, and especially by Archdeacon Paley, whose *Natural Theology* exercised a powerful influence down to recent times. The same tendency appeared in other countries, though various philosophers showed weak points in the argument, and Goethe made sport of it in a noted verse, praising the forethought of the Creator in foreordaining the cork tree to furnish stoppers for wine-bottles.

Shortly before the middle of the nineteenth century the main movement culminated in the *Bridgewater Treatises.* Pursuant to the will of the eighth Earl of Bridgewater, the President of the Royal Society selected eight persons, each to receive a thousand pounds sterling for writing and publishing a treatise on the "power, wisdom, and goodness of God, as manifested in the creation." Of these, the leading essays in regard to animated Nature were those of Thomas Chalmers, on *The Adaptation of External Nature to the Moral and Intellectual Condition of Man*; of Sir Charles Bell, on *The Hand as evincing Design*; of Roget, on *Animal and Vegetable Physiology with reference to Natural Theology*; and of Kirby, on *The Habits and Instincts of Animals with reference to Natural Theology.*

Besides these there were treatises by Whewell, Buckland, Kidd, and Prout. The work was well done. It was a marked advance on all that had appeared before, in matter, method, and spirit. Looking back upon it now we can see that it was provisional, but that it was none the less fruitful in truth, and we may well remember Darwin's remark on the stimulating effect of mistaken *theories*, as compared with the sterilizing effect of mistaken *observations:* mistaken observations lead men astray, mistaken theories suggest true theories.

An effort made in so noble a spirit certainly does not deserve the ridicule that, in our own day, has sometimes been lavished upon it. Curiously, indeed, one of the most contemptuous of these criticisms has been recently made by one of the most strenuous defenders of orthodoxy. No less eminent a standard-bearer of the faith than the Rev. Prof. Zoeckler says of this movement to demonstrate creative purpose and design, and of the men who took part in it, "The earth appeared in their representation of it like a great clothing shop and soup kitchen, and God as a glorified rationalistic professor." Such a statement as this is far from just to the conceptions of such men as Butler, Paley, and Chalmers, no matter how fully the thinking world has now outlived them.*

But, noble as the work of these men was, the foundation of fact on which they reared it became evidently more and more insecure.

For as far back as the seventeenth century acute theologians had begun to discern difficulties more serious than any that had before confronted them. More and more it was seen that the number of different species was far greater than the world had hitherto imagined. Greater and greater had become the old difficulty in conceiving that, of these innumerable species, each had been specially created by the Almighty hand; that each had been brought before Adam by the Almighty to be named; and that each, in couples or in sevens, had been gathered by Noah into the ark. But the difficulties thus suggested were as nothing compared to those raised by the *distribution* of animals.

Even in the first days of the Church this had aroused serious thought, and above all in the great mind of St.

---

* For a very valuable and interesting study on the old idea of the generation of insects from carrion, see Osten-Sacken, *On the Oxen-born Bees of the Ancients*, Heidelberg, 1894; for Ray, see the work cited, London, 1827, p. 153; for Grew, see *Cosmologia Sacra, or a Discourse on the Universe, as it is the Creature and Kingdom of God; chiefly written to demonstrate the Truth and Excellency of the Bible*, by Dr. Nehemiah Grew, Fellow of the College of Physicians and of the Royal Society, London, 1701; for Paley and the Bridgewater Treatises, see the usual editions; also Lange, *History of Rationalism*. Goethe's couplet ran as follows:

> "Welche Verehrung verdient der Weltenerschöpfer, der Gnädig,
> Als er den Korkbaum erschuf, gleich auch die Stopfel erfand."

For the quotation from Zoeckler, see his work already cited, vol. ii, pp. 74, 440.

Augustine. In his *City of God* he had stated the difficulty as follows: "But there is a question about all these kinds of beasts, which are neither tamed by man, nor spring from the earth like frogs, such as wolves and others of that sort, . . . as to how they could find their way to the islands after that flood which destroyed every living thing not preserved in the ark. . . . Some, indeed, might be thought to reach islands by swimming, in case these were very near; but some islands are so remote from continental lands that it does not seem possible that any creature could reach them by swimming. It is not an incredible thing, either, that some animals may have been captured by men and taken with them to those lands which they intended to inhabit, in order that they might have the pleasure of hunting; and it can not be denied that the transfer may have been accomplished through the agency of angels, commanded or allowed to perform this labour by God."

But this difficulty had now assumed a magnitude of which St. Augustine never dreamed. Most powerful of all agencies to increase it were the voyages of Columbus, Vasco da Gama, Magellan, Amerigo Vespucci, and other navigators of the period of discovery. Still more serious did it become as the great islands of the southern seas were explored. Every navigator brought home tidings of new species of animals and of races of men living in parts of the world where the theologians, relying on the statement of St. Paul that the gospel had gone into all lands, had for ages declared there could be none; until finally it overtaxed even the theological imagination to conceive of angels, in obedience to the divine command, distributing the various animals over the earth, dropping the megatherium in South America, the archeopteryx in Europe, the ornithorhynchus in Australia, and the opossum in North America.

The first striking evidence of this new difficulty was shown by the eminent Jesuit missionary, Joseph Acosta. In his *Natural and Moral History of the Indies*, published in 1590, he proved himself honest and lucid. Though entangled in most of the older scriptural views, he broke away from many; but the distribution of animals gave him great trouble. Having shown the futility of St. Augustine's other explanations, he quaintly asks: "Who can imagine that in so long a voyage men woulde take the paines to carrie Foxes to Peru, especially that kinde they call 'Acias,' which is the filthiest I have seene? Who woulde likewise say that they have carried Tygers and Lyons? Truly it were a thing worthy the laughing at to thinke so. It was sufficient, yea, very much, for men driven against their willes by tempest, in so long and unknowne a voyage, to escape with their owne lives, without busying themselves to carrie Woolves and Foxes, and to nourish them at sea."

It was under the impression made by this new array of facts that in 1667 Abraham Milius published at Geneva his book on *The Origin of Animals and the Migrations of Peoples*. This book shows, like that of Acosta, the shock and strain to which the discovery of America subjected the received theological scheme of things. It was issued with the special approbation of the Bishop of Salzburg, and it indicates the possibility that a solution of the whole trouble may be found in the text, "Let the earth bring forth the living creature after his kind." Milius goes on to show that the ancient philosophers agree with Moses, and that "the earth and the waters, and especially the heat of the sun and of the genial sky, together with that slimy and putrid quality which seems to be inherent in the soil, may furnish the origin for fishes, terrestrial animals, and birds." On the other hand, he is very severe against those who imagine that man can have had the same origin with animals. But the subject with which Milius especially grapples is the *distribution* of animals. He is greatly exercised by the many species found in America and in remote islands of the ocean—species entirely unknown in the other continents—and of course he is especially troubled by the fact that these species existing in those exceedingly remote parts of the earth do not exist in the neighbourhood of Mount Ararat. He confesses that to explain the distribution of animals is the most difficult part of the problem. If it be urged that birds could reach America by flying and fishes by swimming, he asks, "What of the beasts which neither fly nor swim?" Yet even as to the birds he asks, "Is there not an infinite variety of winged creatures who fly so slowly and heavily, and have such a horror of the water, that they would not even dare trust themselves to fly over a wide river?" As to fishes, he says, "They are very averse to wandering from their native waters," and he shows that there are now reported many species of American and East Indian fishes entirely unknown on the other continents, whose presence, therefore, can not be explained by any theory of natural dispersion.

Of those who suggest that land animals may have been dispersed over the earth by the direct agency of man for his use or pleasure he asks: "Who would like to get different sorts of lions, bears, tigers, and other ferocious and noxious creatures on board ship? who would trust himself with them? and who would wish to plant colonies of such creatures in new, desirable lands?"

His conclusion is that plants and animals take their origin in the lands wherein they are found; an opinion which he supports by quoting from the two narrations in Genesis passages which imply generative force in earth and water.

But in the eighteenth century matters had become even worse for the theological view. To meet the difficulty the eminent Benedictine, Dom Calmet, in his *Commentary*, expressed the belief that all the species of a genus had originally formed one species, and he dwelt on this view as one which enabled him to explain the possibility of gathering all animals into the ark. This idea, dangerous as it was to the fabric of orthodoxy, and involving a profound separation from the general doctrine of the Church, seems to have been abroad among thinking men, for we find in the latter half of the same century even Linnæus inclining to consider it. It was time, indeed, that some new theological theory be evolved; the great Linnæus himself, in spite of his famous declaration favouring the fixity of species, had dealt a deathblow to the old theory. In his *Systema Naturæ*, published in the middle of the eighteenth century, he had enumerated four thousand species of animals, and the difficulties involved in the naming of each of them by Adam and in bringing them together in the ark appeared to all thinking men more and more insurmountable.

What was more embarrassing, the number of distinct species went on increasing rapidly, indeed enormously, until, as an eminent zoological authority of our own time has declared, "for every one of the species enumerated by Linnæus, more than fifty kinds are known to the naturalist of to-day, and the number of species still unknown doubtless far exceeds the list of those recorded."

Already there were premonitions of the strain made upon Scripture by requiring a hundred and sixty distinct miraculous interventions of the Creator to produce the hundred and sixty species of land shells found in the little island of Madeira alone, and fourteen hundred distinct interventions to produce the actual number of distinct species of a single well-known shell.

Ever more and more difficult, too, became the question of the geographical distribution of animals. As new explorations were made in various parts of the world, this danger to the theological view went on increasing. The sloths in South America suggested painful questions: How could animals so sluggish have got away from the neighbourhood of Mount Ararat so completely and have travelled so far?

The explorations in Australia and neighbouring islands made matters still worse, for there was found in those regions a whole realm of animals differing widely from those of other parts of the earth.

The problem before the strict theologians became, for example, how to explain the fact that the kangaroo can have been in the ark and be now only found in Australia: his saltatory powers are indeed great, but how could he by any series of leaps have sprung across the intervening mountains, plains, and oceans to that remote continent? and, if the theory were adopted that at some period a causeway extended across the vast chasm separating Australia from the nearest mainland, why did not lions, tigers, camels, and camelopards force or find their way across it?

The theological theory, therefore, had by the end of the eighteenth century gone to pieces. The wiser theologians waited; the unwise indulged in exhortations to "root out the wicked heart of unbelief," in denunciation of "science falsely so called," and in frantic declarations that "the Bible is true"—by which they meant that the limited understanding of it which they had happened to inherit is true.

By the middle of the nineteenth century the whole theological theory of creation—though still preached everywhere as a matter of form—was clearly seen by all thinking men to be hopelessly lost: such strong men as Cardinal Wiseman in the Roman Church, Dean Buckland in the Anglican, and Hugh Miller in the Scottish Church, made heroic efforts to save something from it, but all to no purpose. That sturdy Teutonic and Anglo-Saxon honesty, which is the best legacy of the Middle Ages to Christendom, asserted itself in the old strongholds of theological thought, the universities. Neither the powerful logic of Bishop Butler nor the nimble reasoning of Archdeacon Paley availed. Just as the line of astronomical thinkers from Copernicus to Newton had destroyed the old astronomy, in which the earth was the centre, and the Almighty sitting above the firmament the agent in moving the heavenly bodies about it with his own hands, so now a race of biological thinkers had destroyed the old idea of a Creator minutely contriving and fashioning all animals to suit the needs and purposes of man. They had developed a system of a very different sort, and this we shall next consider.*

III. THEOLOGICAL AND SCIENTIFIC THEORIES OF AN EVOLUTION IN ANIMATED NATURE.

WE have seen, thus far, how there came into the thinking of mankind upon the visible universe and its inhabitants the idea of a creation virtually instantaneous and complete, and of a Creator in human form with human attributes, who spoke matter into existence literally by the exercise of his throat and lips, or shaped and placed it with his hands and fingers.

We have seen that this view came from far; that it ex-

---

* For Acosta, see his *Historia natural y moral dè las Indias*, Seville, 1590—the quaint English translation is of London, 1604; for Abraham Milius, see his *De Origine Animalium et Migratione Populorum*, Geneva, 1667; also *Kosmos*, 1877, H. 1, S. 36; for Linnæus's declaration regarding species, see the *Philosophia Botanica*, 99, 157; for Calmet and Linnæus, see Zoeckler, vol. ii, p. 237. As to the enormously increasing numbers of species in zoology and botany, see President D. S. Jordan, *Science Sketches*, pp. 176, 177; also, for pithy statement, Laing's *Problems of the Future*, chap. vi.

isted in the Chaldæo-Babylonian and Egyptian civilizations, and probably in others of the earliest date known to us; that its main features passed thence into the sacred books of the Hebrews and then into the early Christian Church, by whose theologians it was developed through the Middle Ages and maintained during the modern period.

But, while this idea was thus developed by a succession of noble and thoughtful men through thousands of years, another conception, to all appearance equally ancient, was developed, sometimes in antagonism to it, sometimes mingled with it—the conception of all living beings as wholly or in part the result of a growth process—of an evolution.

This idea, in various forms, became a powerful factor in nearly all the greater ancient theologies and philosophies. For very widespread among the early peoples who attained to much thinking power was a conception that, in obedience to the divine fiat, a watery chaos produced the earth, and that the sea and land gave birth to their inhabitants.

This is clearly seen in those records of Chaldæo-Babylonian thought deciphered in these latter years, to which reference has already been made. In these we have a watery chaos which, under divine action, brings forth the earth and its inhabitants; first the sea animals and then the land animals—the latter being separated into three kinds, substantially as recorded afterward in the Hebrew accounts. At the various stages in the work the Chaldean Creator pronounces it "beautiful," just as the Hebrew Creator in our own later account pronounces it "good."

In both accounts there is placed over the whole creation a solid, concave firmament; in both, light is created first, and the heavenly bodies are afterward placed "for signs and for seasons"; in both, the number seven is especially sacred, giving rise to a sacred division of time and to much else. It may be added that, with many other features in the Hebrew legends evidently drawn from the Chaldean, the account of the creation in each is followed by a legend regarding "the fall of man" and a deluge, many details of which clearly passed in slightly modified form from the Chaldean into the Hebrew accounts.

It would have been a miracle indeed if these primitive

conceptions, wrought out with so much poetic vigour in that earlier civilization on the Tigris and Euphrates, had failed to influence the Hebrews, who during the most plastic periods of their development were under the tutelage of their Chaldean neighbours. Since the researches of Layard, George Smith, Oppert, Schrader, Jensen, Sayce, and their compeers, there is no longer a reasonable doubt that this ancient view of the world, elaborated if not originated in that earlier civilization, came thence as a legacy to the Hebrews, who wrought it in a somewhat disjointed but mainly monotheistic form into the poetic whole which forms one of the most precious treasures of ancient thought preserved in the book of Genesis.

Thus it was that, while the idea of a simple material creation literally by the hands and fingers or voice of the Creator became, as we have seen, the starting-point of a powerful stream of theological thought, and while this stream was swollen from age to age by contributions from the fathers, doctors, and learned divines of the Church, Catholic and Protestant, there was poured into it this lesser current, always discernible and at times clearly separated from it—a current of belief in a process of evolution.

The Rev. Prof. Sayce, of Oxford, than whom no English-speaking scholar carries more weight in a matter of this kind, has recently declared his belief that the Chaldæo-Babylonian theory was the undoubted source of the similar theory propounded by the Ionic philosopher Anaximander—the Greek thinkers deriving this view from the Babylonians through the Phœnicians; he also allows that from the same source its main features were adopted into both the accounts given in the first of our sacred books, and in this general view the most eminent Christian Assyriologists concur.

It is true that these sacred accounts of ours contradict each other. In that part of the first or Elohistic account given in the first chapter of Genesis the *waters* bring forth fishes, marine animals, and birds (Genesis, i, 20); but in that part of the second or Jehovistic account given in the second chapter of Genesis both the land animals and birds are declared to have been created not out of the water, but "*out of the ground*" (Genesis, ii, 19).

The dialectic skill of the fathers was easily equal to explaining away this contradiction; but the old current of thought, strengthened by both these legends, arrested their attention, and, passing through the minds of a succession of the greatest men of the Church, influenced theological opinion deeply, if not widely, for ages, in favour of an evolution theory.

But there was still another ancient source of evolution ideas. Thoughtful men of the early civilizations which were developed along the great rivers in the warmer regions of the earth noted how the sun-god as he rose in his fullest might caused the water and the rich soil to teem with the lesser forms of life. In Egypt, especially, men saw how under this divine power the Nile slime brought forth "creeping things innumerable." Hence mainly this ancient belief that the animals and man were produced by lifeless matter at the divine command, "in the beginning," was supplemented by the idea that some of the lesser animals, especially the insects, were produced by a later evolution, being evoked after the original creation from various sources, but chiefly from matter in a state of decay.

This crude, early view aided doubtless in giving germs of a better evolution theory to the early Greeks. Anaximander, Empedocles, Anaxagoras, and, greatest of all, Aristotle, as we have seen, developed them, making their way at times by guesses toward truths since established by observation. Aristotle especially, both by speculation and observation, arrived at some results which, had Greek freedom of thought continued, might have brought the world long since to its present plane of biological knowledge; for he reached something like the modern idea of a succession of higher organizations from lower, and made the fruitful suggestion of "a perfecting principle" in Nature.

With the coming in of Christian theology this tendency toward a yet truer theory of evolution was mainly stopped, but the old crude view remained, and as a typical example of it we may note the opinion of St. Basil the Great in the fourth century. Discussing the work of creation, he declares that, at the command of God, "the waters were gifted with productive power"; "from slime and muddy places frogs, flies, and gnats came into being"; and he finally declares that the same voice which gave this energy and quality of productiveness to earth and water shall be similarly efficacious until the end of the world. St. Gregory of Nyssa held a similar view.

This idea of these great fathers of the Eastern Church took even stronger hold on the great father of the Western Church. For St. Augustine, so fettered usually by the letter of the sacred text, broke from his own famous doctrine as to the acceptance of Scripture and spurned the generally received belief of a creative process like that by which a toymaker brings into existence a box of playthings. In his great treatise on *Genesis* he says: "To suppose that God formed man from the dust with bodily hands is very childish. . . . God neither formed man with bodily hands nor did he breathe upon him with throat and lips."

St. Augustine then suggests the adoption of the old emanation or evolution theory, shows that "certain very small animals may not have been created on the fifth and sixth days, but may have originated later from putrefying matter," argues that, even if this be so, God is still their creator, dwells upon such a potential creation as involved in the actual creation, and speaks of animals "whose numbers the after-time unfolded."

In his great treatise on the *Trinity*—the work to which he devoted the best thirty years of his life—we find the full growth of this opinion. He develops at length the view that in the creation of living beings there was something like a growth—that God is the ultimate author, but works through secondary causes; and finally argues that certain substances are endowed by God with the power of producing certain classes of plants and animals.*

* For the Chaldean view of creation, see George Smith, *Chaldean Account of Genesis*, New York, 1876, pp. 14, 15, and 64–86; also Lukas, as above; also Sayce, *Religion of the Ancient Babylonians*, Hibbert Lectures for 1887, pp. 371 and elsewhere; as to the fall of man, Tower of Babel, sacredness of the number seven, etc., see also Delitzsch, appendix to the German translation of Smith, pp. 305 *et seq.*; as to the almost exact adoption of the Chaldean legends into the Hebrew sacred account, see all these, as also Schrader, *Die Keilinschriften und das Alte Testament*, Giessen, 1883, early chapters; also article *Babylonia* in the *Encyclopædia Britannica*; as to the similar approval of creation by the Creator in both accounts,

This idea of a development by secondary causes apart from the original creation was helped in its growth by a theological exigency. More and more, as the organic world was observed, the vast multitude of petty animals, winged creatures, and "creeping things" was felt to be a strain upon the sacred narrative. More and more it became difficult to reconcile the dignity of the Almighty with his work in bringing each of these creatures before Adam to be named; or to reconcile the human limitations of Adam with his work in naming "every living creature"; or to reconcile the dimensions of Noah's ark with the space required for preserving all of them, and the food of all sorts necessary for their sustenance, whether they were admitted by twos, as stated in one scriptural account, or by sevens, as stated in the other.

The inadequate size of the ark gave especial trouble. Origen had dealt with it by suggesting that the cubit was six times greater than had been supposed. Bede explained Noah's ability to complete so large a vessel by supposing that he worked upon it during a hundred years; and, as to the provision of food taken into it, he declared that there was no need of a supply for more than one day, since God could throw the animals into a deep sleep or otherwise miraculously make one day's supply sufficient; he also les-

---

see George Smith, p. 73; as to the migration of the Babylonian legends to the Hebrews, see Schrader, Whitehouse's translation, pp. 44, 45; as to the Chaldæan belief in a solid firmament, while Schrader in 1883 thought it not proved, Jensen in 1890 has found it clearly expressed—see his *Kosmologie der Babylonier*, pp. 9 *et seq.*, also pp. 304–306, and elsewhere. Dr. Lukas in 1893 also fully accepts this view of a Chaldean record of a "firmament"—see *Kosmologie*, pp. 43, etc.; see also Maspero and Sayce, *The Dawn of Civilization*, and for crude early ideas of evolution in Egypt, see ibid., pp. 156 *et seq.*

For the seven-day week among Chaldeans and rest on the seventh day, and the proof that even the name "Sabbath" is of Chaldean origin, see Delitzsch, *Beigaben zu Smith's Chald. Genesis*, pp. 300 and 306; also Schrader; for St. Basil, see *Hexæmeron* and *Homilies* vii–ix; but, for the steadfastness of Basil's view in regard to the immutability of species, see a Catholic writer on *Evolution and Faith* in the *Dublin Review* for July, 1871, p. 13; for citations of St. Augustine on Genesis, see the *De Genesi contra Manichæos*, lib. ii, cap. 14, in Migne, xxxiv, 188,—lib. v, cap. 5 and cap. 23,—and lib. vii, cap. 1; for the citations from his work on the Trinity, see his *De Trinitate*, lib. iii, cap. 8 and 9, in Migne, xlii, 877, 878; for the general subject very fully and adequately presented, see Osborn, *From the Greeks to Darwin*, New York, 1894, chaps. ii and iii.

sened the strain on faith still more by diminishing the number of animals taken into the ark—supporting his view upon Augustine's theory of the later development of insects out of carrion.

Doubtless this theological necessity was among the main reasons which led St. Isidore of Seville, in the seventh century, to incorporate this theory, supported by St. Basil and St. Augustine, into his great encyclopedic work which gave materials for thought on God and Nature to so many generations. He familiarized the theological world still further with the doctrine of secondary creation, giving such examples of it as that "bees are generated from decomposed veal, beetles from horseflesh, grasshoppers from mules, scorpions from crabs," and, in order to give still stronger force to the idea of such transformations, he dwells on the biblical account of Nebuchadnezzar, which appears to have taken strong hold upon mediæval thought in science, and he declares that other human beings had been changed into animals, especially into swine, wolves, and owls.

This doctrine of after-creations went on gathering strength until, in the twelfth century, Peter Lombard, in his theological summary, *The Sentences*, so powerful in moulding the thought of the Church, emphasized the distinction between animals which spring from carrion and those which are created from earth and water; the former he holds to have been created "potentially," the latter "actually."

In the century following, this idea was taken up by St. Thomas Aquinas and virtually received from him its final form. In the *Summa*, which remains the greatest work of mediæval thought, he accepts the idea that certain animals spring from the decaying bodies of plants and animals, and declares that they are produced by the creative word of God either actually or virtually. He develops this view by saying, "Nothing was made by God, after the six days of creation, absolutely new, but it was in some sense included in the work of the six days"; and that "even new species, if any appear, have existed before in certain native properties, just as animals are produced from putrefaction."

The distinction thus developed between creation "causally" or "potentially," and "materially" or "formally," was made much of by commentators afterward. Cornelius a Lapide spread it by saying that certain animals were created not "absolutely," but only "derivatively," and this thought was still further developed three centuries later by Augustinus Eugubinus, who tells us that, after the first creative energy had called forth land and water, light was made by the Almighty, the instrument of all future creation, and that the light called everything into existence.

All this "science falsely so called," so sedulously developed by the master minds of the Church, and yet so futile that we might almost suppose that the great apostle, in a glow of prophetic vision, had foreseen it in his famous condemnation, seems at this distance very harmless indeed; yet, to many guardians of the "sacred deposit of doctrine" in the Church, even so slight a departure from the main current of thought seemed dangerous. It appeared to them like pressing the doctrine of secondary causes to a perilous extent; and about the beginning of the seventeenth century we have the eminent Spanish Jesuit and theologian Suarez denouncing it, and declaring St. Augustine a heretic for his share in it.

But there was little danger to the older idea just then; the main theological tendency was so strong that the world kept on as of old. Biblical theology continued to spin its own webs out of its own bowels, and all the lesser theological flies continued to be entangled in them; yet here and there stronger thinkers broke loose from this entanglement and helped somewhat to disentangle others.*

---

* For Bede's view of the ark and the origin of insects, see his *Hexæmeron*, i and ii; for Isidore, see the *Etymologiæ*, xi, 4, and xiii, 22; for Peter Lombard, see *Sent.*, lib. ii, dist. xv, 4 (in Migne, cxcii, 682); for St. Thomas Aquinas as to the laws of Nature, see *Summa Theologica*, i, *Quæst.* lxvii, art. iv; for his discussion on Avicenna's theory of the origin of animals, see ibid., *Quæst.* lxxi, vol. i, pp. 1184 and 1185, of Migne's edit.; for his idea as to the word of God being the active producing principle, see ibid., i, *Quæst.* lxxi, art. i; for his remarks on species, see ibid., i, *Quæst.* lxxii, art. i; for his ideas on the necessity of the procreation of man, see ibid., i, *Quæst.* lxxii, art. i; for the origin of animals from putrefaction, see ibid., i, *Quæst.* lxxix, art. i, 3; for Cornelius a Lapide on the derivative creation of animals, see his *In Genesim Comment.*, cap. i, cited by Mivart, *Genesis of Species*, p. 282; for a reference to Suarez's denunciation of the view of St. Augustine, see Huxley's *Essays*.

At the close of the Middle Ages, in spite of the devotion of the Reformed Church to the letter of Scripture, the revival of learning and the great voyages gave an atmosphere in which better thinking on the problems of Nature began to gain strength. On all sides, in every field, men were making discoveries which caused the general theological view to appear more and more inadequate.

First of those who should be mentioned with reverence as beginning to develop again that current of Greek thought which the system drawn from our sacred books by the fathers and doctors of the Church had interrupted for more than a thousand years, was Giordano Bruno. His utterances were indeed vague and enigmatical, but this fault may well be forgiven him, for he saw but too clearly what must be his reward for any more open statements. His reward indeed came—even for his faulty utterances—when, toward the end of the nineteenth century, thoughtful men from all parts of the world united in erecting his statue on the spot where he had been burned by the Roman Inquisition nearly three hundred years before.

After Bruno's death, during the first half of the seventeenth century, Descartes seemed about to take the leadership of human thought: his theories, however superseded now, gave a great impulse to investigation then. His genius in promoting an evolution doctrine as regards the mechanical formation of the solar system was great, and his mode of thought strengthened the current of evolutionary doctrine generally; but his constant dread of persecution, both from Catholics and Protestants, led him steadily to veil his thoughts and even to suppress them. The execution of Bruno had occurred in his childhood, and in the midst of his career he had watched the Galileo struggle in all its stages. He had seen his own works condemned by university after university under the direction of theologians, and placed upon the Roman *Index*. Although he gave new and striking arguments to prove the existence of God, and humbled himself before the Jesuits, he was condemned by Catholics and Protestants alike. Since Roger Bacon, perhaps, no great thinker had been so completely abased and thwarted by theological oppression.

Near the close of the same century another great thinker, Leibnitz, though not propounding any full doctrine on evolution, gave it an impulse by suggesting a view contrary to the sacrosanct belief in the immutability of species—that is, to the pious doctrine that every species in the animal kingdom now exists as it left the hands of the Creator, the naming process by Adam, and the door of Noah's ark.

His punishment at the hands of the Church came a few years later, when, in 1712, the Jesuits defeated his attempt to found an Academy of Science at Vienna. The imperial authorities covered him with honours, but the priests—ruling in the confessionals and pulpits—would not allow him the privilege of aiding his fellow-men to ascertain God's truths revealed in Nature.

Spinoza, Hume, and Kant may also be mentioned as among those whose thinking, even when mistaken, might have done much to aid in the development of a truer theory had not the theologic atmosphere of their times been so unpropitious; but a few years after Leibnitz's death came in France a thinker in natural science of much less influence than any of these, who made a decided step forward.

Early in the eighteenth century Benoist de Maillet, a man of the world, but a wide observer and close thinker upon Nature, began meditating especially upon the origin of animal forms, and was led into the idea of the transformation of species and so into a theory of evolution, which in some important respects anticipated modern ideas. He definitely, though at times absurdly, conceived the production of existing species by the modification of their predecessors, and he plainly accepted one of the fundamental maxims of modern geology—that the structure of the globe must be studied in the light of the present course of Nature.

But he fell between two ranks of adversaries. On one side, the Church authorities denounced him as a freethinker; on the other, Voltaire ridiculed him as a devotee. Feeling that his greatest danger was from the orthodox theologians, De Maillet endeavoured to protect himself by disguising his name in the title of his book, and by so wording its preface and dedication that, if persecuted, he could declare it a mere sport of fancy; he therefore announced it as the reverie of a

Hindu sage imparted to a Christian missionary. But this strategy availed nothing: he had allowed his Hindu sage to suggest that the days of creation named in Genesis might be long periods of time; and this, with other ideas of equally fearful import, was fatal. Though the book was in type in 1735, it was not published till 1748—three years after his death.

On the other hand, the heterodox theology of Voltaire was also aroused; and, as De Maillet had seen in the presence of fossils on high mountains a proof that these mountains were once below the sea, Voltaire, recognising in this an argument for the deluge of Noah, ridiculed the new thinker without mercy. Unfortunately, some of De Maillet's vagaries lent themselves admirably to Voltaire's sarcasm; better material for it could hardly be conceived than the theory, seriously proposed, that the first human being was born of a mermaid.

Hence it was that, between these two extremes of theology, De Maillet received no recognition until, very recently, the greatest men of science in England and France have united in giving him his due. But his work was not lost, even in his own day; Robinet and Bonnet pushed forward victoriously on helpful lines.

In the second half of the eighteenth century a great barrier was thrown across this current—the authority of Linnæus. He was the most eminent naturalist of his time, a wide observer, a close thinker; but the atmosphere in which he lived and moved and had his being was saturated with biblical theology, and this permeated all his thinking.

He who visits the tomb of Linnæus to-day, entering the beautiful cathedral of Upsala by its southern porch, sees above it, wrought in stone, the Hebrew legend of creation. In a series of medallions, the Almighty—in human form—accomplishes the work of each creative day. In due order he puts in place the solid firmament with the waters above it, the sun, moon, and stars within it, the beasts, birds, and plants below it, and finishes his task by taking man out of a little hillock of "the earth beneath," and woman out of man's side. Doubtless Linnæus, as he went to his devotions, often smiled at this childlike portrayal. Yet he was never able to break away from the idea it embodied. At times, in face of the difficulties which beset the orthodox theory, he ventured to favour some slight concessions. Toward the end of his life he timidly advanced the hypothesis that all the species of one genus constituted at the creation one species; and from the last edition of his *Systema Naturæ* he quietly left out the strongly orthodox statement of the fixity of each species, which he had insisted upon in his earlier works. But he made no adequate declaration. What he might expect if he openly and decidedly sanctioned a newer view he learned to his cost; warnings came speedily both from the Catholic and Protestant sides.

At a time when eminent prelates of the older Church were eulogizing debauched princes like Louis XV, and using the unspeakably obscene casuistry of the Jesuit Sanchez in the education of the priesthood as to the relations of men to women, the modesty of the Church authorities was so shocked by Linnæus's proofs of a sexual system in plants that for many years his writings were prohibited in the Papal States and in various other parts of Europe where clerical authority was strong enough to resist the new scientific current. Not until 1773 did one of the more broad-minded cardinals —Zelanda—succeed in gaining permission that Prof. Minasi should discuss the Linnæan system at Rome.

And Protestantism was quite as oppressive. In a letter to Eloius, Linnæus tells of the rebuke given to science by one of the great Lutheran prelates of Sweden, Bishop Svedberg. From various parts of Europe detailed statements had been sent to the Royal Academy of Science that water had been turned into blood, and well-meaning ecclesiastics had seen in this an indication of the wrath of God, certainly against the regions in which these miracles had occurred and possibly against the whole world. A miracle of this sort appearing in Sweden, Linnæus looked into it carefully and found that the reddening of the water was caused by dense masses of minute insects. News of this explanation having reached the bishop, he took the field against it; he denounced this scientific discovery as "a Satanic abyss" (*abyssum Satanæ*), and declared "The reddening of the water is *not* natural," and "when God allows such a miracle to take place Satan endeavours, and so do his ungodly, self-reliant, self-sufficient, and worldly tools, to make it signify nothing." In face of this onslaught Linnæus retreated; he tells his correspondent that "it is difficult to say anything in this matter," and shields himself under the statement "It is certainly a miracle that so many millions of creatures can be so suddenly propagated," and "it shows undoubtedly the all-wise power of the Infinite."

The great naturalist, grown old and worn with labours for science, could no longer resist the contemporary theology; he settled into obedience to it, and while the modification of his early orthodox view was, as we have seen, quietly imbedded in the final edition of his great work, he made no special effort to impress it upon the world. To all appearance he continued to adhere to the doctrine that all existing species had been created by the Almighty "in the beginning," and that since "the beginning" no new species had appeared.

Yet even his great authority could not arrest the swelling tide; more and more vast became the number of species, more and more incomprehensible under the old theory became the newly ascertained facts in geographical distribution, more and more it was felt that the universe and animated beings had come into existence by some process other than a special creation "in the beginning," and the question was constantly pressing, "By *what* process?"

Throughout the whole of the eighteenth century one man was at work on natural history who might have contributed much toward an answer to this question: this man was Buffon. His powers of research and thought were remarkable, and his gift in presenting results of research and thought showed genius. He had caught the idea of an evolution in Nature by the variation of species, and was likely to make a great advance with it; but he, too, was made to feel the power of theology.

As long as he gave pleasing descriptions of animals the Church petted him, but when he began to deduce truths of philosophical import the batteries of the Sorbonne were opened upon him; he was made to know that "the sacred deposit of truth committed to the Church" was, that "in the beginning God made the heavens and the earth"; and that "all things were made at the beginning of the world." For his simple statement of truths in natural science which are to-day truisms, he was, as we have seen, dragged forth by the theological faculty, forced to recant publicly, and to print his recantation. In this he announced, "I abandon everything in my book respecting the formation of the earth, and generally all which may be contrary to the narrative of Moses." *

But all this triumph of the Chaldeo-Babylonian creation legends which the Church had inherited availed but little.

For about the end of the eighteenth century fruitful suggestions and even clear presentations of this or that part of a large evolutionary doctrine came thick and fast, and from the most divergent quarters. Especially remarkable were those which came from Erasmus Darwin in England, from Maupertuis in France, from Oken in Switzerland, and from Herder, and, most brilliantly of all, from Goethe in Germany.

Two men among these thinkers must be especially mentioned—Treviranus in Germany and Lamarck in France; each independently of the other drew the world more completely than ever before in this direction.

From Treviranus came, in 1802, his work on biology, and in this he gave forth the idea that from forms of life originally simple had arisen all higher organizations by gradual development; that every living creature has a capacity for

* For Descartes in his relation to the Copernican theory, see Saisset, *Descartes et ses Précurseurs*; also Fouillée, *Descartes*, Paris, 1893, chaps. ii and iii; also other authorities cited in my chapter on Astronomy; for his relation to the theory of evolution, see the *Principes de Philosophie*, 3ème partie, § 45. For De Maillet, see Quatrefages, *Darwin et ses Précurseurs français*, chap. i, citing D'Archiac, *Paléontologie, Stratigraphie*, vol. i; also, Perrier, *La Philosophie zoologique avant Darwin*, chap. vi; also the admirable article, *Evolution*, by Huxley, in *Encyc. Brit.* The title of De Maillet's book is, *Telliamed, ou Entretiens d'un Philosophe indien avec un Missionnaire français sur la Diminution de la Mer*, 1748 and 1756. For Buffon, see the authorities previously given, also the chapter on Geology in this work. For the resistance of both Catholic and Protestant authorities to the Linnæan system and ideas, see Alberg, *Life of Linnæus*, London, 1888, pp. 143–147, and 237. As to the creation medallions at the Cathedral of Upsala, it is a somewhat curious coincidence that the present writer came upon them while visiting that edifice during the preparation of this chapter.

receiving modifications of its structure from external influences; and that no species had become really extinct, but that each had passed into some other species. From Lamarck came about the same time his *Researches*, and a little later his *Zoölogical Philosophy*, which introduced a new factor into the process of evolution—the action of the animal itself in its efforts toward a development to suit new needs—and he gave as his principal conclusions the following:

1. Life tends to increase the volume of each living body and of all its parts up to a limit determined by its own necessities.
2. New wants in animals give rise to new organs.
3. The development of these organs is in proportion to their employment.
4. New developments may be transmitted to offspring.

His well-known examples to illustrate these views, such as that of successive generations of giraffes lengthening their necks by stretching them to gather high-growing foliage, and of successive generations of kangaroos lengthening and strengthening their hind legs by the necessity of keeping themselves erect while jumping, provoked laughter, but the very comicality of these illustrations aided to fasten his main conclusion in men's memories.

In both these statements, imperfect as they were, great truths were embodied—truths which were sure to grow.

Lamarck's declaration, especially, that the development of organs is in ratio to their employment, and his indications of the reproduction in progeny of what is gained or lost in parents by the influence of circumstances, entered as a most effective force into the development of the evolution theory.

The next great successor in the apostolate of this idea of the universe was Geoffroy Saint-Hilaire. As early as 1795 he had begun to form a theory that species are various modifications of the same type, and this theory he developed, testing it at various stages as Nature was more and more displayed to him. It fell to his lot to bear the brunt in a struggle against heavy odds which lasted many years.

For the man who now took up the warfare, avowedly for science but unconsciously for theology, was the foremost naturalist then living—Cuvier. His scientific eminence was deserved; the highest honours of his own and other countries were given him, and he bore them worthily. An Imperial Councillor under Napoleon; President of the Council of Public Instruction and Chancellor of the University under the restored Bourbons; Grand Officer of the Legion of Honour, a Peer of France, Minister of the Interior, and President of the Council of State under Louis Philippe; he was eminent in all these capacities, and yet the dignity given by such high administrative positions was as nothing compared to his leadership in natural science. Science throughout the world acknowledged in him its chief contemporary ornament, and to this hour his fame rightly continues. But there was in him, as in Linnæus, a survival of certain theological ways of looking at the universe and certain theological conceptions of a plan of creation; it must be said, too, that while his temperament made him distrust new hypotheses, of which he had seen so many born and die, his environment as a great functionary of state, honoured, admired, almost adored by the greatest, not only in the state but in the Church, his solicitude lest science should receive some detriment by openly resisting the Church, which had recaptured Europe after the French Revolution, and had made of its enemies its footstool—all these considerations led him to oppose the new theory. Amid the plaudits, then, of the foremost churchmen he threw across the path of the evolution doctrines the whole mass of his authority in favour of the old theory of catastrophic changes and special creations.

Geoffroy Saint-Hilaire stoutly withstood him, braving non-recognition, ill-treatment, and ridicule. Treviranus, afar off in his mathematical lecture-room at Bremen, seemed simply forgotten.

But the current of evolutionary thought could not thus be checked: dammed up for a time, it broke out in new channels and in ways and places least expected; turned away from France, it appeared especially in England, where great paleontologists and geologists arose whose work culminated in that of Lyell. Specialists throughout all the world now became more vigorous than ever, gathering facts and thinking upon them in a way which caused the special creation theory to shrink more and more. Broader and more full became these various rivulets, soon to unite in one great stream of thought.

In 1813 Dr. Wells developed a theory of evolution by natural selection to account for varieties in the human race. About 1820 Dean Herbert, eminent as an authority in horticulture, avowed his conviction that species are but fixed varieties. In 1831 Patrick Matthews stumbled upon and stated the main doctrine of natural selection in evolution; and others here and there, in Europe and America, caught an inkling of it.

But no one outside of a circle apparently uninfluential cared for these things: the Church was serene: on the Continent it had obtained reactionary control of courts, cabinets, and universities; in England, Dean Cockburn was denouncing Mary Somerville and the geologists to the delight of churchmen; and the Rev. Mellor Brown was doing the same thing for the edification of dissenters.

In America the mild suggestions of Silliman and his compeers were met by the protestations of the Andover theologians headed by Moses Stuart. Neither of the great English universities, as a rule, took any notice of the innovators save by sneers.

To this current of thought there was joined a new element when, in 1844, Robert Chambers published his *Vestiges of Creation*. The book was attractive and was widely read. In Chambers's view the several series of animated beings, from the simplest and oldest up to the highest and most recent, were the result of two distinct impulses, each given once and for all time by the Creator. The first of these was an impulse imparted to forms of life, lifting them gradually through higher grades; the second was an impulse tending to modify organic substances in accordance with external circumstances; in fact, the doctrine of the book was evolution tempered by miracle—a stretching out of the creative act through all time—a pious version of Lamarck.

Two results followed, one mirth-provoking, the other leading to serious thought. The amusing result was that the theologians were greatly alarmed by the book: it was loudly insisted that it promoted atheism. Looking back along the line of thought which has since been developed, one feels that the older theologians ought to have put up thanksgivings for Chambers's theory, and prayers that it might prove true. The more serious result was that it accustomed men's minds to a belief in evolution as in some form possible or even probable. In this way it was provisionally of service.

Eight years later Herbert Spencer published an essay contrasting the theories of creation and evolution—reasoning with great force in favour of the latter, showing that species had undoubtedly been modified by circumstances; but still only few and chosen men saw the significance of all these lines of reasoning which had been converging during so many years toward one conclusion.

On July 1, 1858, there were read before the Linnæan Society at London two papers—one presented by Charles Darwin, the other by Alfred Russel Wallace—and with the reading of these papers the doctrine of evolution by natural selection was born. Then and there a fatal breach was made in the great theological barrier of the continued fixity of species since the creation.

The story of these papers the scientific world knows by heart: how Charles Darwin, having been sent to the University of Cambridge to fit him for the Anglican priesthood, left it in 1831 to go upon the scientific expedition of the Beagle; how for five years he studied with wonderful vigour and acuteness the problems of life as revealed on land and at sea—among volcanoes and coral reefs, in forests and on the sands, from the tropics to the arctic regions; how, in the Cape Verde and the Galapagos Islands, and in Brazil, Patagonia, and Australia he interrogated Nature with matchless persistency and skill; how he returned unheralded, quietly settled down to his work, and soon set the world thinking over its first published results, such as his book on *Coral Reefs*, and the monograph on the *Cirripedia*; and, finally, how he presented his paper, and followed it up with treatises which made him one of the great leaders in the history of human thought.

The scientific world realizes, too, more and more, the power of character shown by Darwin in all this great career; the faculty of silence, the reserve of strength seen in keeping his great thought—his idea of evolution by natural selection—under silent study and meditation for nearly twenty years, giving no hint of it to the world at large, but working in every field to secure proofs or disproofs, and accumulating masses of precious material for the solution of the questions involved.

To one man only did he reveal his thought—to Dr. Joseph Hooker, to whom in 1844, under the seal of secrecy, he gave a summary of his conclusions. Not until fourteen years later occurred the event which showed him that the fulness of time had come—the letter from Alfred Russel Wallace, to whom, in brilliant researches during the decade from 1848 to 1858, in Brazil and in the Malay Archipelago, the same truth of evolution by natural selection had been revealed. Among the proofs that scientific study does no injury to the more delicate shades of sentiment is the well-known story of this letter. With it Wallace sent Darwin a memoir, asking him to present it to the Linnæan Society: on examining it, Darwin found that Wallace had independently arrived at conclusions similar to his own—possibly had deprived him of fame; but Darwin was loyal to his friend, and his friend remained ever loyal to him. He publicly presented the paper from Wallace, with his own conclusions; and the date of this presentation—July 1, 1858—separates two epochs in the history, not merely of natural science, but of human thought.

In the following year, 1859, came the first instalment of his work in its fuller development—his book on *The Origin of Species*. In this book one at least of the main secrets at the heart of the evolutionary process, which had baffled the long line of investigators and philosophers from the days of Aristotle, was more broadly revealed. The effective mechanism of evolution was shown at work in three ascertained facts: in the struggle for existence among organized beings; in the survival of the fittest; and in heredity. These facts were presented with such minute research, wide observation, patient collation, transparent honesty, and judicial fairness, that they at once commanded the world's attention. It was the outcome of thirty years' work and thought by a worker and thinker of genius, but it was yet more than that

—it was the outcome, also, of the work and thought of another man of genius fifty years before. The book of Malthus on the *Principle of Population*, mainly founded on the fact that animals increase in a geometrical ratio, and therefore, if unchecked, must encumber the earth, had been generally forgotten, and was only recalled with a sneer. But the genius of Darwin recognised in it a deeper meaning, and now the thought of Malthus was joined to the new current. Meditating upon it in connection with his own observations of the luxuriance of Nature, Darwin had arrived at his doctrine of natural selection and survival of the fittest.

As the great dogmatic barrier between the old and new views of the universe was broken down, the flood of new thought pouring over the world stimulated and nourished strong growths in every field of research and reasoning: edition after edition of the book was called for; it was translated even into Japanese and Hindustani; the stagnation of scientific thought, which Buckle, only a few years before, had so deeply lamented, gave place to a widespread and fruitful activity; masses of accumulated observations, which had seemed stale and unprofitable, were made alive; facts formerly without meaning now found their interpretation. Under this new influence an army of young men took up every promising line of scientific investigation in every land. Epoch-making books appeared in all the great nations. Spencer, Wallace, Huxley, Galton, Tyndall, Tylor, Lubbock, Bagehot, Lewes, in England, and a phalanx of strong men in Germany, Italy, France, and America gave forth works which became authoritative in every department of biology. If some of the older men in France held back, overawed perhaps by the authority of Cuvier, the younger and more vigorous pressed on.

One source of opposition deserves to be especially mentioned—Louis Agassiz.

A great investigator, an inspired and inspiring teacher, a noble man, he had received and elaborated a theory of animated creation which he could not readily change. In his heart and mind still prevailed the atmosphere of the little Swiss parsonage in which he was born, and his religious and moral nature, so beautiful to all who knew him, was especially repelled by sundry evolutionists, who, in their zeal as neophytes, made proclamations seeming to have a decidedly irreligious if not immoral bearing. In addition to this was the direction his thinking had received from Cuvier. Both these influences combined to prevent his acceptance of the new view.

He was the third great man who had thrown his influence as a barrier across the current of evolutionary thought. Linnæus in the second half of the eighteenth century, Cuvier in the first half, and Agassiz in the second half of the nineteenth—all made the same effort. Each remains great; but not all of them together could arrest the current. Agassiz's strong efforts throughout the United States, and indeed throughout Europe, to check it, really promoted it. From the great museum he had founded at Cambridge, from his summer school at Penikese, from his lecture rooms at Harvard and Cornell, his disciples went forth full of love and admiration for him, full of enthusiasm which he had stirred and into fields which he had indicated; but their powers, which he had aroused and strengthened, were devoted to developing the truth he failed to recognise; Shaler, Verrill, Packard, Hartt, Wilder, Jordan, with a multitude of others, and especially the son who bore his honoured name, did justice to his memory by applying what they had received from him to research under inspiration of the new revelation.

Still another man deserves especial gratitude and honour in this progress—Edward Livingston Youmans. He was perhaps the first in America to recognise the vast bearings of the truths presented by Darwin, Wallace, and Spencer. He became the apostle of these truths, sacrificing the brilliant career on which he had entered as a public lecturer, subordinating himself to the three leaders, and giving himself to editorial drudgery in the stimulation of research and the announcement of results.

In support of the new doctrine came a world of new proofs; those which Darwin himself added in regard to the cross-fertilization of plants, and which he had adopted from embryology, led the way, and these were followed by the discoveries of Wallace, Bates, Huxley, Marsh, Cope, Leidy, Haeckel, Müller, Gaudry, and a multitude of others in all lands.*

IV. THE FINAL EFFORT OF THEOLOGY.

DARWIN'S *Origin of Species* had come into the theological world like a plough into an ant-hill. Everywhere those thus rudely awakened from their old comfort and repose had swarmed forth angry and confused. Reviews, sermons, books light and heavy, came flying at the new thinker from all sides.

The keynote was struck at once in the *Quarterly Review* by Wilberforce, Bishop of Oxford. He declared that Darwin was guilty of "a tendency to limit God's glory in creation"; that "the principle of natural selection is absolutely incompatible with the word of God"; that it "contradicts the revealed relations of creation to its Creator"; that it is "inconsistent with the fulness of his glory"; that it is "a dishonouring view of Nature"; and that there is "a simpler explanation of the presence of these strange forms among the works of God": that explanation being—"the fall of Adam." Nor did the bishop's efforts end here; at the meeting of the British Association for the Advancement of Science he again disported himself in the tide of popular applause. Referring to the ideas of Darwin, who was absent on account of illness, he congratulated himself in a public speech that he was not descended from a monkey. The reply came from Huxley, who said in substance: "If I had to choose, I would prefer to be a descendant of a humble monkey rather than of a man who employs his knowledge

* For Agassiz's opposition to evolution, see the *Essay on Classification*, vol. i, 1857, as regards Lamarck, and vol. iii, 1860, as regards Darwin; also *Silliman's Journal*, July, 1860; also the *Atlantic Monthly*, January, 1874; also his *Life and Correspondence*, vol. ii, p. 647; also Asa Gray, *Scientific Papers*, vol ii, p. 484. A reminiscence of my own enables me to appreciate his deep ethical and religious feeling. I was passing the day with him at Nahant in 1868, consulting him regarding candidates for various scientific chairs at the newly established Cornell University, in which he took a deep interest. As we discussed one after another of the candidates he suddenly said: "Who is to be your Professor of Moral Philosophy? That is a far more important position than all the others."

and eloquence in misrepresenting those who are wearing out their lives in the search for truth."

This shot reverberated through England, and indeed through other countries.

The utterances of this the most brilliant prelate of the Anglican Church received a sort of antiphonal response from the leaders of the English Catholics. In an address before the "Academia," which had been organized to combat "science falsely so called," Cardinal Manning declared his abhorrence of the new view of Nature, and described it as "a brutal philosophy—to wit, there is no God, and the ape is our Adam."

These attacks from such eminent sources set the clerical fashion for several years. One distinguished clerical reviewer, in spite of Darwin's thirty years of quiet labour, and in spite of the powerful summing up of his book, prefaced a diatribe by saying that Darwin "might have been more modest had he given some slight reason for dissenting from the views generally entertained." Another distinguished clergyman, vice-president of a Protestant institute to combat "dangerous" science, declared Darwinism "an attempt to dethrone God." Another critic spoke of persons accepting the Darwinian views as "under the frenzied inspiration of the inhaler of mephitic gas," and of Darwin's argument as "a jungle of fanciful assumption." Another spoke of Darwin's views as suggesting that "God is dead," and declared that Darwin's work "does open violence to everything which the Creator himself has told us in the Scriptures of the methods and results of his work." Still another theological authority asserted: "If the Darwinian theory is true, Genesis is a lie, the whole framework of the book of life falls to pieces, and the revelation of God to man, as we Christians know it, is a delusion and a snare." Another, who had shown excellent qualities as an observing naturalist, declared the Darwinian view "a huge imposture from the beginning."

Echoes came from America. One review, the organ of the most widespread of American religious sects, declared that Darwin was "attempting to befog and to pettifog the whole question"; another denounced Darwin's views as "infidelity"; another, representing the American branch of the Anglican Church, poured contempt over Darwin as "sophistical and illogical," and then plunged into an exceedingly dangerous line of argument in the following words: "If this hypothesis be true, then is the Bible an unbearable fiction; . . . then have Christians for nearly two thousand years been duped by a monstrous lie. . . . Darwin requires us to disbelieve the authoritative word of the Creator." A leading journal representing the same church took pains to show the evolution theory to be as contrary to the explicit declarations of the New Testament as to those of the Old, and said: "If we have all, men and monkeys, oysters and eagles, developed from an original germ, then is St. Paul's grand deliverance—'All flesh is not the same flesh; there is one kind of flesh of men, another of beasts, another of fishes, and another of birds'—untrue."

Another echo came from Australia, where Dr. Perry, Lord Bishop of Melbourne, in a most bitter book on *Science and the Bible*, declared that the obvious object of Chambers, Darwin, and Huxley is "to produce in their readers a disbelief of the Bible."

Nor was the older branch of the Church to be left behind in this chorus. Bayma, in the *Catholic World*, declared, "Mr. Darwin is, we have reason to believe, the mouthpiece or chief trumpeter of that infidel clique whose well-known object is to do away with all idea of a God."

Worthy of especial note as showing the determination of the theological side at that period was the foundation of sacro-scientific organizations to combat the new ideas. First to be noted is the "Academia," planned by Cardinal Wiseman. In a circular letter the cardinal, usually so moderate and just, sounded an alarm and summed up by saying, "Now it is for the Church, which alone possesses divine certainty and divine discernment, to place itself at once in the front of a movement which threatens even the fragmentary remains of Christian belief in England." The necessary permission was obtained from Rome, the Academia was founded, and the "divine discernment" of the Church was seen in the utterances which came from it, such as those of Cardinal Manning, which every thoughtful Catholic would now desire to recall, and in the diatribes of Dr. Laing, which only aroused laughter on all sides. A similar effort was seen in Protestant quarters; the "Victoria Institute" was created, and perhaps the most noted utterance which ever came from it was the declaration of its vice-president, the Rev. Walter Mitchell, that "Darwinism endeavours to dethrone God." *

In France the attack was even more violent. Fabre d'Envieu brought out the heavy artillery of theology, and in a long series of elaborate propositions demonstrated that any other doctrine than that of the fixity and persistence of species is absolutely contrary to Scripture. The Abbé Désorges, a former Professor of Theology, stigmatized Darwin as a "pedant," and evolution as "gloomy"; Monseigneur Ségur, referring to Darwin and his followers, went into hysterics and shrieked: "These infamous doctrines have for their only support the most abject passions. Their father is pride, their mother impurity, their offspring revolutions. They come from hell and return thither, taking with them the gross creatures who blush not to proclaim and accept them."

In Germany the attack, if less declamatory, was no less severe. Catholic theologians vied with Protestants in bitterness. Prof. Michelis declared Darwin's theory "a caricature of creation." Dr. Hagermann asserted that it "turned the Creator out of doors." Dr. Schund insisted that "every

---

* For Wilberforce's article, see *Quarterly Review*, July, 1860. For the reply of Huxley to the bishop's speech I have relied on the account given in *Quatrefages*, who had it from Carpenter; a somewhat different version is given in the *Life and Letters of Darwin*. For Cardinal Manning's attack, see *Essays on Religion and Literature*, London, 1865. For the review articles, see the *Quarterly* already cited, and that for July, 1874; also the *North British Review*, May, 1860; also, F. O. Morris's letter in the *Record*, reprinted at Glasgow, 1870; also the *Addresses of Rev. Walter Mitchell* before the Victoria Institute, London, 1867; also Rev. B. G. Johns, *Moses not Darwin, a Sermon*, March 31, 1871. For the earlier American attacks, see *Methodist Quarterly Review*, April, 1871; *The American Church Review*, July and October, 1865, and January, 1866. For the Australian attack, see *Science and the Bible*, by the Right Reverend Charles Perry, D. D., Bishop of Melbourne, London, 1869. For Bayma, see the *Catholic World*, vol. xxvi, p. 782. For the Academia, see *Essays* edited by Cardinal Manning, above cited; and for the Victoria Institute, see *Scientia Scientiarum*, by a member of the Victoria Institute, London, 1865.

idea of the Holy Scriptures, from the first to the last page, stands in diametrical opposition to the Darwinian theory"; and, "if Darwin be right in his view of the development of man out of a brutal condition, then the Bible teaching in regard to man is utterly annihilated." Rougemont in Switzerland called for a crusade against the obnoxious doctrine. Luthardt, Professor of Theology at Leipsic, declared: "The idea of creation belongs to religion and not to natural science; the whole superstructure of personal religion is built upon the doctrine of creation"; and he showed the evolution theory to be in direct contradiction to Holy Writ.

But in 1863 came an event which brought serious confusion to the theological camp: Sir Charles Lyell, the most eminent of living geologists, a man of deeply Christian feeling and of exceedingly cautious temper, who had opposed the evolution theory of Lamarck and declared his adherence to the idea of successive creations, then published his work on the *Antiquity of Man*, and in this and other utterances showed himself a complete though unwilling convert to the fundamental ideas of Darwin. The blow was serious in many ways, and especially so in two—first, as withdrawing all foundation in fact from the scriptural chronology, and secondly, as discrediting the creation theory. The blow was not unexpected; in various review articles against the Darwinian theory there had been appeals to Lyell, at times almost piteous, "not to flinch from the truths he had formerly proclaimed." But Lyell, like the honest man he was, yielded unreservedly to the mass of new proofs arrayed on the side of evolution against that of creation.

At the same time came Huxley's *Man's Place in Nature*, giving new and most cogent arguments in favour of evolution by natural selection.

In 1871 was published Darwin's *Descent of Man*. Its doctrine had been anticipated by critics of his previous books, but it made, none the less, a great stir; again the opposing army trooped forth, though evidently with much less heart than before. A few were very violent. The *Dublin University Magazine*, after the traditional Hibernian fashion, charged Mr. Darwin with seeking "to displace God by the unerring action of vagary," and with being "resolved to hunt God out of the world." But most notable from the side of the older Church was the elaborate answer to Darwin's book by the eminent French Catholic physician, Dr. Constantin James. In his work, *On Darwinism, or the Man-Ape*, published at Paris in 1877, Dr. James not only refuted Darwin scientifically but poured contempt on his book, calling it "a fairy tale," and insisted that a work "so fantastic and so burlesque" was, doubtless, only a huge joke, like Erasmus's *Praise of Folly*, or Montesquieu's *Persian Letters*. The princes of the Church were delighted. The Cardinal Archbishop of Paris assured the author that the book had become his "spiritual reading," and begged him to send a copy to the Pope himself. His Holiness, Pope Pius IX, acknowledged the gift in a remarkable letter. He thanked his dear son, the writer, for the book in which he "refutes so well the aberrations of Darwinism." "A system," His Holiness adds, "which is repugnant at once to history, to the tradition of all peoples, to exact science, to observed facts, and even to Reason herself, would seem to need no refutation, did not alienation from God and the leaning toward materialism, due to depravity, eagerly seek a support in all this tissue of fables. . . . And, in fact, pride, after rejecting the Creator of all things and proclaiming man independent, wishing him to be his own king, his own priest, and his own God—pride goes so far as to degrade man himself to the level of the unreasoning brutes, perhaps even of lifeless matter, thus unconsciously confirming the Divine declaration, *When pride cometh, then cometh shame*. But the corruption of this age, the machinations of the perverse, the danger of the simple, demand that such fancies, altogether absurd though they are, should—since they borrow the mask of science—be refuted by true science." Wherefore the Pope thanked Dr. James for his book, "so opportune and so perfectly appropriate to the exigencies of our time," and bestowed on him the apostolic benediction. Nor was this brief all. With it there came a second, creating the author an officer of the Papal Order of St. Sylvester. The cardinal archbishop assured the delighted physician that such a double honour of brief and brevet was perhaps unprecedented, and suggested only that in a new edition of his book he should "insist a little more on the relation existing between the narratives of Genesis and the discoveries of modern science, in such fashion as to convince the most incredulous of their perfect agreement." The prelate urged also a more dignified title. The proofs of this new edition were accordingly all submitted to His Eminence, and in 1882 it appeared as *Moses and Darwin: the Man of Genesis compared with the Man-Ape, or Religious Education opposed to Atheistic.* No wonder the cardinal embraced the author, thanking him in the name of science and religion. "We have at last," he declared, "a handbook which we can safely put into the hands of youth."

Scarcely less vigorous were the champions of English Protestant orthodoxy. In an address at Liverpool, Mr. Gladstone remarked: "Upon the grounds of what is termed evolution God is relieved of the labour of creation; in the name of unchangeable laws he is discharged from governing the world"; and, when Herbert Spencer called his attention to the fact that Newton with the doctrine of gravitation and with the science of physical astronomy is open to the same charge, Mr. Gladstone retreated in the *Contemporary Review* under one of his characteristic clouds of words. The Rev. Dr. Coles, in the *British and Foreign Evangelical Review*, declared that the God of evolution is not the Christian's God. Burgon, Dean of Chichester, in a sermon preached before the University of Oxford, pathetically warned the students that "those who refuse to accept the history of the creation of our first parents according to its obvious literal intention, and are for substituting the modern dream of evolution in its place, cause the entire scheme of man's salvation to collapse." Dr. Pusey also came into the fray with most earnest appeals against the new doctrine, and the Rev. Gavin Carlyle was perfervid on the same side. The Society for Promoting Christian Knowledge published a book by the Rev. Mr. Birks, in which the evolution doctrine was declared to be "flatly opposed to the fundamental doctrine of creation." Even the *London Times* admitted a review stigmatizing Darwin's *Descent of Man* as an "utterly unsupported hypothesis," full of "unsubstantiated premises, cursory investigations, and disintegrating speculations," and Darwin himself as "reckless and unscientific." *

But it was noted that this second series of attacks, on the *Descent of Man*, differed in one remarkable respect—so far as England was concerned—from those which had been made over ten years before on the *Origin of Species*. While everything was done to discredit Darwin, to pour contempt upon him, and even, of all things in the world, to make him—the gentlest of mankind, only occupied with the scientific side of the problem—"a persecutor of Christianity," while his followers were represented more and more as charlatans or dupes, there began to be in the most influential quarters careful avoidance of the old argument that evolution—even by natural selection—contradicts Scripture. It began to be felt that this was dangerous ground. The defection of Lyell had, perhaps, more than anything else, started the question among theologians who had preserved some equanimity, "*What if, after all, the Darwinian theory should prove to be true?*" Recollections of the position in which the Roman Church found itself after the establishment of the doctrines

* For the French theological opposition to the Darwinian theory, see Pozzy, *La Terre et le Récit Biblique de la Création*, 1874, especially pp 353, 363 ; also, Félix Ducane, *Études sur le Transformisme*, 1876, especially pp. 107 to 119. As to Fabre d'Envieu, see especially his Proposition xliii. For the Abbé Désorges, "former Professor of Philosophy and Theology," see his *Erreurs Modernes*, Paris, 1878, pp. 677 and 595 to 598. For Monseigneur Ségur, see his *La Foi devant la Science Moderne*, sixth ed., Paris, 1874, pp. 23, 34, etc. For Herbert Spencer's reply to Mr. Gladstone, see his *Study of Sociology* ; for the passage in the *Dublin Review*, see the issue for July, 1871. For the review in the *London Times*, see *Nature* for April 20, 1871. For Gavin Carlyle, see *The Battle of Unbelief*, 1870, pp. 86 and 171. For the attacks by Michelis and Hagermann, see *Natur und Offenbarung*, Münster, 1861 to 1869. For Schund, see his *Darwin's Hypothese und ihr Verhältniss zu Religion und Moral*, Stuttgart, 1869. For Luthardt, see *Fundamental Truths of Christianity*, translated by Sophia Taylor, second ed., Edinburgh, 1869. For Rougemont, see his *L'Homme et le Singe*, Neuchâtel, 1863 (also in German trans.). For Constantin James, see his *Mes Entretiens avec l'Empéreur Don Pédro sur le Darwinisme*, Paris, 1888, where the papal briefs are printed in full. For the English attacks on Darwin's *Descent of Man*, see the *Edinburgh Review* July, 1871, and elsewhere ; the *Dublin Review*, July, 1871 ; the *British and Foreign Evangelical Review*, April, 1886. See also *The Scripture Doctrine of Creation*, by the Rev. T. R. Birks, London, 1873, published by the S. P. C. K. For Dr. Pusey's attack, see his *Unscience, not Science, adverse to Faith*, 1878 ; also, *Darwin's Life and Letters*, vol. ii, pp. 411, 412.

of Copernicus and Galileo naturally came into the minds of the more thoughtful. In Germany this consideration does not seem to have occurred at quite so early a day. One eminent Lutheran clergyman at Magdeburg called on his hearers to choose between Darwin and religion; Delitszch, in his new commentary on Genesis, attempted to bring science back to recognise human sin as an important factor in creation; Prof. Heinrich Ewald, while carefully avoiding any sharp conflict between the scriptural doctrine and evolution, comforted himself by covering Darwin and his followers with contempt; Christlieb, in his address before the Evangelical Alliance at New York in 1873, simply took the view that the tendencies of the Darwinian theory were "toward infidelity," but declined to make any serious battle on biblical grounds; the Jesuit, Father Pesch, in Holland, drew up in Latin, after the old scholastic manner, a sort of general indictment of evolution, of which one may say that it was interesting—as interesting as the display of a troop in chain armour and with cross-bows on a nineteenth-century battlefield.

From America there came new echoes. Among the myriad attacks on the Darwinian theory by Protestants and Catholics two should be especially mentioned. The first of these was by Dr. Noah Porter, President of Yale College, an excellent scholar, an interesting writer, a noble man, broadly tolerant, combining in his thinking a curious mixture of radicalism and conservatism. While giving great latitude to the evolutionary teaching in the university under his care, he felt it his duty upon one occasion to avow his disbelief in it; but he was too wise a man to suggest any necessary antagonism between it and the Scriptures. He confined himself mainly to pointing out the tendency of the evolution doctrine in this form toward agnosticism and pantheism. To those who knew and loved him, and had noted the genial way in which by wise neglect he had allowed scientific studies to flourish at Yale, there was an amusing side to all this. Within a stone's throw of his college rooms was the Museum of Paleontology, in which Prof. Marsh had laid side by side, among other evidences of the new truth, that wonderful series of specimens showing the evolution of the horse from the earliest form of the animal, "not larger than a fox, with five toes," through the whole series up to his present form and size—that series which Huxley declared an absolute proof of the existence of natural selection as an agent in evolution. In spite of the veneration and love which all Yale men felt for President Porter, it was hardly to be expected that these particular arguments of his would have much permanent effect upon them when there was constantly before their eyes so convincing a refutation.

But a far more determined opponent was the Rev. Dr. Hodge, of Princeton; his anger toward the evolution doctrine was bitter: he denounced it as thoroughly "atheistic"; he insisted that Christians "have a right to protest against the arraying of probabilities against the clear evidence of the Scriptures"; he even censured so orthodox a writer as the Duke of Argyll, and declared that the Darwinian theory of natural selection is "utterly inconsistent with the Scriptures," and that "an absent God, who does nothing, is to us no God"; that "to ignore design as manifested in God's creation is to dethrone God"; that "a denial of design in Nature is virtually a denial of God"; and that "no teleologist can be a Darwinian." Even more uncompromising was another of the leading authorities at the same university—the Rev. Dr. Duffield. He declared war not only against Darwin but even against men like Asa Gray, Le Conte, and others, who had attempted to reconcile the new theory with the Bible: he insisted that "evolutionism and the scriptural account of the origin of man are irreconcilable"—that the Darwinian theory is "in direct conflict with the teaching of the apostle, 'All scripture is given by inspiration of God'"; he pointed out, in his opposition to Darwin's *Descent of Man* and Lyell's *Antiquity of Man*, that in the Bible "the genealogical links which connect the Israelites in Egypt with Adam and Eve in Eden are explicitly given." These utterances of Prof. Duffield culminated in a declaration which deserves to be cited as showing that a Presbyterian minister can "deal damnation round the land" *ex cathedra* in a fashion quite equal to that of popes and bishops. It is as follows: "If the development theory of the origin of man," wrote Dr. Duffield in the *Princeton Review*, "shall in a little while take its place—as doubtless it will—with other exploded scientific speculations, then they who accept it with its proper logical consequences will in the life to come have their portion with those who in this life 'know not God and obey not the gospel of his Son.'"

Fortunately, at about the time when Darwin's *Descent of Man* was published, there had come into Princeton University a "*deus ex machina*" in the person of Dr. James McCosh. Called to the presidency, he at once took his stand against teachings so dangerous to Christianity as those of Drs. Hodge, Duffield, and their associates. In one of his personal confidences he has let us into the secret of this matter. With that hard Scotch sense which Thackeray had applauded in his well-known verses, he saw that the most dangerous thing which could be done to Christianity at Princeton was to reiterate in the university pulpit, week after week, solemn declarations that if evolution by natural selection, or indeed evolution at all, be true, the Scriptures are false. He tells us that he saw that this was the certain way to make the students unbelievers; he therefore not only checked this dangerous preaching but preached an opposite doctrine. With him began the inevitable compromise, and, in spite of mutterings against him as a Darwinian, he carried the day. Whatever may be thought of his general system of philosophy, no one can deny his great service in neutralizing the teachings of his predecessors and colleagues—so dangerous to all that is essential in Christianity.

Other divines of strong sense in other parts of the country began to take similar ground—namely, that men could be Christians and at the same time Darwinians. There appeared, indeed, here and there, curious discrepancies: thus in 1873 the *Monthly Religious Magazine* of Boston congratulated its readers that the Rev. Mr. Burr had "demolished the evolution theory, knocking the breath of life out of it and throwing it to the dogs." This amazing performance by the Rev. Mr. Burr was repeated in a very striking way by Bishop Keener before the Œcumenical Council of Methodism at Washington in 1891. In what the newspapers described as an "admirable speech," he refuted evolution doctrines by saying that evolutionists had "only to make a journey of twelve hours from the place where he was then standing to find together the bones of the muskrat, the opossum, the coprolite, and the ichthyosaurus." He asserted that Agassiz—whom the good bishop, like so many others, seemed to think an evolutionist—when he visited these beds near Charleston, declared: "These old beds have set me crazy; they have destroyed the work of a lifetime." And the Methodist prelate ended by saying: "Now, gentlemen, brethren, take these facts home with you; get down and look at them. This is the watch that was under the steam hammer—the doctrine of evolution; and this steam hammer is the wonderful deposit of the Ashley beds."

Exhibitions like these availed little. While the good bishop amid vociferous applause thus made comically evident his belief that Agassiz was a Darwinian and a coprolite an animal, scientific men were recording in all parts of the world facts confirming the dreaded theory of an evolution by natural selection. While the Rev. Mr. Burr was so loudly praised for "throwing Darwinism to the dogs," Marsh was completing his series leading from the five-toed ungulates to the horse. While Dr. Tayler Lewis at Union, and Drs. Hodge and Duffield at Princeton, were showing that if evolution be true the biblical accounts must be false, the indefatigable Yale professor was showing his cretaceous birds, and among them *Hesperornis* and *Ichthyornis* with teeth. While in Germany Luthardt, Schund, and their compeers were demonstrating that Scripture requires a belief in special and separate creations, the *Archæopteryx*, showing a most remarkable connection between birds and reptiles, was discovered. While in France Monseigneur Ségur and others were indulging in diatribes against "a certain Darwin," Gaudry and Filhol were discovering a striking series of "missing links" among the carnivora.

In view of the proofs accumulating in favour of the new evolutionary hypothesis, the change in the tone of controlling theologians was now rapid. From all sides came evidences of desire to compromise with the theory. Strict adherents of the biblical text pointed significantly to the verses in Genesis in which the earth and sea were made to bring forth birds and fishes, and man was created out of the dust of the ground. Men of larger mind like Kingsley and Farrar, with English and American broad churchmen generally, took ground directly in Darwin's favour. Even Whewell took pains to show that there might be such a thing as a Darwinian argument for design in Nature; and the Rev. Samuel Houghton, of the Royal Society, gave interesting suggestions of a divine design in evolution.

Both the great English universities received the new teaching as a leaven: at Oxford, in the very front of the High Church party at Keble College, was elaborated a statement that the evolution doctrine is "an advance in our theological thinking." And Temple, Bishop of London, perhaps the most influential thinker then in the Anglican episcopate, accepted the new revelation in the following words: "It seems something more majestic, more befitting him to whom a thousand years are as one day, thus to impress his will once for all on his creation, and provide for all the countless varieties by this one original impress, than by special acts of creation to be perpetually modifying what he had previously made."

In Scotland the Duke of Argyll, head and front of the orthodox party, dissenting in many respects from Darwin's full conclusions, made concessions which badly shook the old position.

Curiously enough, from the Roman Catholic Church, bitter as some of its writers had been, now came argument to prove that the Catholic faith does not prevent any one from holding the Darwinian theory, and especially a declaration from an authority eminent among American Catholics —a declaration which has a very curious sound, but which it would be ungracious to find fault with—that "the doctrine of evolution is no more in opposition to the doctrine of the Catholic Church than is the Copernican theory or that of Galileo."

Here and there, indeed, men of science like Dawson, Mivart, and Wigand, in view of theological considerations, sought to make conditions; but the current was too strong, and eminent theologians in every country accepted natural selection as at least a very important part in the mechanism of evolution.

At the death of Darwin it was felt that there was but one place in England where his body should be laid, and that this place was next the grave of Sir Isaac Newton in Westminster Abbey. The noble address of Canon Farrar at his funeral was echoed from many pulpits in Europe and America, and theological opposition as such was ended. Occasionally appeared, it is true, a survival of the old feeling: the Rev. Dr. Laing referred to the burial of Darwin in Westminster Abbey as "a proof that England is no longer a Christian country," and added that this burial was a desecration—that this honour was given him because he had been "the chief promoter of the mock doctrine of evolution of the species and the ape descent of man."

Still another of these belated prophets was, of all men, Thomas Carlyle. Soured and embittered, in the same spirit which led him to find more heroism in a marauding Viking or in one of Frederick the Great's generals than in Washington, or Lincoln, or Grant, and which caused him to see in the American civil war only the burning out of a foul chimney, he, with the petulance natural to a dyspeptic eunuch, railed at Darwin as an "apostle of dirt worship."

The last echoes of these utterances reverberated between Scotland and America. In the former country, in 1885, the Rev. Dr. Lee issued a volume declaring that, if the Darwinian view be true, "there is no place for God"; that "by no method of interpretation can the language of Holy Scripture be made wide enough to re-echo the orang-outang theory of man's natural history"; that "Darwinism reverses the revelation of God" and "implies utter blasphemy against the divine and human character of our Incarnate Lord"; and he was pleased to call Darwin and his followers "gospellers of the gutter." In one of the intellectual centres of America the editor of a periodical called *The Christian* urged frantically that "the battle be set in array, and that men find out who is on the Lord's side and who is on the side of the devil and the monkeys."

To the honour of the Church of England it should be recorded that a considerable number of her truest men opposed such utterances as these, and that one of them—Farrar, Archdeacon of Westminster—made a protest worthy to be held in perpetual remembrance. While confessing his own inability to accept fully the new scientific belief, he said: "We should consider it disgraceful and humiliating to try to shake it by an *ad captandum* argument, or by a clap-trap platform appeal to the unfathomable ignorance and unlimited arrogance of a prejudiced assembly. We should blush to meet it with an anathema or a sneer."

All opposition had availed nothing; Darwin's work and fame were secure. As men looked back over his beautiful life—simple, honest, tolerant, kindly—and thought upon his great labours in the search for truth, all the attacks faded into nothingness.

There were indeed some dark spots, which as time goes on appear darker. At Trinity College, Cambridge, Whewell, the "omniscient," author of the *History of the Inductive Sciences*, refused to allow a copy of the *Origin of Species* to be placed in the library. At multitudes of institutions under theological control—Protestant as well as Catholic—attempts were made to stamp out or to stifle evolutionary teaching. Especially was this true for a time in America, and the case of the American College at Beyrout, where nearly all the younger professors were dismissed for adhering to Darwin's views, is worthy of remembrance. The treatment of Dr. Winchell at the Vanderbilt University in Tennessee showed the same spirit; one of the truest of men, devoted to science but of deeply Christian feeling, he was driven forth for views which centred in the Darwinian theory.

Still more striking was the case of Dr. Woodrow. He had, about 1857, been appointed to a professorship of Natural Science as connected with Revealed Religion, in the Presbyterian Seminary at Columbia, South Carolina. He was a devoted Christian man, and his training had led him to accept the Presbyterian standards of faith. With great gifts for scientific study he visited Europe, made a most conscientious examination of the main questions under discussion, and adopted the chief points in the doctrine of evolution by natural selection. A struggle soon began. A movement hostile to him grew more and more determined, and at last, in spite of the efforts made in his behalf by the directors of the seminary and by a large and broad-minded minority in the representative bodies controlling it, an orthodox storm, raised by the delegates from various Presbyterian bodies, drove him from his post. Fortunately, he was received into a professorship at the University of South Carolina, where he has since taught with more power than ever before.

This testimony to the faith by American provincial Protestantism was very properly echoed from Spanish provincial Catholicism. In the year 1878 a Spanish colonial man of science, Dr. Chil y Marango, published a work on the Canary Islands. But Dr. Chil had the imprudence to sketch, in his introduction, the modern hypothesis of evolution, and to exhibit some proofs, found in the Canary Islands, of the barbarism of primitive man. The ecclesiastical authorities, under the lead of Bishop Urquinaona y Bidot, at once grappled with this new idea. By a solemn act they declared it "*falsa, impia, scandalosa*"; all persons possessing copies of the work were ordered to surrender them at once to the proper ecclesiastics, and the author was placed under the major excommunication.

But all this opposition may be reckoned among the last expiring convulsions of the old theologic theory. Even from the new Catholic University at Washington has come an utterance in favour of the new doctrine, and in other universities in the Old World and in the New the doctrine of evolution by natural selection has asserted its right to full and honest consideration. More than this, it is clearly evident that the stronger men in the Church have, in these latter days, not only relinquished the struggle against science in this field, but have determined frankly and manfully to make an alliance with it. In two very remarkable lectures given in 1892 at the parish church of Rochdale, Wilson, Archdeacon of Manchester, not only accepted Darwinism as true, but wrought it with great argumentative power into a higher view of Christianity; and what is of great significance, these sermons were published by the same Society for the Promotion of Christian Knowledge which only a few years before had published the most bitter attacks against the Darwinian theory. So, too, during the year 1893, Prof. Henry Drummond, whose praise is in all the dissenting churches, developed a similar view most brilliantly in a series of lectures delivered before the American Chautauqua schools, and published in one of the most widespread of English orthodox newspapers.

Whatever additional factors may be added to natural selection—and Darwin himself fully admitted that there might be others—the theory of an evolution process in the formation of the universe and of animated nature is established, and the old theory of direct creation is gone forever. In place of it science has given us conceptions far more noble, and opened the way to an argument for design infinitely more beautiful than any ever developed by theology.*

---

* For causes of the bitterness shown regarding the Darwinian hypothesis, see Reusch, *Bibel und Natur*, vol. ii, pp. 46 *et seq.* For hostility in the United States toward the Darwinian theory, see, among a multitude of writers, the following: Dr. Charles Hodge, of Princeton, monograph, *What is Darwinism?* New York, 1874; also his *Systematic Theology*, New York, 1872, vol. ii, part 2, *Anthropology*; also *The Light by which we see Light, or Nature and the Scriptures*, Vedder Lectures, 1875, Rutgers College, New York, 1875; also *Positivism and Evolutionism*, in the *American Catholic Quarterly*, October, 1877, pp. 607, 619; and, in the same number, *Professor Huxley and Evolution*, by Rev. A. M. Kirsch, pp. 662, 664; *The Logic of Evolution*, by Prof. Edward F. X. McSweeney, D. D., July, 1879, p. 561; *Das Hexæmeron und die Geologie*, von P. Eirich, Pastor in Albany, N. Y., Lutherischer Concordia-Verlag, St. Louis, Mo., 1878, pp. 81, 82, 84, 92–94; *Evolutionism respecting Man and the Bible*, by John T. Duffield, of Princeton, January, 1878, *Princeton Review*, pp. 151, 153, 154, 158, 159, 160, 188; *A Lecture on Evolution*, before the Nineteenth Century Club of New York, May 25, 1886, by ex-President Noah Porter, pp. 4, 26–29. For the laudatory notice of the Rev. E. F. Burr's demolition of evolution in his book *Pater Mundi*, see *Monthly Religious Magazine*, Boston, May, 1873, p. 492. Concerning the removal of Rev. Dr. James Woodrow, Professor of Natural Science in the Columbia Theological Seminary, see *Evolution or Not*, art. in the *New York. Weekly Sun*, October 24, 1888. For the dealings of Spanish ecclesiastics with Dr. Chil and his Darwinian exposition, see the *Revue d'Anthropologie*, cited in the *Academy* for April 6, 1878; see also the *Catholic World*, xix, 433, *A Discussion with an Infidel*, directed against Dr. Louis Büchner and his Kraft und Stoff; also *Mind and Matter*, by Rev. James Tait, of Canada, p. 66 (in the third edition the author bemoans the "horrible plaudits" that "have accompanied every effort to establish man's brutal descent)"; also *The Church Journal*, New York, May 28, 1874. For the effort in favour of a teleological evolution, see Rev. Samuel Houghton, F. R. S., *Principles of Animal Mechanics*, London, 1873, preface and p. 156 and elsewhere. For details of the persecution of Drs. Winchell and Woodrow, and of the Beyrout professors, with authorities cited, see my chapter on *The Fall of Man and Anthropology.* For more liberal views among religious thinkers regarding the Darwinian theory, and for efforts to mitigate and adapt it to theological views, see, among the great mass of utterances, the following: Charles Kingsley's letters to Darwin, November 18, 1859, in *Dar-*

win's Life and Letters, vol. ii, p. 82; Adam Sedgwick to Charles Darwin, December 24, 1859, see ibid., vol. ii, pp. 356–359; the same to Miss Gerard, January 2, 1860, see *Sedgwick's Life and Letters*, vol. ii, pp. 359, 360; the same in *The Spectator*, London, March 24, 1860; *The Rambler*, March, 1860, cited by Mivart, *Genesis of Species*, p. 30; *The Dublin Review*, May, 1860; *The Christian Examiner*, May, 1860; Charles Kingsley to F. D. Maurice in 1863, in Kingsley's *Life*, vol. ii, p. 171; Adam Sedgwick to Livingstone (the explorer), March 16, 1865, in *Life and Letters of Sedgwick*, vol. ii, pp. 410–412; the Duke of Argyll, *The Reign of Law*, New York, pp. 16, 18, 31, 116, 117, 120, 159; Joseph P. Thompson, D. D., LL. D., *Man in Genesis and Geology*, New York, 1870, pp. 48, 49, 82; Canon H. P. Liddon, *Sermons preached before the University of Oxford*, 1871, Sermon III; St. George Mivart, *Evolution and its Consequences*, *Contemporary Review*, January, 1872; *British and Foreign Evangelical Review*, 1872, article on *The Theory of Evolution*; *The Lutheran Quarterly*, Gettysburg, Pa., April, 1872, article by Rev. Cyrus Thomas, Assistant United States Geological Survey, on *The Descent of Man*, pp. 214, 239, 372–376; *The Lutheran Quarterly*, July, 1873, article on *Some Assumptions against Christianity*, by Rev. C. A. Stork, Baltimore, Md., pp. 325, 326; also, in the same number, see a review of Dr. Burr's *Pater Mundi*, pp. 474, 475, and contrast with the review in the *Andover Review* of that period; an article in the *Religious Magazine and Monthly Review*, Boston, on *Religion and Evolution*, by Rev. S. R. Calthrop, September, 1873, p. 200; *The Popular Science Monthly*, January, 1874, article *Genesis, Geology, and Evolution*, by Rev. George Henslow—this article first appeared in his book *Evolution and Religion*; article by Asa Gray, *Nature*, London, June 4, 1874; *Materialism*, by Rev. W. Streissguth, *Lutheran Quarterly*, July, 1875, originally written in German, and translated by J. G. Morris, D. D., pp. 406, 408; *Darwinismus und Christenthum*, von R. Steck, Ref. Pfarrer in Dresden, Berlin, 1875, pp. 5, 6, and 26, reprinted from the *Protestantische Kirchenzeitung*, and issued as a tract by the Protestantenverein; Rev. W. E. Adams, article in the *Lutheran Quarterly*, April, 1879, on *Evolution: Shall it be Atheistic?* John Wood, *Bible Anticipations of Modern Science*, 1880, pp. 18, 19, 22; *Lutheran Quarterly*, January, 1881, *Some Postulates of the New Ethics*, by Rev. C. A. Stork, D. D.; *Lutheran Quarterly*, January, 1882, *The Religion of Evolution as against the Religion of Jesus*, by Prof. W. H. Wynn, Iowa State Agricultural College—this article was republished as a pamphlet; Canon Liddon, prefatory note to sermon on *The Recovery of St. Thomas*, pp. 4, 11, 12, 13, and 26, preached in St. Paul's Cathedral, April 23, 1882; *Lutheran Quarterly*, January, 1882, *Evolution and the Scripture*, by Rev. John A. Earnest, pp. 101, 105; *Glimpses in the Twilight*, by Rev. F. G. Lee, D. D., Edinburgh, 1885, especially pp. 18 and 19; the *Hibbert Lectures* for 1883, by Rev. Charles Beard, pp. 392, 393, *et seq.*; F. W. Farrar, D. D., Canon of Westminster, *The History of Interpretation*, being the *Bampton Lectures* for 1885, pp. 426, 427; Bishop Temple, *Bampton Lectures*, pp. 184–186; article *Evolution*, in the *Dictionary of Religion*, edited by Rev. William Benham, 1887; Prof. Huxley, *An Episcopal Trilogy*, *Nineteenth Century*, November, 1887—this article discusses three sermons delivered by the Bishops of Carlisle, Bedford, and Manchester, in Manchester Cathedral, during the meeting of the British Association, September, 1887—these sermons were afterward published in pamphlet form under the title *The Advance of Science*; John Fiske, *Darwinism, and other Essays*, Boston, 1888; Harriet Mackenzie, *Evolution illuminating the Bible*, London, 1891, dedicated to Prof. Huxley; H. E. Ryle, Hulsean Professor of Divinity at Cambridge, *The Early Narratives of Genesis*, London, 1892, preface, pp. vii–ix, pp. 7, 9, 11; Rev. G. M. Searle, of the Catholic University, Washington, article in the *Catholic World*, November, 1892, pp. 223, 227, 229, 231. For the statement from Keble College, see Rev. Mr. Illingworth, in *Lux Mundi*. For Bishop Temple, see citation in Laing. For a complete and admirable acceptance of the evolution theory as lifting Christian doctrine and practice to a higher plane, with suggestions for a new theology, see two *Sermons* by Archdeacon Wilson, of Manchester, S. P. C. K., London, and Young & Co., New York, 1893; and for a characteristically lucid statement of the most recent development of evolution doctrines, and the relations of Spencer, Weismann, Galton, and others to them, see Lester F. Ward's *Address* as President of the Biological Society, Washington, 1891; also, recent articles in the leading English reviews. For a brilliant glorification of evolution by natural selection as a doctrine necessary to the highest and truest view of Christianity, see Prof. Drummond's *Chautauqua Lectures*, published in *The British Weekly*, London, from April 20 to May 11, 1893.

CHAPTER II.

GEOGRAPHY.

I. THE FORM OF THE EARTH.

AMONG various rude tribes we find survivals of a primitive idea that the earth is a flat table or disk, ceiled, domed, or canopied by the sky, and that the sky rests upon the mountains as pillars. Such a belief is entirely natural; it conforms to the appearance of things, and hence at a very early period entered into various theologies.

In the civilizations of Chaldea and Egypt it was very fully developed. The Assyrian inscriptions deciphered in these latter years represent the god Marduk as in the beginning creating the heavens and the earth: the earth rests upon the waters; within it is the realm of the dead; above it is spread "the firmament"—a solid dome coming down to the horizon on all sides and resting upon foundations laid in the "great waters" which extend around the earth.

On the east and west sides of this domed firmament are doors, through which the sun enters in the morning and departs at night; above it extends another ocean, which goes down to the ocean surrounding the earth at the horizon on all sides, and which is supported and kept away from the earth by the firmament. Above the firmament and the upper ocean which it supports is the interior of heaven.

The Egyptians considered the earth as a table, flat and oblong, the sky being its ceiling—a huge "firmament" of metal. At the four corners of the earth were the pillars supporting this firmament, and on this solid sky were the "waters above the heavens." They believed that, when chaos was taking form, one of the gods by main force raised the waters on high and spread them out over the firmament;

that on the under side of this solid vault, or ceiling, or firmament, the stars were suspended to light the earth, and that the rains were caused by the letting down of the waters through its windows. This idea and others connected with it seem to have taken strong hold of the Egyptian priestly caste, entering into their theology and sacred science: ceilings of great temples, with stars, constellations, planets, and signs of the zodiac figured upon them, remain to-day as striking evidences of this.

In Persia we have theories of geography based upon similar conceptions and embalmed in sacred texts.

From these and doubtless from earlier sources common to them all came geographical legacies to the Hebrews. Various passages in their sacred books, many of them noble in conception and beautiful in form, regarding "the foundation of the earth upon the waters," "the fountains of the great deep," "the compass upon the face of the depth," the "firmament," the "corners of the earth," the "pillars of heaven," the "waters above the firmament," the "windows of heaven," and "doors of heaven," point us back to both these ancient springs of thought.*

---

* For survivals of the early idea, among the Eskimos, of the sky as supported by mountains, and, among sundry Pacific islanders, of the sky as a firmament or vault of stone, see Tylor, *Early History of Mankind*, second edition, London, 1870, chap. xi; Spencer, *Sociology*, vol. i, chap. viii; also Andrew Lang, *La Mythologie*, Paris, 1886, pp. 68–73. For the Babylonian theories, see George Smith's *Chaldean Genesis*, and especially the German translation by Delitzsch, Leipsic, 1876; also, Jensen, *Die Kosmogonie der Babylonier*, Strasburg, 1890; see especially in the appendices, pp. 9 and 10, a drawing representing the whole Babylonian scheme so closely followed in the Hebrew book Genesis. See also Lukas, *Die Grundbegriffe in den Kosmogonien der alten Völker*, Leipsic, 1893, for a most thorough summing up of the whole subject, with texts showing the development of Hebrew out of Chaldean and Egyptian conceptions, pp. 44, etc.; also pp. 127 *et seq.* For the early view in India and Persia, see citations from the Vedas and the Zend-Avesta in Lethaby, *Architecture, Mysticism, and Myth*, chap. i. For the Egyptian view, see Champollion; also, Lenormant, *Histoire Ancienne*, Maspero, and others. As to the figures of the heavens upon the ceilings of Egyptian temples, see Maspero, *Archéologie Egyptienne*, Paris, 1890; and for engravings of them, see Lepsius, *Denkmäler*, vol. i, Bl. 41, and vol. ix, Abth. iv, Bl. 35; also the *Déscription de l'Égypte*, published by order of Napoleon, tome ii, Pl. 14; also Prisse d'Avennes, *Art Égyptien*, Atlas, tome i, Pl. 35; and especially for a survival at the Temple of Denderah, see Denon, *Voyage en Égypte*, Planches 129, 130. For the Egyptian idea of "pillars of heaven," as alluded to on the stele of victory of Thotmes III,

But, as civilization was developed, there were evolved, especially among the Greeks, ideas of the earth's sphericity. The Pythagoreans, Plato, and Aristotle especially cherished them. These ideas were vague, they were mixed with absurdities, but they were germ ideas, and even amid the luxuriant growth of theology in the early Christian Church these germs began struggling into life in the minds of a few thinking men, and these men renewed the suggestion that the earth is a globe.*

A few of the larger-minded fathers of the Church, influenced possibly by Pythagorean traditions, but certainly by Aristotle and Plato, were willing to accept this view, but the majority of them took fright at once. To them it seemed fraught with dangers to Scripture, by which, of course, they meant *their interpretation* of Scripture. Among the first who took up arms against it was Eusebius. In view of the New Testament texts indicating the immediately approaching end of the world, he endeavoured to turn off this idea by bringing scientific studies into contempt. Speaking of investigators, he said, "It is not through ignorance of the things admired by them, but through contempt of their use-

---

in the Cairo Museum, see Ebers, *Uarda*, vol. ii, p. 175, note, Leipsic, 1877. For a similar Babylonian belief, see Sayce's *Herodotus*, Appendix, p. 403. For the belief of Hebrew scriptural writers in a solid "firmament," see especially Job, xxxviii, 18; also Smith's *Bible Dictionary*. For engravings showing the earth and heaven above it as conceived by Egyptians and Chaldeans, with "pillars of heaven" and "firmament," see Maspero and Sayce, *Dawn of Civilization*, London, 1894, pp. 17 and 543.

* The agency of the Pythagoreans in first spreading the doctrine of the earth's sphericity is generally acknowledged, but the first clear and full utterance of it to the world was by Aristotle. Very fruitful, too, was the statement of the new theory given by Plato in the *Timæus*; see Jowett's translation, 62, c. Also the *Phædo*, pp. 449 *et seq.* See also Grote on Plato's doctrine of the sphericity of the earth; also Sir G. C. Lewis's *Astronomy of the Ancients*, London, 1862, chap. iii, section i, and note. Cicero's mention of the antipodes, and his reference to the passage in the Timæus, are even more remarkable than the latter, in that they much more clearly foreshadow the modern doctrine. See his *Academic Questions*, ii; also *Tusc. Quest.*, i and v, 24. For a very full summary of the views of the ancients on the sphericity of the earth, see Kretschmer, *Die physische Erdkunde im christlichen Mittelalter*, Wien, 1889, pp. 35 *et seq.*; also, Eicken, *Geschichte der mittelalterlichen Weltanschauung*, Stuttgart, 1887, Dritter Theil, chap. vi. For citations and summaries, see Whewell, *Hist. Induct. Sciences*, vol. i, p. 189, and St. Martin, *Hist. de la Géog.*, Paris, 1873, p. 96; also, Leopardi, *Saggio sopra gli errori popolari degli antichi*, Firenze, 1851, chap. xii, pp. 184 *et seq.*

less labour, that we think little of these matters, turning our souls to better things." Basil of Cæsarea declared it "a matter of no interest to us whether the earth is a sphere or a cylinder or a disk, or concave in the middle like a fan." Lactantius referred to the ideas of those studying astronomy as "bad and senseless," and opposed the doctrine of the earth's sphericity both from Scripture and reason. St. John Chrysostom also exerted his influence against this scientific belief; and Ephraem Syrus, the greatest man of the old Syrian Church, widely known as the "lute of the Holy Ghost," opposed it no less earnestly.

But the strictly biblical men of science, such eminent fathers and bishops as Theophilus of Antioch in the second century, and Clement of Alexandria in the third, with others in centuries following, were not content with merely opposing what they stigmatized as an old heathen theory; they drew from their Bibles a new Christian theory, to which one Church authority added one idea and another another, until it was fully developed. Taking the survival of various early traditions, given in the seventh verse of the first chapter of Genesis, they insisted on the clear declarations of Scripture that the earth was, at creation, arched over with a solid vault, "a firmament," and to this they added the passages from Isaiah and the Psalms, in which it declared that the heavens are stretched out "like a curtain," and again "like a tent to dwell in." The universe, then, is like a house: the earth is its ground floor, the firmament its ceiling, under which the Almighty hangs out the sun to rule the day and the moon and stars to rule the night. This ceiling is also the floor of the apartment above, and in this is a cistern, shaped, as one of the authorities says, "like a bathing-tank," and containing "the waters which are above the firmament." These waters are let down upon the earth by the Almighty and his angels through the "windows of heaven." As to the movement of the sun, there was a citation of various passages in Genesis, mixed with metaphysics in various proportions, and this was thought to give ample proofs from the Bible that the earth could not be a sphere.*

* For Eusebius, see the *Præp. Ev.*, xv, 61. For Basil, see the *Hexæmeron*,

In the sixth century this development culminated in what was nothing less than a complete and detailed system of the universe, claiming to be based upon Scripture, its author being the Egyptian monk Cosmas Indicopleustes. Egypt was a great treasure-house of theologic thought to various religions of antiquity, and Cosmas appears to have urged upon the early Church this Egyptian idea of the construction of the world, just as another Egyptian ecclesiastic, Athanasius, urged upon the Church the Egyptian idea of a triune deity ruling the world. According to Cosmas, the earth is a parallelogram, flat, and surrounded by four seas. It is four hundred days' journey long and two hundred broad. At the outer edges of these four seas arise massive walls closing in the whole structure and supporting the firmament or vault of the heavens, whose edges are cemented to the walls. These walls inclose the earth and all the heavenly bodies.

The whole of this theologico-scientific structure was built most carefully and, as was then thought, most scripturally. Starting with the expression applied in the ninth chapter of Hebrews to the tabernacle in the desert, Cosmas insists, with other interpreters of his time, that it gives the key to the whole construction of the world. The universe is, therefore, made on the plan of the Jewish tabernacle—boxlike and oblong. Going into details, he quotes the sublime words of Isaiah: "It is He that sitteth upon the circle of the earth; . . . that stretcheth out the heavens like a curtain, and spreadeth them out like a tent to dwell in"; and the passage in Job which speaks of the "pillars of heaven." He works all this into his system, and reveals, as he thinks, treasures of science.

This vast box is divided into two compartments, one above the other. In the first of these, men live and stars move; and it extends up to the first solid vault, or firmament, above which live the angels, a main part of whose business it is to push and pull the sun and planets to and

---

Hom. ix. For Lactantius, see his *Inst. Div.*, lib. iii, cap. 3; also, citations in Whewell, *Hist. Induct. Sciences*, London, 1857, vol. i, p. 194, and in St. Martin, *Histoire de la Géographie*, pp. 216, 217. For the views of St. John Chrysostom, Ephraem Syrus, and other great churchmen, see Kretschmer as above, chap. i.

fro. Next, he takes the text, "Let there be a firmament in the midst of the waters, and let it divide the waters from the waters," and other texts from Genesis; to these he adds the text from the Psalms, "Praise him, ye heaven of heavens, and ye waters that be above the heavens"; then casts all these growths of thought into his crucible together, and finally brings out the theory that over this first vault is a vast cistern containing "the waters." He then takes the expression in Genesis regarding the "windows of heaven" and establishes a doctrine regarding the regulation of the rain, to the effect that the angels not only push and pull the heavenly bodies to light the earth, but also open and close the heavenly windows to water it.

To understand the surface of the earth, Cosmas, following the methods of interpretation which Origen and other early fathers of the Church had established, studies the table of shew-bread in the Jewish tabernacle. The surface of this table proves to him that the earth is flat, and its dimensions prove that the earth is twice as long as broad; its four corners symbolize the four seasons; the twelve loaves of bread, the twelve months; the hollow about the table proves that the ocean surrounds the earth. To account for the movement of the sun, Cosmas suggests that at the north of the earth is a great mountain, and that at night the sun is carried behind this; but some of the commentators ventured to express a doubt here: they thought that the sun was pushed into a pit at night and pulled out in the morning.

Nothing can be more touching in its simplicity than Cosmas's summing up of his great argument. He declares, "We say therefore with Isaiah that the heaven embracing the universe is a vault, with Job that it is joined to the earth, and with Moses that the length of the earth is greater than its breadth." The treatise closes with rapturous assertions that not only Moses and the prophets, but also angels and apostles, agree to the truth of his doctrine, and that at the last day God will condemn all who do not accept it.

Although this theory was drawn from Scripture, it was also, as we have seen, the result of an evolution of theological thought begun long before the scriptural texts on which it rested were written. It was not at all strange that Cosmas,

Egyptian as he was, should have received this old Nile-born doctrine, as we see it indicated to-day in the structure of Egyptian temples, and that he should have developed it by the aid of the Jewish Scriptures; but the theological world knew nothing of this more remote evolution from pagan germs; it was received as virtually inspired, and was soon regarded as a fortress of scriptural truth. Some of the foremost men in the Church devoted themselves to buttressing it with new texts and throwing about it new outworks of theological reasoning; the great body of the faithful considered it a direct gift from the Almighty. Even in the later centuries of the Middle Ages John of San Geminiano made a desperate attempt to save it. Like Cosmas, he takes the Jewish tabernacle as his starting-point, and shows how all the newer ideas can be reconciled with the biblical accounts of its shape, dimensions, and furniture.*

---

* For a notice of the views of Cosmas in connection with those of Lactantius, Augustine, St John Chrysostom, and others, see Schoell, *Histoire de la Littérature Grecque*, vol. vii, p. 37. The main scriptural passages referred to are as follows: (1) Isaiah xl, 22; (2) Genesis i, 6; (3) Genesis vii, 11; (4) Exodus xxiv, 10; (5) Job xxvi, 11, and xxxvii, 18; (6) Psalm cxlviii, 4, and civ, 9; (7) Ezekiel i, 22–26. For Cosmas's theory, see Montfaucon, *Collectio Nova Patrum*, Paris, 1706, vol. ii, p. 188; also pp. 298, 299. The text is illustrated with engravings showing walls and solid vault (firmament), with the whole apparatus of "fountains of the great deep," "windows of heaven," angels, and the mountain behind which the sun is drawn. For reduction of one of them, see Peschel, *Geschichte der Erdkunde*, p. 98; also article *Maps*, in Knight's *Dictionary of Mechanics*, New York, 1875. For curious drawings showing Cosmas's scheme in a different way from that given by Montfaucon, see extracts from a Vatican codex of the ninth century in Garucci, *Storia de l'Arte Christiana*, vol. iii, pp. 70 *et seq*. For a good discussion of Cosmas's ideas, see Santarem, *Hist. de la Cosmographie*, vol. ii, pp. 8 *et seq*., and for a very thorough discussion of its details, Kretschmer, as above. For still another theory, very droll, and thought out on similar principles, see Mungo Park, cited in De Morgan, *Paradoxes*, p. 309. For Cosmas's joyful summing up, see Montfaucon, *Collectio Nova Patrum*, vol. ii, p. 255. For a curious survival in the thirteenth century of the old idea of the "waters above the heavens," see the story in Gervase of Tilbury, how in his time some people coming out of church in England found an anchor let down by a rope out of the heavens, how there came voices from sailors above trying to loose the anchor, and, finally, how a sailor came down the rope, who, on reaching the earth, died as if drowned in water. See Gervase of Tilbury, *Otia Imperialia*, edit. Liebrecht, Hanover, 1856, Prima Decisio, cap. xiii. The work was written about 1211. For John of San Geminiano, see his *Summa de Exemplis*, lib. ix, cap. 43. For the Egyptian Trinitarian views, see Sharpe, *History of Egypt*, vol. i, pp. 94, 102.

From this old conception of the universe as a sort of house, with heaven as its upper story and the earth as its ground floor, flowed important theological ideas into heathen, Jewish, and Christian mythologies. Common to them all are legends regarding attempts of mortals to invade the upper apartment from the lower. Of such are the Greek legends of the Aloidae, who sought to reach heaven by piling up mountains, and were cast down; the Chaldean and Hebrew legends of the wicked who at Babel sought to build "a tower whose top may reach heaven," which Jehovah went down from heaven to see, and which he brought to naught by the "confusion of tongues"; the Hindu legend of the tree which sought to grow into heaven and which Brahma blasted; and the Mexican legend of the giants who sought to reach heaven by building the Pyramid of Cholula, and who were overthrown by fire from above.

Myths having this geographical idea as their germ developed in luxuriance through thousands of years. Ascensions to heaven and descents from it, "translations," "assumptions," "annunciations," mortals "caught up" into it and returning, angels flying between it and the earth, thunderbolts hurled down from it, mighty winds issuing from its corners, voices speaking from the upper floor to men on the lower, temporary openings of the floor of heaven to reveal the blessedness of the good, "signs and wonders" hung out from it to warn the wicked, interventions of every kind—from the heathen gods coming down on every sort of errand, and Jehovah coming down to walk in Eden in the cool of the day, to St. Mark swooping down into the market-place of Venice to break the shackles of a slave—all these are but features in a vast evolution of myths arising largely from this geographical germ.

Nor did this evolution end here. Naturally, in this view of things, if heaven was a loft, hell was a cellar; and if there were ascensions into one, there were descents into the other. Hell being so near, interferences by its occupants with the dwellers of the earth just above were constant, and form a vast chapter in mediæval literature. Dante made this conception of the location of hell still more vivid, and we find some forms of it serious barriers to geographical investigation. Many a bold navigator, who was quite ready to brave pirates and tempests, trembled at the thought of tumbling with his ship into one of the openings into hell which a widespread belief placed in the Atlantic at some unknown distance from Europe. This terror among sailors was one of the main obstacles in the great voyage of Columbus. In a mediæval text-book, giving science the form of a dialogue, occur the following question and answer: "Why is the sun so red in the evening?" "Because he looketh down upon hell."

But the ancient germ of scientific truth in geography—the idea of the earth's sphericity—still lived. Although the great majority of the early fathers of the Church, and especially Lactantius, had sought to crush it beneath the utterances attributed to Isaiah, David, and St. Paul, the better opinion of Eudoxus and Aristotle could not be forgotten. Clement of Alexandria and Origen had even supported it. Ambrose and Augustine had tolerated it, and, after Cosmas had held sway a hundred years, it received new life from a great churchman of southern Europe, Isidore of Seville, who, however fettered by the dominant theology in many other things, braved it in this. In the eighth century a similar declaration was made in the north of Europe by another great Church authority, Bede. Against the new life thus given to the old truth, the sacred theory struggled long and vigorously but in vain. Eminent authorities in later ages, like Albert the Great, St. Thomas Aquinas, Dante, and Vincent of Beauvais, felt obliged to accept the doctrine of the earth's sphericity, and as we approach the modern period we find its truth acknowledged by the vast majority of thinking men. The Reformation did not at first yield fully to this better theory. Luther, Melanchthon, and Calvin were very strict in their adherence to the exact letter of Scripture. Even Zwingli, broad as his views generally were, was closely bound down in this matter, and held to the opinion of the fathers that a great firmament, or floor, separated the heavens from the earth; that above it were the waters and angels, and below it the earth and man.

The main scope given to independent thought on this general subject among the Reformers was in a few minor speculations regarding the universe which encompassed Eden, the exact character of the conversation of the serpent with Eve, and the like.

In the times immediately following the Reformation matters were even worse. The interpretations of Scripture by Luther and Calvin became as sacred to their followers as the Scripture itself. When Calixt ventured, in interpreting the Psalms, to question the accepted belief that "the waters above the heavens" were contained in a vast receptacle upheld by a solid vault, he was bitterly denounced as heretical.

In the latter part of the sixteenth century Musæus interpreted the accounts in Genesis to mean that first God made the heavens for the roof or vault, and left it there on high swinging until three days later he put the earth under it. But the new scientific thought as to the earth's form had gained the day. The most sturdy believers were obliged to adjust their biblical theories to it as best they could.*

II. THE DELINEATION OF THE EARTH.

Every great people of antiquity, as a rule, regarded its own central city or most holy place as necessarily the centre of the earth.

The Chaldeans held that their "holy house of the gods" was the centre. The Egyptians sketched the world under the form of a human figure, in which Egypt was the heart, and the centre of it Thebes. For the Assyrians, it was Babylon; for the Hindus, it was Mount Meru; for the Greeks, so far as the civilized world was concerned, Olympus or the temple at Delphi; for the modern Mohammedans, it is Mecca and its sacred stone; the Chinese, to this day, speak of their empire as the "middle kingdom." It was in accordance, then, with a simple tendency of human

---

* For a discussion of the geographical views of Isidore and Bede, see Santarem, *Cosmographie*, vol. i, pp 22–24. For the gradual acceptance of the idea of the earth's sphericity after the eighth century, see Kretschmer, pp. 51 *et seq.*, where citations from a multitude of authors are given. For the views of the Reformers, see Zöckler, vol. i, pp. 679 and 693. For Calixt, Musæus, and others, ibid., pp. 673–677 and 761.

thought that the Jews believed the centre of the world to be Jerusalem.

The book of Ezekiel speaks of Jerusalem as in the middle of the earth, and all other parts of the world as set around the holy city. Throughout the "ages of faith" this was very generally accepted as a direct revelation from the Almighty regarding the earth's form. St. Jerome, the greatest authority of the early Church upon the Bible, declared, on the strength of this utterance of the prophet, that Jerusalem could be nowhere but at the earth's centre; in the ninth century Archbishop Rabanus Maurus reiterated the same argument; in the eleventh century Hugh of St. Victor gave to the doctrine another scriptural demonstration; and Pope Urban, in his great sermon at Clermont urging the Franks to the crusade, declared, "Jerusalem is the middle point of the earth"; in the thirteenth century an ecclesiastical writer much in vogue, the monk Cæsarius of Heisterbach, declared, "As the heart in the midst of the body, so is Jerusalem situated in the midst of our inhabited earth,"—"so it was that Christ was crucified at the centre of the earth." Dante accepted this view of Jerusalem as a certainty, wedding it to immortal verse; and in the pious book of travels ascribed to Sir John Mandeville, so widely read in the Middle Ages, it is declared that Jerusalem is at the centre of the world, and that a spear standing erect at the Holy Sepulchre casts no shadow at the equinox.

Ezekiel's statement thus became the standard of orthodoxy to early map-makers. The map of the world at Hereford Cathedral, the maps of Andrea Bianco, Marino Sanuto, and a multitude of others fixed this view in men's minds, and doubtless discouraged during many generations any scientific statements tending to unbalance this geographical centre revealed in Scripture.*

* For the beliefs of various nations of antiquity that the earth's centre was in their most sacred place, see citations from Maspero, Charton, Sayce, and others in Lethaby, *Architecture, Mysticism, and Myth*, chap. iv. As to the Greeks, we have typical statements in the *Eumenides* of Æschylus, where the stone on the altar at Delphi is repeatedly called "the earth's navel"—which is precisely the expression used regarding Jerusalem in the Septuagint translation of Ezekiel (see below). The proof texts on which the mediæval geographers mainly relied as to the form

Nor did mediæval thinkers rest with this conception. In accordance with the dominant view that physical truth must be sought by theological reasoning, the doctrine was evolved that not only the site of the cross on Calvary marked the geographical centre of the world, but that on this very spot had stood the tree which bore the forbidden fruit in Eden. Thus was geography made to reconcile all parts of the great theologic plan. This doctrine was hailed with joy by multitudes; and we find in the works of mediæval pilgrims to Palestine, again and again, evidence that this had become precious truth to them, both in theology and geography. Even as late as 1664 the eminent French priest Eugène Roger, in his published travels in Palestine, dwelt upon the thirty-eighth chapter of Ezekiel, coupled with a text from Isaiah, to prove that the exact centre of the earth is a spot marked on the pavement of the Church of the Holy Sepulchre, and that on this spot once stood the tree which bore the forbidden fruit and the cross of Christ.*

---

of the earth were Ezekiel v, 5, and xxxviii, 12. The progress of geographical knowledge evidently caused them to be softened down somewhat in our King James's version; but the first of them reads, in the Vulgate, "*Ista est Hierusalem, in medio gentium posui eam et in circuitu ejus terræ*"; and the second reads, in the Vulgate, "*in medio terræ,*" and in the Septuagint, ἐπὶ τὸν ὀμφαλὸν τῆς γῆς. That the literal centre of the earth was understood, see proof in St. Jerome, *Commentar. in Ezekiel*, lib. ii; and for general proof, see Leopardi, *Saggio sopra gli errori popolari degli antichi*, pp. 207, 208. For Rabanus Maurus, see his *De Universo*, lib. xii, cap. 4, in Migne, tome cxi, p. 339. For Hugh of St. Victor, see his *De Situ Terrarum*, cap. ii. For Dante's belief, see *Inferno*, canto xxxiv, 112–115:

"E se' or sotto l'emisperio giunto,
Ch' è opposito a quel che la gran secca
Coverchia, e sotto il cui colmo consunto
Fu l'uom che nacque e visse senza pecca."

For orthodox geography in the Middle Ages, see Wright's *Essays on Archæology*, vol. ii, chapter on the map of the world in Hereford Cathedral; also the rude maps in Cardinal d'Ailly's *Ymago Mundi*; also copies of maps of Marino Sanuto and others in Peschel, *Erdkunde*, p. 210; also Münster, *Fac Simile dell' Atlante di Andrea Bianco*, Venezia, 1869. And for discussions of the whole subject, see Santarem, vol. ii, p. 295, vol. iii, pp. 71, 183, 184, and elsewhere. For a brief summary with citations, see Eicken, *Geschichte*, etc., pp. 622, 623.

* For the site of the cross on Calvary, as the point where stood "the tree of the knowledge of good and evil" in Eden, at the centre of the earth, see various Eastern travellers cited in Tobler; but especially the travels of Bishop Arculf in the Holy Land, in Wright's *Early Travels in Palestine*, p. 8; also *Travels of Saewulf*, ibid., p. 38; also, Sir John Mandeville, ibid., pp. 166, 167. For Roger,

Nor was this the only misconception which forced its way from our sacred writings into mediæval map-making: two others were almost as marked.

First of these was the vague terror inspired by Gog and Magog. Few passages in the Old Testament are more sublime than the denunciation of these great enemies by Ezekiel; and the well-known statement in the Apocalypse fastened the Hebrew feeling regarding them with a new meaning into the mind of the early Church: hence it was that the mediæval map-makers took great pains to delineate these monsters and their habitations on the maps. For centuries no map was considered orthodox which did not show them.

The second conception was derived from the mention in our sacred books of the "four winds." Hence came a vivid belief in their real existence, and their delineation on the maps, generally as colossal heads with distended cheeks, blowing vigorously toward Jerusalem.

After these conceptions had mainly disappeared we find here and there evidences of the difficulty men found in giving up the scriptural idea of direct personal interference by agents of Heaven in the ordinary phenomena of Nature: thus, in a noted map of the sixteenth century representing the earth as a sphere, there is at each pole a crank, with an angel laboriously turning the earth by means of it: and, in another map, the hand of the Almighty, thrust forth from the clouds, holds the earth suspended by a rope and spins it with his thumb and fingers. Even as late as the middle of the seventeenth century Heylin, the most authoritative English geographer of the time, shows a like tendency to mix science and theology. He warps each to help the other, as follows: "Water, making but one globe with the earth, is

---

see his *La Terre Saincte*, Paris, 1664, pp. 89–218, etc.; see also Quaresmio, *Terræ Sanctæ Elucidatio*, 1639, for similar view; and, for one narrative in which the idea was developed into an amazing mass of pious myths, see *Pilgrimage of the Russian Abbot Daniel*, edited by Sir C. W. Wilson, London, 1885, p. 14. (The passage deserves to be quoted as an example of myth-making; it is as follows: "At the time of our Lord's crucifixion, when he gave up the ghost on the cross, the veil of the temple was rent, and the rock above Adam's skull opened, and the blood and water which flowed from Christ's side ran down through the fissure upon the skull, thus washing away the sins of men.")

yet higher than it. This appears, first, because it is a body not so heavy; secondly, it is observed by sailors that their ships move faster to the shore than from it, whereof no reason can be given but the height of the water above the land; thirdly, to such as stand on the shore the sea seems to swell into the form of a round hill till it puts a bound upon our sight. Now that the sea, hovering thus over and above the earth, doth not overwhelm it, can be ascribed only to his Providence who 'hath made the waters to stand on an heap that they turn not again to cover the earth.'" *

III. THE INHABITANTS OF THE EARTH.

Even while the doctrine of the sphericity of the earth was undecided, another question had been suggested which theologians finally came to consider of far greater importance. The doctrine of the sphericity of the earth naturally led to thought regarding its inhabitants, and another ancient germ was warmed into life—the idea of antipodes: of human beings on the earth's opposite sides.

In the Greek and Roman world this idea had found supporters and opponents, Cicero and Pliny being among the former, and Epicurus, Lucretius, and Plutarch among the latter. Thus the problem came into the early Church unsolved.

Among the first churchmen to take it up was, in the East, St. Gregory Nazianzen, who showed that to sail be-

---

* For Gog and Magog, see Ezekiel xxxviii and xxxix, and Rev. xx, 8; and for the general subject, Toy, *Judaism and Christianity*, Boston, 1891, pp. 373, 374. For maps showing these two great terrors, and for geographical discussion regarding them, see Lelewel, *Géog. du Moyen Age*, Bruxelles, 1850, Atlas; also Ruge, *Gesch. des Zeitalters der Entdeckungen*, Berlin, 1881, pp. 78, 79; also Peschel's *Abhandlungen*, pp. 28–35, and *Gesch. der Erdkunde*, p. 210. For representations on maps of the "Four Winds," see Charton, *Voyageurs*, tome ii, p. 11; also Ruge, as above, pp. 324, 325; also, for a curious mixture of the scriptural four winds with the classical winds issuing from the bags of Æolus, see a map of the twelfth century in Léon Gautier, *La Chevalerie*, p. 153; and for maps showing additional winds, see various editions of Ptolemy. For a map with angels turning the earth by means of cranks at the poles, see Grynæus, *Novus Orbis*, Basileæ, 1537. For the globe kept spinning by the Almighty, see J. Hondius's map, 1589; and for Heylin, his first folio, 1652, p. 27.

yond Gibraltar was impossible; and, in the West, Lactantius, who asked: "Is there any one so senseless as to believe that there are men whose footsteps are higher than their heads? . . . that the crops and trees grow downward? . . . that the rains and snow and hail fall upward toward the earth? . . . I am at a loss what to say of those who, when they have once erred, steadily persevere in their folly and defend one vain thing by another."

In all this contention by Gregory and Lactantius there was nothing to be especially regretted, for, whatever their motive, they simply supported their inherited belief on grounds of natural law and probability.

Unfortunately, the discussion was not long allowed to rest on these scientific and philosophical grounds; other Christian thinkers followed, who in their ardour adduced texts of Scripture, and soon the question had become theological; hostility to the belief in antipodes became dogmatic. The universal Church was arrayed against it, and in front of the vast phalanx stood, to a man, the fathers.

To all of them this idea seemed dangerous; to most of them it seemed damnable. St. Basil and St. Ambrose were tolerant enough to allow that a man might be saved who thought the earth inhabited on its opposite sides; but the great majority of the fathers doubted the possibility of salvation to such misbelievers.

The great champion of the orthodox view was St. Augustine. Though he seemed inclined to yield a little in regard to the sphericity of the earth, he fought the idea that men exist on the other side of it, saying that "Scripture speaks of no such descendants of Adam." He insists that men could not be allowed by the Almighty to live there, since if they did they could not see Christ at his second coming descending through the air. But his most cogent appeal, one which we find echoed from theologian to theologian during a thousand years afterward, is to the nineteenth Psalm, and to its confirmation in the Epistle to the Romans; to the words, "Their line is gone out through all the earth, and their words to the end of the world." He dwells with great force on the fact that St. Paul based one of his most powerful arguments upon this declaration regarding the preachers of the gospel, and that he declared even more explicitly that "Verily, their sound went into all the earth, and their words unto the ends of the world." Thenceforth we find it constantly declared that, as those preachers did not go to the antipodes, no antipodes can exist; and hence that the supporters of this geographical doctrine "give the lie direct to King David and to St. Paul, and therefore to the Holy Ghost." Thus the great Bishop of Hippo taught the whole world for over a thousand years that, as there was no preaching of the gospel on the opposite side of the earth, there could be no human beings there.

The great authority of Augustine, and the cogency of his scriptural argument, held the Church firmly against the doctrine of the antipodes; all schools of interpretation were now agreed—the followers of the allegorical tendencies of Alexandria, the strictly literal exegetes of Syria, the more eclectic theologians of the West. For over a thousand years it was held in the Church, "always, everywhere, and by all," that there could not be human beings on the opposite sides of the earth, even if the earth had opposite sides; and, when attacked by gainsayers, the great mass of true believers, from the fourth century to the fifteenth, simply used that opiate which had so soothing an effect on John Henry Newman in the nineteenth century—*securus judicat orbis terrarum.*

Yet gainsayers still appeared. That the doctrine of the antipodes continued to have life, is shown by the fact that in the sixth century Procopius of Gaza attacks it with a tremendous argument. He declares that, if there be men on the other side of the earth, Christ must have gone there and suffered a second time to save them; and, therefore, that there must have been there, as necessary preliminaries to his coming, a duplicate Eden, Adam, serpent, and deluge.

Cosmas Indicopleustes also attacked the doctrine with especial bitterness, citing a passage from St. Luke to prove that antipodes are theologically impossible.

At the end of the sixth century came a man from whom much might be expected—St. Isidore of Seville. He had pondered over ancient thought in science, and, as we have seen, had dared proclaim his belief in the sphericity of the earth; but with that he stopped. As to the antipodes, the authority of the Psalmist, St. Paul, and St. Augustine silences him; he shuns the whole question as unlawful, subjects reason to faith, and declares that men can not and ought not to exist on opposite sides of the earth.*

Under such pressure this scientific truth seems to have disappeared for nearly two hundred years; but by the eighth century the sphericity of the earth had come to be generally accepted among the leaders of thought, and now the doctrine of the antipodes was again asserted by a bishop, Virgil of Salzburg.

There then stood in Germany, in those first years of the eighth century, one of the greatest and noblest of men—St. Boniface. His learning was of the best then known. In labours he was a worthy successor of the apostles; his genius for Christian work made him unwillingly primate of Germany; his devotion to duty led him willingly to martyrdom. There sat, too, at that time, on the papal throne a great Christian statesman—Pope Zachary. Boniface immediately declared against the revival of such a heresy as the doctrine of the antipodes; he stigmatized it as an assertion that there are men beyond the reach of the appointed means of salvation; he attacked Virgil, and called on Pope Zachary for aid.

---

* For the opinions of Basil, Ambrose, and others, see Lecky, *History of Rationalism in Europe*, New York, 1872, vol. i, p. 279, note. Also Letronne, in *Revue des Deux Mondes*, March, 1834. For Lactantius, see citations already given. For St. Augustine's opinion, see the *De Civitate Dei*, xvi, 9, where this great father of the Church shows that the existence of the antipodes "*nulla ratione credendum est.*" For the unanimity of the fathers against the antipodes, see Zöckler, vol. i, p. 127. For a very naïve summary, see Joseph Acosta, *Natural and Moral History of the Indies*, Grimston's translation, republished by the Hakluyt Soc., chaps. vii and viii; also citations in Buckle's *Posthumous Works*, vol. ii, p. 645. For Procopius of Gaza, see Kretschmer, p. 55. See also, on the general subject, Peschel, *Geschichte der Erdkunde*, pp. 96, 97. For Isidore, see citations already given. To understand the embarrassment caused by these utterances of the fathers to scientific men of a later period, see letter of Agricola to Joachim Vadianus in 1514. Agricola asks Vadianus to give his views regarding the antipodes, saying that he himself does not know what to do, between the fathers on the one side and the learned men of modern times on the other. On the other hand, for the embarrassment caused to the Church by this mistaken zeal of the fathers, see Kepler's references and Fromund's replies; also De Morgan, *Paradoxes*, p. 58. Kepler appears to have taken great delight in throwing the views of Lactantius into the teeth of his adversaries.

The Pope, as the infallible teacher of Christendom, made a strong response. He cited passages from the book of Job and the Wisdom of Solomon against the doctrine of the antipodes; he declared it "perverse, iniquitous, and against Virgil's own soul," and indicated a purpose of driving him from his bishopric. Whether this purpose was carried out or not, the old theological view, by virtue of the Pope's divinely ordered and protected "inerrancy," was re-established, and the doctrine that the earth has inhabitants on but one of its sides became more than ever orthodox, and precious in the mind of the Church.*

This decision seems to have been regarded as final, and five centuries later the great encyclopedist of the Middle Ages, Vincent of Beauvais, though he accepts the sphericity of the earth, treats the doctrine of the antipodes as disproved, because contrary to Scripture. Yet the doctrine still lived. Just as it had been previously revived by William of Conches and then laid to rest, so now it is somewhat timidly brought out in the thirteenth century by no less a personage than Albert the Great, the most noted man of science in that time. But his utterances are perhaps purposely obscure. Again it disappears beneath the theological wave, and a hundred years later Nicolas d'Oresme, geographer of the King of France, a light of science, is forced to yield to the clear teaching of the Scripture as cited by St. Augustine.

Nor was this the worst. In Italy, at the beginning of the fourteenth century, the Church thought it necessary to deal with questions of this sort by rack and fagot. In 1316 Peter of Abano, famous as a physician, having promulgated

---

* For Virgil of Salzburg, see Neander's *History of the Christian Church*, Torrey's translation, vol. iii, p. 63; also Herzog, *Real-Encyklopädie*, etc., recent edition by Prof. Hauck, s. v. *Virgilius*; also Kretschmer, pp. 56–58; also Whewell, vol. i, p. 197; also De Morgan, *Budget of Paradoxes*, pp. 24–26. For very full notes as to pagan and Christian advocates of the doctrine of the sphericity of the earth and of the antipodes, and for extract from Zachary's letter, see Migne, *Patrologia*, vol. vi, p. 426, and vol. xli, p. 487. For St. Boniface's part, see *Bonifacii Epistolæ*, ed. Giles, i, 173. Berger de Xivrey, *Traditions Tératologiques*, pp. 186–188, makes a curious attempt to show that Pope Zachary denounced the wrong man; that the real offender was the Roman poet—in the sixth book of the *Æneid* and the first book of the *Georgics*.

this with other obnoxious doctrines in science, only escaped the Inquisition by death; and in 1327 Cecco d'Ascoli, noted as an astronomer, was for this and other results of thought, which brought him under suspicion of sorcery, driven from his professorship at Bologna and burned alive at Florence. Nor was this all his punishment: Orcagna, whose terrible frescoes still exist on the walls of the Campo Santo at Pisa, immortalized Cecco by representing him in the flames of hell.*

Years rolled on, and there came in the fifteenth century one from whom the world had a right to expect much. Pierre d'Ailly, by force of thought and study, had risen to be Provost of the College of St. Dié in Lorraine; his ability had made that little village a centre of scientific thought for all Europe, and finally made him Archbishop of Cambray and a cardinal. Toward the end of the fifteenth century was printed what Cardinal d'Ailly had written long before as a summing up of his best thought and research—the collection of essays known as the *Ymago Mundi.* It gives us one of the most striking examples in history of a great man in theological fetters. As he approaches this question he states it with such clearness that we expect to hear him assert the truth; but there stands the argument of St. Augustine; there, too, stand the biblical texts on which it is founded—the text from the Psalms and the explicit declaration of St. Paul to the Romans, "Their sound went into all the earth, and their words unto the ends of the world." D'Ailly attempts to reason, but he is overawed, and gives to the world virtually nothing.

---

* For Vincent of Beauvais and the antipodes, see his *Speculum Naturale*, Book VII, with citations from St. Augustine, *De Civitate Dei*, cap. xvi. For Albert the Great's doctrine regarding the antipodes, compare Kretschmer, as above, with Eicken, *Geschichte*, etc., p. 621. Kretschmer finds that Albert supports the doctrine, and Eicken finds that he denies it—a fair proof that Albert was not inclined to state his views with dangerous clearness. For D'Oresme, see Santarem, *Histoire de la Cosmographie*, vol. i, p. 142. For Peter of Abano, or Apono, as he is often called, see Tiraboschi; also Ginguené, vol. ii, p. 293; also Naudé, *Histoire des Grands Hommes soupçonnés de Magie.* For Cecco d'Ascoli, see Montucla, *Histoire des Mathématiques*, i, 528; also Daunou, *Études Historiques*, vol. vi, p. 320; also Kretschmer, p. 59. Concerning Orcagna's representation of Cecco in the flames of hell, see Renan, *Averroes et l'Averroisme*, Paris, 1867, p. 328.

Still, the doctrine of the antipodes lived and moved: so much so that the eminent Spanish theologian Tostatus, even as late as the age of Columbus, felt called upon to protest against it as "unsafe." He had shaped the old missile of St. Augustine into the following syllogism: "The apostles were commanded to go into all the world and to preach the gospel to every creature; they did not go to any such part of the world as the antipodes; they did not preach to any creatures there: *ergo*, no antipodes exist."

The warfare of Columbus the world knows well: how the Bishop of Ceuta worsted him in Portugal; how sundry wise men of Spain confronted him with the usual quotations from the Psalms, from St. Paul, and from St. Augustine; how, even after he was triumphant, and after his voyage had greatly strengthened the theory of the earth's sphericity, with which the theory of the antipodes was so closely connected, the Church by its highest authority solemnly stumbled and persisted in going astray. In 1493 Pope Alexander VI, having been appealed to as an umpire between the claims of Spain and Portugal to the newly discovered parts of the earth, issued a bull laying down upon the earth's surface a line of demarcation between the two powers. This line was drawn from north to south a hundred leagues west of the Azores; and the Pope in the plenitude of his knowledge declared that all lands discovered east of this line should belong to the Portuguese, and all west of it should belong to the Spaniards. This was hailed as an exercise of divinely illuminated power by the Church; but difficulties arose, and in 1506 another attempt was made by Pope Julius II to draw the line three hundred and seventy leagues west of the Cape Verde Islands. This, again, was supposed to bring divine wisdom to settle the question; but, shortly, overwhelming difficulties arose; for the Portuguese claimed Brazil, and, of course, had no difficulty in showing that they could reach it by sailing to the east of the line, provided they sailed long enough. The lines laid down by Popes Alexander and Julius may still be found upon the maps of the period, but their bulls have quietly passed into the catalogue of ludicrous errors.

Yet the theological barriers to this geographical truth

yielded but slowly. Plain as it had become to scholars, they hesitated to declare it to the world at large. Eleven hundred years had passed since St. Augustine had proved its antagonism to Scripture, when Gregory Reysch gave forth his famous encyclopædia, the *Margarita Philosophica.* Edition after edition was issued, and everywhere appeared in it the orthodox statements; but they were evidently strained to the breaking point; for while, in treating of the antipodes, Reysch refers respectfully to St. Augustine as objecting to the scientific doctrine, he is careful not to cite Scripture against it, and not less careful to suggest geographical reasoning in favour of it.

But in 1519 science gains a crushing victory. Magellan makes his famous voyage. He proves the earth to be round, for his expedition circumnavigates it; he proves the doctrine of the antipodes, for his shipmates see the peoples of the antipodes. Yet even this does not end the war. Many conscientious men oppose the doctrine for two hundred years longer. Then the French astronomers make their measurements of degrees in equatorial and polar regions, and add to their proofs that of the lengthened pendulum. When this was done, when the deductions of science were seen to be established by the simple test of measurement, beautifully and perfectly, and when a long line of trustworthy explorers, including devoted missionaries, had sent home accounts of the antipodes, then, and then only, this war of twelve centuries ended.

Such was the main result of this long war; but there were other results not so fortunate. The efforts of Eusebius, Basil, and Lactantius to deaden scientific thought; the efforts of Augustine to combat it; the efforts of Cosmas to crush it by dogmatism; the efforts of Boniface and Zachary to crush it by force, conscientious as they all were, had resulted simply in impressing upon many leading minds the conviction that science and religion are enemies.

On the other hand, what was gained by the warriors of science for religion? Certainly a far more worthy conception of the world, and a far more ennobling conception of that power which pervades and directs it. Which is more consistent with a great religion, the cosmography of Cosmas or that of Isaac Newton? Which presents a nobler field for religious thought, the diatribes of Lactantius or the calm statements of Humboldt?*

IV. THE SIZE OF THE EARTH.

But at an early period another subject in geography had stirred the minds of thinking men—*the earth's size.* Various ancient investigators had by different methods reached measurements more or less near the truth; these methods were continued into the Middle Ages, supplemented by new thought, and among the more striking results were those obtained by Roger Bacon and Gerbert, afterward Pope Sylvester II. They handed down to after-time the torch of

---

* For D'Ailly's acceptance of St. Augustine's argument, see the *Ymago Mundi*, cap. vii. For Tostatus, see Zöckler, vol. i, pp. 467, 468. He based his opposition on Romans x, 18. For Columbus, see Winsor, Fiske, and Adams; also Humboldt, *Histoire de la Géographie du Nouveau Continent.* For the bull of Alexander VI, see Daunou, *Études Historiques*, vol. ii, p. 417; also Peschel, *Zeitalter der Entdeckungen*, Book II, chap. iv. The text of the bull is given with an English translation in Arber's reprint of *The First Three English Books on America*, etc., Birmingham, 1885, pp. 201–204; also especially Peschel, *Die Theilung der Erde unter Papst Alexander VI und Julius II*, Leipsic, 1871, pp. 14 *et seq.* For remarks on the power under which the line was drawn by Alexander VI, see Mamiani, *Del Papato nei Tre Ultimi Secoli*, p. 170. For maps showing lines of division, see Kohl, *Die beiden ältesten General-Karten von Amerika*, Weimar, 1860, where maps of 1527 and 1529 are reproduced; also Mercator, *Atlas*, tenth edition, Amsterdam, 1628, pp. 70, 71. For latest discussion on *The Demarcation Line of Alexander VI*, see E. G. Bourne in *Yale Review*, May, 1892. For the *Margarita Philosophica*, see the editions of 1503, 1509, 1517, lib. vii, cap. 48. For the effect of Magellan's voyages, and the reluctance to yield to proof, see Henri Martin, *Histoire de France*, vol. xiv, p. 395; St. Martin's *Histoire de la Géographie*, p. 369; Peschel, *Geschichte des Zeitalters der Entdeckungen*, concluding chapters; and for an admirable summary, Draper, *Hist. Int. Devel. of Europe*, pp. 451–453; also an interesting passage in Sir Thomas Browne's *Vulgar and Common Errors*, Book I, chap. vi; also a striking passage in Acosta, chap. ii. For general statement as to supplementary proof by measurement of degrees and by pendulum, see Somerville, *Phys. Geog.*, chap. i, par. 6, note; also Humboldt, *Cosmos*, vol. ii, p. 736, and vol. v, pp. 16, 32; also Montucla, iv, 138. As to the effect of travel, see Acosta's history above cited. The good missionary says, in Grimston's quaint translation, "Whatsoever Lactantius saieth, wee that live now at Peru, and inhabite that parte of the worlde which is opposite to Asia and their Antipodes, finde not ourselves to bee hanging in the aire, our heades downward and our feete on high."

knowledge, but, as their reward among their contemporaries, they fell under the charge of sorcery.

Far more consonant with the theological spirit of the Middle Ages was a solution of the problem from Scripture, and this solution deserves to be given as an example of a very curious theological error, chancing to result in the establishment of a great truth. The second book of Esdras, which among Protestants is placed in the Apocrypha, was held by many of the foremost men of the ancient Church as fully inspired: though Jerome looked with suspicion on this book, it was regarded as prophetic by Clement of Alexandria, Tertullian, and Ambrose, and the Church acquiesced in that view. In the Eastern Church it held an especially high place, and in the Western Church, before the Reformation, was generally considered by the most eminent authorities to be part of the sacred canon. In the sixth chapter of this book there is a summary of the works of creation, and in it occur the following verses:

"Upon the third day thou didst command that the waters should be gathered in the seventh part of the earth; six parts hast thou dried up and kept them to the intent that of these some, being planted of God and tilled, might serve thee."

"Upon the fifth day thou saidst unto the seventh part where the waters were gathered, that it should bring forth living creatures, fowls and fishes, and so it came to pass."

These statements were reiterated in other verses, and were naturally considered as of controlling authority.

Among the scholars who pondered on this as on all things likely to increase knowledge was Cardinal Pierre d'Ailly. As we have seen, this great man, while he denied the existence of the antipodes, as St. Augustine had done, believed firmly in the sphericity of the earth, and, interpreting these statements of the book of Esdras in connection with this belief, he held that, as only one seventh of the earth's surface was covered by water, the ocean between the west coast of Europe and the east coast of Asia could not be very wide. Knowing, as he thought, the extent of the land upon the globe, he felt that in view of this divinely authorized statement the globe must be much smaller, and the land of "Zipango," reached by Marco Polo, on the extreme east coast of Asia, much nearer than had been generally believed.

On this point he laid stress in his great work, the *Ymago Mundi*, and an edition of it having been published in the days when Columbus was thinking most closely upon the problem of a westward voyage, it naturally exercised much influence upon his reasonings. Among the treasures of the library at Seville, there is nothing more interesting than a copy of this work annotated by Columbus himself: from this very copy it was that Columbus obtained confirmation of his belief that the passage across the ocean to Marco Polo's land of Zipango in Asia was short. But for this error, based upon a text supposed to be inspired, it is unlikely that Columbus could have secured the necessary support for his voyage. It is a curious fact that this single theological error thus promoted a series of voyages which completely destroyed not only this but every other conception of geography based upon the sacred writings.*

V. THE CHARACTER OF THE EARTH'S SURFACE.

It would be hardly just to dismiss the struggle for geographical truth without referring to one passage more in the history of the Protestant Church, for it shows clearly the difficulties in the way of the simplest statement of geographical truth which conflicted with the words of the sacred books.

In the year 1553 Michael Servetus was on trial for his life at Geneva on the charge of Arianism. Servetus had rendered many services to scientific truth, and one of these

---

* For this error, so fruitful in discovery, see D'Ailly, *Ymago Mundi*; the passage referred to is fol. 12 verso. For the passage from Esdras, see chap. vi, verses 42, 47, 50, and 52; see also Zöckler, *Geschichte der Beziehungen zwischen Theologie und Naturwissenschaft*, vol. i, p. 461. For one of the best recent statements, see Ruge, *Gesch. des Zeitalters der Entdeckungen*, Berlin, 1882, pp. 221 *et seq.* For a letter of Columbus acknowledging his indebtedness to this mistake in Esdras, see Navarrete, *Viajes y Descubrimientos*, Madrid, 1825, tome i, pp. 242, 264; also Humboldt, *Hist. de la Géographie du Nouveau Continent*, vol. i, pp. 68, 69.

was an edition of Ptolemy's *Geography*, in which Judea was spoken of, not as "a land flowing with milk and honey," but, in strict accordance with the truth, as, in the main, meagre, barren, and inhospitable. In his trial this simple statement of geographical fact was used against him by his arch-enemy John Calvin with fearful power. In vain did Servetus plead that he had simply drawn the words from a previous edition of Ptolemy; in vain did he declare that this statement was a simple geographical truth of which there were ample proofs; it was answered that such language "necessarily inculpated Moses, and grievously outraged the Holy Ghost." *

In summing up the action of the Church upon geography, we must say, then, that the dogmas developed in strict adherence to Scripture and the conceptions held in the Church during many centuries "always, everywhere, and by all," were, on the whole, steadily hostile to truth; but it is only just to make a distinction here between the religious and the theological spirit. To the religious spirit are largely due several of the noblest among the great voyages of discovery. A deep longing to extend the realms of Christianity influenced the minds of Prince John of Portugal, in his great series of efforts along the African coast; of Vasco da Gama, in his circumnavigation of the Cape of Good Hope; of Magellan, in his voyage around the world; and doubtless found a place among the more worldly motives of Columbus.†

Thus, in this field, from the supremacy accorded to theology, we find resulting that tendency to dogmatism which has shown itself in all ages the deadly foe not only of scientific inquiry but of the higher religious spirit itself, while from the love of truth for truth's sake, which has been the inspiration of all fruitful work in science, nothing but advantage has ever resulted to religion.

---

* For Servetus's geographical offense, see Rilliet, *Rélation du Procès criminel contre Michel Servet d'après les Documents originaux*, Geneva, 1844, pp. 42, 43; also Willis, *Servetus and Calvin*, London, 1877, p. 325. The passage condemned is in the Ptolemy of 1535, fol. 41. It was discreetly retrenched in a reprint of the same edition.

† As to the mixture in the motives of Columbus, it may be well to compare with the earlier biographies the recent ones by Dr. Winsor and President Adams.

CHAPTER III.

ASTRONOMY.

I. THE OLD SACRED THEORY OF THE UNIVERSE.

THE next great series of battles was fought over the relations of the visible heavens to the earth.

In the early Church, in view of the doctrine so prominent in the New Testament, that the earth was soon to be destroyed, and that there were to be "new heavens and a new earth," astronomy, like other branches of science, was generally looked upon as futile. Why study the old heavens and the old earth, when they were so soon to be replaced with something infinitely better? This feeling appears in St. Augustine's famous utterance, "What concern is it to me whether the heavens as a sphere inclose the earth in the middle of the world or overhang it on either side?"

As to the heavenly bodies, theologians looked on them as at best only objects of pious speculation. Regarding their nature the fathers of the Church were divided. Origen, and others with him, thought them living beings possessed of souls, and this belief was mainly based upon the scriptural vision of the morning stars singing together, and upon the beautiful appeal to the "stars and light" in the song of the three children—the *Benedicite*—which the Anglican communion has so wisely retained in its Liturgy.

Other fathers thought the stars abiding-places of the angels, and that stars were moved by angels. The Gnostics thought the stars spiritual beings governed by angels, and appointed not to cause earthly events but to indicate them.

As to the heavens in general, the prevailing view in the Church was based upon the scriptural declarations that a solid vault—a "firmament"—was extended above the earth, and that the heavenly bodies were simply lights hung within it. This was for a time held very tenaciously. St. Philastrius, in his famous treatise on heresies, pronounced it a heresy to deny that the stars are brought out by God from his treasure-house and hung in the sky every evening; any other view he declared "false to the Catholic faith." This view also survived in the sacred theory established so firmly by Cosmas in the sixth century. Having established his plan of the universe upon various texts in the Old and New Testaments, and having made it a vast oblong box, covered by the solid "firmament," he brought in additional texts from Scripture to account for the planetary movements, and developed at length the theory that the sun and planets are moved and the "windows of heaven" opened and shut by angels appointed for that purpose.

How intensely real this way of looking at the universe was, we find in the writings of St. Isidore, the greatest leader of orthodox thought in the seventh century. He affirms that since the fall of man, and on account of it, the sun and moon shine with a feebler light; but he proves from a text in Isaiah that when the world shall be fully redeemed these "great lights" will shine again in all their early splendour. But, despite these authorities and their theological finalities, the evolution of scientific thought continued, its main germ being the geocentric doctrine—the doctrine that the earth is the centre, and that the sun and planets revolve about it.*

This doctrine was of the highest respectability: it had been developed at a very early period, and had been elabo-

---

* For passage cited from Clement of Alexandria, see English translation, Edinburgh, 1869, vol. ii, p. 368; also the *Miscellanies*, Book V, cap. vi. For typical statements by St. Augustine, see *De Genesi*, ii, cap. ix, in Migne, *Patr. Lat.*, tome xxxiv, pp. 270, 271. For Origen's view, see the *De Principiis*, lib. i, cap. vii; see also Leopardi's *Errori Popolari*, cap. xi; also Wilson's *Selections from the Prophetic Scriptures* in *Ante-Nicene Library*, p. 132. For Philo Judæus, see *On the Creation of the World*, chaps. xviii and xix, and *On Monarchy*, chap. i. For St. Isidore, see the *De Ordine Creaturarum*, cap. v, in Migne, *Patr. Lat.*, lxxxiii, pp. 923–925; also, 1000, 1001. For Philastrius, see the *De Hæresibus*, chap. cxxxiii, in Migne, tome xii, p. 1264. For Cosmas's view, see his *Topographia Christiana*, in Montfauçon, *Col. Nov. Patrum*, ii, p. 150, and elsewhere as cited in my chapter on Geography.

rated until it accounted well for the apparent movements of the heavenly bodies; its final name, "Ptolemaic theory," carried weight; and, having thus come from antiquity into the Christian world, St. Clement of Alexandria demonstrated that the altar in the Jewish tabernacle was "a symbol of the earth placed in the middle of the universe": nothing more was needed; the geocentric theory was fully adopted by the Church and universally held to agree with the letter and spirit of Scripture.*

Wrought into this foundation, and based upon it, there was developed in the Middle Ages, mainly out of fragments of Chaldean and other early theories preserved in the Hebrew Scriptures, a new sacred system of astronomy, which became one of the great treasures of the universal Church —the last word of revelation.

Three great men mainly reared this structure. First was the unknown who gave to the world the treatises ascribed to Dionysius the Areopagite. It was unhesitatingly believed that these were the work of St. Paul's Athenian convert, and therefore virtually of St. Paul himself. Though now known to be spurious, they were then considered a treasure of inspiration, and an emperor of the East sent them to an emperor of the West as the most worthy of gifts. In the ninth century they were widely circulated in western Europe, and became a fruitful source of thought, especially on the whole celestial hierarchy. Thus the old ideas of astronomy were vastly developed, and the heavenly hosts were classed and named in accordance with indications scattered through the sacred Scriptures.

The next of these three great theologians was Peter Lombard, professor at the University of Paris. About the middle of the twelfth century he gave forth his collection of

---

* As to the respectability of the geocentric theory, etc., see Grote's *Plato*, vol. iii, p. 257; also Sir G. C. Lewis's *Astronomy of the Ancients*, chap. iii, sec. 1, for a very thoughtful statement of Plato's view, and differing from ancient statements. For plausible elaboration of it, and for supposed agreement of Scripture with it, see Fromundus, *Anti-Aristarchus*, Antwerp, 1631; also Melanchthon's *Initia Doctrinæ Physicæ*. For an admirable statement of the theological view of the geocentric theory, antipodes, etc., see Eicken, *Geschichte und System der mittelalterlichen Weltanschauung*, pp. 618 *et seq.*

Sentences, or Statements by the Fathers, and this remained until the end of the Middle Ages the universal manual of theology. In it was especially developed the theological view of man's relation to the universe. The author tells the world: "Just as man is made for the sake of God—that is, that he may serve Him,—so the universe is made for the sake of man—that is, that it may serve *him*; therefore is man placed at the middle point of the universe, that he may both serve and be served."

The vast significance of this view, and its power in resisting any real astronomical science, we shall see, especially in the time of Galileo.

The great triad of thinkers culminated in St. Thomas Aquinas—the sainted theologian, the glory of the mediæval Church, the "Angelic Doctor," the most marvellous intellect between Aristotle and Newton; he to whom it was believed that an image of the Crucified had spoken words praising his writings. Large of mind, strong, acute, yet just—even more than just—to his opponents, he gave forth, in the latter half of the thirteenth century, his Cyclopædia of Theology, the *Summa Theologica*. In this he carried the sacred theory of the universe to its full development. With great power and clearness he brought the whole vast system, material and spiritual, into its relations to God and man.*

Thus was the vast system developed by these three leaders of mediæval thought; and now came the man who wrought it yet more deeply into European belief, the poet divinely inspired who made the system part of the world's *life*. Pictured by Dante, the empyrean and the concentric heavens, paradise, purgatory, and hell, were seen of all men;

---

* For the beliefs of Chaldean astronomers in revolving spheres carrying sun, moon, and planets, in a solid firmament supporting the celestial waters, and in angels as giving motion to the planets, see Lenormant; also Lethaby, 13–21; also Schröder, Jensen, Lukas, *et al.* For the contribution of the pseudo-Dionysius to mediæval cosmology, see Dion. Areopagita, *De Cælesti Hierarchia*, vers. Joan. Scoti, in Migne, *Patr. Lat.*, cxxii. For the contribution of Peter Lombard, see Pet. Lomb., *Libr. Sent.*, II, i, 8,—IV, i, 6, 7, in Migne, tome 192. For the citations from St. Thomas Aquinas, see the *Summa*, ed. Migne, especially Pars I, Qu. 70, (tome i, pp. 1174–1184); also Quæstio 47, Art. iii. For good general statement, see Milman, *Latin Christianity*, iv, 191 *et seq.*; and for relation of Cosmas to these theologians of western Europe, see Milman, as above, viii, 228, note.

the God Triune, seated on his throne upon the circle of the heavens, as real as the Pope seated in the chair of St. Peter; the seraphim, cherubim, and thrones, surrounding the Almighty, as real as the cardinals surrounding the Pope; the three great orders of angels in heaven, as real as the three great orders, bishops, priests, and deacons, on earth; and the whole system of spheres, each revolving within the one above it, and all moving about the earth, subject to the *primum mobile*, as real as the feudal system of western Europe, subject to the Emperor.*

Let us look into this vast creation—the highest achievement of theology—somewhat more closely.

Its first feature shows a development out of earlier theological ideas. The earth is no longer a flat plain inclosed by four walls and solidly vaulted above, as theologians of previous centuries had believed it, under the inspiration of Cosmas; it is no longer a mere flat disk, with sun, moon, and stars hung up to give it light, as the earlier cathedral sculptors had figured it; it has become a globe at the centre of the universe. Encompassing it are successive transparent spheres, rotated by angels about the earth, and each carrying one or more of the heavenly bodies with it: that nearest the earth carrying the moon; the next, Mercury; the next, Venus; the next, the sun; the next three, Mars, Jupiter, and Saturn; the eighth carrying the fixed stars. The ninth was the *primum mobile*, and inclosing all was the tenth heaven—the Empyrean. This was immovable—the boundary between creation and the great outer void; and here, in a light which no one can enter, the Triune God sat enthroned, the "music of the spheres" rising to Him as they moved. Thus was the old heathen doctrine of the spheres made Christian.

In attendance upon the Divine Majesty, thus enthroned,

---

* For the central sun, hierarchy of angels, and concentric circles, see Dante, *Paradiso*, canto xxviii. For the words of St. Thomas Aquinas, showing to Virgil and Dante the great theologians of the Middle Ages, see canto x, and in Dean Plumptre's translation, vol. ii, pp. 56 *et seq.*; also Botta, *Dante*, pp. 350, 351. As to Dante's deep religious feeling and belief in his own divine mission, see J. R. Lowell, *Among my Books*, vol. i, p. 36. For a remarkable series of coloured engravings showing Dante's whole cosmology, see *La Materia della Divina Commedia di Dante dichiarata* in vi tavole, da Michelangelo Caetani, published by the monks of Monte Cassino, to whose kindness I am indebted for my copy.

are vast hosts of angels, who are divided into three hierarchies, one serving in the empyrean, one in the heavens, between the empyrean and the earth, and one on the earth.

Each of these hierarchies is divided into three choirs, or orders; the first, into the orders of Seraphim, Cherubim, and Thrones; and the main occupation of these is to chant incessantly—to "continually cry" the divine praises.

The order of Thrones conveys God's will to the second hierarchy, which serves in the movable heavens. This second hierarchy is also made up of three orders. The first of these, the order of Dominions, receives the divine commands; the second, the order of Powers, moves the heavens, sun, moon, planets, and stars, opens and shuts the "windows of heaven," and brings to pass all other celestial phenomena; the third, the order of Empire, guards the others.

The third and lowest hierarchy is also made up of three orders. First of these are the Principalities, the guardian spirits of nations and kingdoms. Next come Archangels; these protect religion, and bear the prayers of the saints to the foot of God's throne. Finally come Angels; these care for earthly affairs in general, one being appointed to each mortal, and others taking charge of the qualities of plants, metals, stones, and the like. Throughout the whole system, from the great Triune God to the lowest group of angels, we see at work the mystic power attached to the triangle and sacred number three—the same which gave the triune idea to ancient Hindu theology, which developed the triune deities in Egypt, and which transmitted this theological gift to the Christian world, especially through the Egyptian Athanasius.

Below the earth is hell. This is tenanted by the angels who rebelled under the lead of Lucifer, prince of the seraphim—the former favourite of the Trinity; but, of these rebellious angels, some still rove among the planetary spheres, and give trouble to the good angels; others pervade the atmosphere about the earth, carrying lightning, storm, drought, and hail; others infest earthly society, tempting men to sin; but Peter Lombard and St. Thomas Aquinas take pains to show that the work of these devils is, after all, but to discipline man or to mete out deserved punishment.

All this vast scheme had been so riveted into the Ptolemaic view by the use of biblical texts and theological reasonings that the resultant system of the universe was considered impregnable and final. To attack it was blasphemy.

It stood for centuries. Great theological men of science, like Vincent of Beauvais and Cardinal d'Ailly, devoted themselves to showing not only that it was supported by Scripture, but that it supported Scripture. Thus was the geocentric theory embedded in the beliefs and aspirations, in the hopes and fears, of Christendom down to the middle of the sixteenth century.*

II. THE HELIOCENTRIC THEORY.

But, on the other hand, there had been planted, long before, the germs of a heliocentric theory. In the sixth century before our era, Pythagoras, and after him Philolaus, had suggested the movement of the earth and planets about a central fire; and, three centuries later, Aristarchus had restated the main truth with striking precision. Here comes in a proof that the antagonism between theological and sci-

---

* For the earlier sacred cosmology of Cosmas, with citations from Montfauçon, see the chapter on *Geography* in this work. For the views of the mediæval theologians, see foregoing notes in this chapter. For the passages of Scripture on which the theological part of this structure was developed, see especially Romans viii, 38; Ephesians i, 21; Colossians i, 16, and ii, 15; and innumerable passages in the Old Testament. As to the music of the spheres, see Dean Plumptre's *Dante*, vol. ii, p. 4, note. For an admirable summing up of the mediæval cosmology in its relation to thought in general, see Rydberg, *Magic of the Middle Ages*, chap. i, whose summary I have followed in the main. For striking woodcuts showing the view taken of the successive heavens with their choirs of angels, the earth being at the centre and the spheres about it, and the Almighty on his throne above all, see the *Nuremberg Chronicle*, ff. iv and v; its date is 1493. For charts showing the continuance of this general view down to the beginning of the sixteenth century, see the various editions of the *Margarita Philosophica*, from that of 1503 onward, astronomical part. For interesting statements regarding the trinities of gods in ancient Egypt, see Sharpe, *History of Egypt*, vol. i, pp. 94 and 101. The present writer once heard a lecture in Cairo, from an eminent Scotch Doctor of Medicine, to account for the ancient Hindu and Egyptian sacred threes and trinities. The lecturer's theory was that, when Jehovah came down into the garden of Eden and walked with Adam in "the cool of the day," he explained his triune character to Adam, and that from Adam it was spread abroad to the various ancient nations.

entific methods is not confined to Christianity; for this statement brought upon Aristarchus the charge of blasphemy, and drew after it a cloud of prejudice which hid the truth for six hundred years. Not until the fifth century of our era did it timidly appear in the thoughts of Martianus Capella: then it was again lost to sight for a thousand years, until in the fifteenth century, distorted and imperfect, it appeared in the writings of Cardinal Nicholas de Cusa.

But in the shade cast by the vast system which had grown from the minds of the great theologians and from the heart of the great poet there had come to this truth neither bloom nor fruitage.

Quietly, however, the soil was receiving enrichment and the air warmth. The processes of mathematics were constantly improved, the heavenly bodies were steadily observed, and at length appeared, far from the centres of thought, on the borders of Poland, a plain, simple-minded scholar, who first fairly uttered to the modern world the truth—now so commonplace, then so astounding—that the sun and planets do not revolve about the earth, but that the earth and planets revolve about the sun: this man was Nicholas Copernicus.

Copernicus had been a professor at Rome, and even as early as 1500 had announced his doctrine there, but more in the way of a scientific curiosity or paradox, as it had been previously held by Cardinal de Cusa, than as the statement of a system representing a great fact in Nature. About thirty years later one of his disciples, Widmanstadt, had explained it to Clement VII; but it still remained a mere hypothesis, and soon, like so many others, disappeared from the public view. But to Copernicus, steadily studying the subject, it became more and more a reality, and as this truth grew within him he seemed to feel that at Rome he was no longer safe. To announce his discovery there as a theory or a paradox might amuse the papal court, but to announce it as a truth—as *the* truth—was a far different matter. He therefore returned to his little town in Poland.

To publish his thought as it had now developed was evidently dangerous even there, and for more than thirty years it lay slumbering in the mind of Copernicus and of the friends to whom he had privately intrusted it.

At last he prepared his great work on the *Revolutions of the Heavenly Bodies*, and dedicated it to the Pope himself. He next sought a place of publication. He dared not send it to Rome, for there were the rulers of the older Church ready to seize it; he dared not send it to Wittenberg, for there were the leaders of Protestantism no less hostile; he therefore intrusted it to Osiander, at Nuremberg.*

---

* For germs of heliocentric theory planted long before, see Sir G. C. Lewis; and for a succinct statement of the claims of Pythagoras, Philolaus, Aristarchus, and Martianus Capella, see Hoefer, *Histoire de l'Astronomie*, 1873, p. 107 *et seq.*; also Heller, *Geschichte der Physik*, Stuttgart, 1882, vol. i, pp. 12, 13; also pp. 99 *et seq.* For germs among thinkers of India, see Whewell, vol. i, p. 277; also Whitney, *Oriental and Linguistic Studies*, New York, 1874; *Essay on the Lunar Zodiac*, p. 345. For the views of Vincent of Beauvais, see his *Speculum Naturale*, lib. xvi, cap. 21. For Cardinal d'Ailly's view, see his treatise *De Concordia Astronomicæ Veritatis cum Theologia* (in his *Ymago Mundi* and separately). For general statement of De Cusa's work, see Draper, *Intellectual Development of Europe*, p. 512. For skilful use of De Cusa's view in order to mitigate censure upon the Church for its treatment of Copernicus's discovery, see an article in the *Catholic World* for January, 1869. For a very exact statement, in a spirit of judicial fairness, see Whewell, *History of the Inductive Sciences*, p. 275 and pp. 379, 380. In the latter, Whewell cites the exact words of De Cusa in the *De Docta Ignorantia*, and sums up in these words: "This train of thought might be a preparation for the reception of the Copernican system; but it is very different from the doctrine that the sun is the centre of the planetary system." Whewell says: "De Cusa propounded the doctrine of the motion of the earth more as a paradox than as a reality. We can not consider this as any distinct anticipation of a profound and consistent view of the truth." On De Cusa, see also Heller, vol. i, p. 216. For Aristotle's views, and their elaboration by St. Thomas Aquinas, see the *De Cœlo et Mundo*, sec. xx, and elsewhere in the latter. It is curious to see how even such a biographer as Archbishop Vaughan slurs over the Angelic Doctor's errors. See Vaughan's *Life and Labours of St. Thomas of Aquin*, pp. 459, 460.

As to Copernicus's danger at Rome, the *Catholic World* for January, 1869, cites a speech of the Archbishop of Mechlin before the University of Louvain, to the effect that Copernicus defended his theory at Rome, in 1500, before two thousand scholars; also, that another professor taught the system in 1528, and was made apostolic notary by Clement VIII. All this, even if the doctrines taught were identical with those of Copernicus as finally developed—which is simply not the case—avails nothing against the overwhelming testimony that Copernicus felt himself in danger—testimony which the after-history of the Copernican theory renders invincible. The very title of Fromundus's book, already cited, published within a few miles of the archbishop's own cathedral, and sanctioned expressly by the theological faculty of that same University of Louvain in 1630, utterly refutes the archbishop's idea that the Church was inclined to treat Copernicus kindly. The

But Osiander's courage failed him: he dared not launch the new thought boldly. He wrote a grovelling preface, endeavouring to excuse Copernicus for his novel idea, and in this he inserted the apologetic lie that Copernicus had propounded the doctrine of the earth's movement not as a fact, but as a hypothesis. He declared that it was lawful for an astronomer to indulge his imagination, and that this was what Copernicus had done.

Thus was the greatest and most ennobling, perhaps, of scientific truths—a truth not less ennobling to religion than to science—forced, in coming before the world, to sneak and crawl.*

On the 24th of May, 1543, the newly printed book arrived at the house of Copernicus. It was put into his hands; but he was on his deathbed. A few hours later he was be-

---

title is as follows: *Ant-Aristarchus sive Orbis-Terræ Immobilis, in quo decretum S. Congregationis S. R. E. Cardinal. an. M.DC.XVI adversus Pythagorico-Copernicanos editum defenditur*, Antverpiæ, MDCXXXI. L'Épinois, *Galilée*, Paris, 1867, lays stress, p. 14, on the broaching of the doctrine by De Cusa in 1435, and by Widmanstadt in 1533, and their kind treatment by Eugenius IV and Clement VII; but this is absolutely worthless in denying the papal policy afterward. Lange, *Geschichte des Materialismus*, vol. i, pp. 217, 218, while admitting that De Cusa and Widmanstadt sustained this theory and received honours from their respective popes, shows that, when the Church gave it serious consideration, it was condemned. There is nothing in this view unreasonable. It would be a parallel case to that of Leo X, at first inclined toward Luther and others, in their "squabbles with the envious friars," and afterward forced to oppose them. That Copernicus felt the danger, is evident, among other things, by the expression in the preface; "*Statim me explodendum cum tali opinione clamitant.*" For dangers at Wittenberg, see Lange, as above, vol. i, p. 217.

* Osiander, in a letter to Copernicus, dated April 20, 1541, had endeavoured to reconcile him to such a procedure, and ends by saying, "*Sic enim placidiores reddideris peripatheticos et theologos quos contradicturos metuis.*" See *Apologia Tychonis* in Kepler's *Opera Omnia*, Frisch's edition, vol. i, p. 246. Kepler holds Osiander entirely responsible for this preface. Bertrand, in his *Fondateurs de l'Astronomie moderne*, gives its text, and thinks it possible that Copernicus may have yielded "in pure condescension toward his disciple." But this idea is utterly at variance with expressions in Copernicus's own dedicatory letter to the Pope, which follows the preface. For a good summary of the argument, see Figuier, *Savants de la Renaissance*, pp. 378, 379; see also citation from Gassendi's *Life of Copernicus*, in Flammarion, *Vie de Copernic*, p. 124. Mr. John Fiske, accurate as he usually is, in his *Outlines of Cosmic Philosophy* appears to have followed Laplace, Delambre, and Petit into the error of supposing that Copernicus, and not Osiander, is responsible for the preface. For the latest proofs, see Menzer's translation of Copernicus's work, Thorn, 1879, notes on pp. 3 and 4 of the appendix.

yond the reach of the conscientious men who would have blotted his reputation and perhaps have destroyed his life.

Yet not wholly beyond their reach. Even death could not be trusted to shield him. There seems to have been fear of vengeance upon his corpse, for on his tombstone was placed no record of his lifelong labours, no mention of his great discovery; but there was graven upon it simply a prayer: "I ask not the grace accorded to Paul; not that given to Peter; give me only the favour which Thou didst show to the thief on the cross." Not till thirty years after did a friend dare write on his tombstone a memorial of his discovery.*

The preface of Osiander, pretending that the book of Copernicus suggested a hypothesis instead of announcing a truth, served its purpose well. During nearly seventy years the Church authorities evidently thought it best not to stir the matter, and in some cases professors like Calganini were allowed to present the new view purely as a hypothesis. There were, indeed, mutterings from time to time on the theological side, but there was no great demonstration against the system until 1616. Then, when the Copernican doctrine was upheld by Galileo as a *truth*, and proved to be a truth by his telescope, the book was taken in hand by the Roman curia. The statements of Copernicus were condemned, "until they should be corrected"; and the corrections required were simply such as would substitute for his conclusions the old Ptolemaic theory.

That this was their purpose was seen in that year when Galileo was forbidden to teach or discuss the Copernican theory, and when were forbidden "all books which affirm the motion of the earth." Henceforth to read the work of Copernicus was to risk damnation, and the world accepted the decree.† The strongest minds were thus held fast. If

---

* See Flammarion, *Vie de Copernic*, p. 190.

† The authorities deciding this matter in accordance with the wishes of Pope Paul V and Cardinal Bellarmine were the Congregation of the Index, or cardinals having charge of the *Index Librorum Prohibitorum*. Recent desperate attempts to fasten the responsibility on them as individuals seem ridiculous in view of the simple fact that their work was sanctioned by the highest Church authority, and required to be universally accepted by the Church. Eleven different editions of

they could not believe the old system, they must *pretend* that they believed it;—and this, even after the great circumnavigation of the globe had done so much to open the eyes of the world! Very striking is the case of the eminent Jesuit missionary Joseph Acosta, whose great work on the *Natural and Moral History of the Indies*, published in the last quarter of the sixteenth century, exploded so many astronomical and geographical errors. Though at times curiously credulous, he told the truth as far as he dared; but as to the movement of the heavenly bodies he remained orthodox—declaring, "I have seen the two poles, whereon the heavens turn as upon their axletrees."

There was, indeed, in Europe one man who might have done much to check this current of unreason which was to sweep away so many thoughtful men on the one hand from scientific knowledge, and so many on the other from Christianity. This was Peter Apian. He was one of the great mathematical and astronomical scholars of the time. His brilliant abilities had made him the astronomical teacher of the Emperor Charles V; his work on geography had brought him a world-wide reputation; his work on astronomy brought him a patent of nobility; his improvements in mathematical processes and astronomical instruments brought him the praise of Kepler and a place in the history of science: never had a true man better opportunity to do a great deed. When Copernicus's work appeared, Apian

---

the *Index* in my own possession prove this. Nearly all of these declare on their title-pages that they are issued by order of the pontiff of the period, and each is prefaced by a special papal bull or letter. See especially the *Index* of 1664, issued under order of Alexander VII, and that of 1761, under Benedict XIV. Copernicus's statements were prohibited in the *Index* "*donec corrigantur*." Kepler said that it ought to be worded "*donec explicetur*." See Bertrand, *Fondateurs de l'Astronomie moderne*, p. 57. De Morgan, pp. 57–60, gives the corrections required by the *Index* of 1620. Their main aim seems to be to reduce Copernicus to the grovelling level of Osiander, making of his discovery a mere hypothesis; but occasionally they require a virtual giving up of the whole Copernican doctrine—e. g., "correction" insisted upon for chap. viii, p. 6. For a scholarly account of the relation of the Prohibitory and Expurgatory Indexes to each other, see Mendham, *Literary Policy of the Church of Rome*; also Reusch, *Index der verbotenen Bücher*, Bonn, 1855, vol. ii, chaps. i and ii. For a brief but very careful statement, see Gebler, *Galileo Galilei*, English translation, London, 1879, chap. i; see also Addis and Arnold's *Catholic Dictionary*, article *Galileo*, p. 8.

was at the height of his reputation and power: a quiet, earnest plea from him, even if it had been only for ordinary fairness and a suspension of judgment, must have carried much weight. His devoted pupil, Charles V, who sat on the thrones of Germany and Spain, must at least have given a hearing to such a plea. But, unfortunately, Apian was a professor in an institution of learning under the strictest Church control—the University of Ingolstadt. His foremost duty was to teach *safe* science—to keep science within the line of scriptural truth as interpreted by theological professors. His great opportunity was lost. Apian continued to maunder over the Ptolemaic theory and astrology in his lecture-room. The attacks on the Copernican theory he neither supported nor opposed; he was silent; and the cause of his silence should never be forgotten so long as any Church asserts its title to control university instruction.*

Doubtless many will exclaim against the Roman Catholic Church for this; but the simple truth is that Protestantism was no less zealous against the new scientific doctrine. All branches of the Protestant Church—Lutheran, Calvinist, Anglican—vied with each other in denouncing the Copernican doctrine as contrary to Scripture; and, at a later period, the Puritans showed the same tendency.

Said Martin Luther: "People gave ear to an upstart astrologer who strove to show that the earth revolves, not the heavens or the firmament, the sun and the moon. Whoever wishes to appear clever must devise some new system, which of all systems is of course the very best. This fool wishes to reverse the entire science of astronomy; but sacred Scripture tells us that Joshua commanded the sun to stand still, and not the earth." Melanchthon, mild as he was, was not behind Luther in condemning Copernicus. In his treatise on the *Elements of Physics*, published six years after Copernicus's death, he says: "The eyes are witnesses that

---

* For Joseph Acosta's statement, see the translation of his *History*, published by the Hakluyt Society, chap. ii. For Peter Apian, see Mädler, *Geschichte der Astronomie*, Braunschweig, 1873, vol. i, p. 141. For evidences of the special favour of Charles V, see Delambre, *Histoire de l'Astronomie au Moyen Age*, p. 390; also Brühns, in the *Allgemeine deutsche Biographie*. For an attempted apology for him, see Günther, *Peter and Philipp Apian*, Prag, 1882, p. 62.

the heavens revolve in the space of twenty-four hours. But certain men, either from the love of novelty, or to make a display of ingenuity, have concluded that the earth moves; and they maintain that neither the eighth sphere nor the sun revolves. . . . Now, it is a want of honesty and decency to assert such notions publicly, and the example is pernicious. It is the part of a good mind to accept the truth as revealed by God and to acquiesce in it." Melanchthon then cites the passages in the Psalms and Ecclesiastes, which he declares assert positively and clearly that the earth stands fast and that the sun moves around it, and adds eight other proofs of his proposition that "the earth can be nowhere if not in the centre of the universe." So earnest does this mildest of the Reformers become, that he suggests severe measures to restrain such impious teachings as those of Copernicus.*

While Lutheranism was thus condemning the theory of the earth's movement, other branches of the Protestant Church did not remain behind. Calvin took the lead, in his *Commentary on Genesis*, by condemning all who asserted that the earth is not at the centre of the universe. He clinched the matter by the usual reference to the first verse of the ninety-third Psalm, and asked, "Who will venture to place the authority of Copernicus above that of the Holy Spirit?" Turretin, Calvin's famous successor, even after Kepler and Newton had virtually completed the theory of Copernicus and Galileo, put forth his compendium of theology, in which he proved, from a multitude of scriptural texts, that the heavens, sun, and moon move about the earth, which stands still in the centre. In England we see similar theological efforts, even after they had become evidently futile. Hutchinson's *Moses's Principia*, Dr. Samuel Pike's *Sacred Philosophy*, the writings of Horne, Bishop Horsley, and President Forbes contain most earnest attacks upon the ideas of New-

* See the *Tischreden* in the Walsch edition of Luther's *Works*, 1743, vol. xxii, p. 2260; also Melanchthon's *Initia Doctrinæ Physicæ*. This treatise is cited under a mistaken title by the *Catholic World*, September, 1870. The correct title is as given above; it will be found in the *Corpus Reformatorum*, vol. xiii (ed. Bretschneider, Halle, 1846), pp. 216, 217. See also Mädler, vol. i, p. 176; also Lange, *Geschichte des Materialismus*, vol. i, p. 217; also Prowe, *Ueber die Abhängigkeit des Copernicus*, Thorn, 1865, p. 4; also note, pp. 5, 6, where text is given in full.

ton, such attacks being based upon Scripture. Dr. John Owen, so famous in the annals of Puritanism, declared the Copernican system a "delusive and arbitrary hypothesis, contrary to Scripture"; and even John Wesley declared the new ideas to "tend toward infidelity." *

And Protestant peoples were not a whit behind Catholic in following out such teachings. The people of Elbing made themselves merry over a farce in which Copernicus was the main object of ridicule. The people of Nuremberg, a Protestant stronghold, caused a medal to be struck with inscriptions ridiculing the philosopher and his theory.

Why the people at large took this view is easily understood when we note the attitude of the guardians of learning, both Catholic and Protestant, in that age. It throws great light upon sundry claims by modern theologians to take charge of public instruction and of the evolution of science. So important was it thought to have "sound learning" guarded and "safe science" taught, that in many of the universities, as late as the end of the seventeenth century, professors were forced to take an oath not to hold the "Pythagorean"—that is, the Copernican—idea as to the movement of the heavenly bodies. As the contest went on, professors were forbidden to make known to students the facts revealed by the telescope. Special orders to this effect were issued by the ecclesiastical authorities to the universities and colleges of Pisa, Innspruck, Louvain, Douay, Salamanca, and others. During generations we find the authorities of these universities boasting that these godless doctrines were kept away from their students. It is touching to hear such boasts made then, just as it is touching now to hear sundry excellent university authorities boast that they discourage the reading of Mill, Spencer, and Darwin. Nor were such attempts to keep the truth from students confined to the Roman Catholic institutions of learning. Strange as it may seem, nowhere were the facts confirming the Copernican theory more carefully kept out of sight than at Wit-

---

* On the teachings of Protestantism as regards the Copernican theory, see citations in Canon Farrar's *History of Interpretation*, preface, xviii; also Rev. Dr. Shields, of Princeton, *The Final Philosophy*, pp. 60, 61.

tenberg—the university of Luther and Melanchthon. About the middle of the sixteenth century there were at that centre of Protestant instruction two astronomers of a very high order, Rheticus and Reinhold; both of these, after thorough study, had convinced themselves that the Copernican system was true, but neither of them was allowed to tell this truth to his students. Neither in his lecture announcements nor in his published works did Rheticus venture to make the new system known, and he at last gave up his professorship and left Wittenberg, that he might have freedom to seek and tell the truth. Reinhold was even more wretchedly humiliated. Convinced of the truth of the new theory, he was obliged to advocate the old; if he mentioned the Copernican ideas, he was compelled to overlay them with the Ptolemaic. Even this was not thought safe enough, and in 1571 the subject was intrusted to Peucer. He was eminently "sound," and denounced the Copernican theory in his lectures as "absurd, and unfit to be introduced into the schools."

To clinch anti-scientific ideas more firmly into German Protestant teaching, Rector Hensel wrote a text-book for schools entitled *The Restored Mosaic System of the World*, which showed the Copernican astronomy to be unscriptural.

Doubtless this has a far-off sound; yet its echo comes very near modern Protestantism in the expulsion of Dr. Woodrow by the Presbyterian authorities in South Carolina; the expulsion of Prof. Winchell by the Methodist Episcopal authorities in Tennessee; the expulsion of Prof. Toy by Baptist authorities in Kentucky; the expulsion of the professors at Beyrout under authority of American Protestant divines—all for holding the doctrines of modern science, and in the last years of the nineteenth century.*

But the new truth could not be concealed; it could neither be laughed down nor frowned down. Many minds

---

* For treatment of Copernican ideas by the people, see *The Catholic World*, as above; also Melanchthon, *ubi supra*; also Prowe, *Copernicus*, Berlin, 1883, vol. i, p. 269, note; also pp. 279, 280; also Mädler, i, p. 167. For Rector Hensel, see Rev. Dr. Shield's *Final Philosophy*, p. 60. For details of recent Protestant efforts against evolution doctrines, see the chapter on *The Fall of Man and Anthropology* in this work.

had received it, but within the hearing of the papacy only one tongue appears to have dared to utter it clearly. This new warrior was that strange mortal, Giordano Bruno. He was hunted from land to land, until at last he turned on his pursuers with fearful invectives. For this he was entrapped at Venice, imprisoned during six years in the dungeons of the Inquisition at Rome, then burned alive, and his ashes scattered to the winds. Still, the new truth lived on. Ten years after the martyrdom of Bruno the truth of Copernicus's doctrine was established by the telescope of Galileo.*

Herein was fulfilled one of the most touching of prophecies. Years before, the opponents of Copernicus had said to him, "If your doctrines were true, Venus would show phases like the moon." Copernicus answered: "You are right; I know not what to say; but God is good, and will in time find an answer to this objection." The God-given answer came when, in 1611, the rude telescope of Galileo showed the phases of Venus.†

III. THE WAR UPON GALILEO.

On this new champion, Galileo, the whole war was at last concentrated. His discoveries had clearly taken the Copernican theory out of the list of hypotheses, and had placed it before the world as a truth. Against him, then, the war was long and bitter. The supporters of what was called "sound learning" declared his discoveries deceptions and his announcements blasphemy. Semi-scientific profes-

* For Bruno, see Bartholmèss, *Vie de Jordano Bruno*, Paris, 1846, vol. i, p. 121 and pp. 212 *et seq.*; also Berti, *Vita di Giordano Bruno*, Firenze, 1868, chap. xvi; also Whewell, vol. i, pp. 272, 273. That Whewell is somewhat hasty in attributing Bruno's punishment entirely to the *Spaccio della Bestia Trionfante* will be evident, in spite of Montucla, to any one who reads the account of the persecution in Bartholmèss or Berti; and, even if Whewell be right, the *Spaccio* would never have been written but for Bruno's indignation at ecclesiastical oppression. See Tiraboschi, vol. vii, pp. 466 *et seq.*

† For the relation of these discoveries to Copernicus's work, see Delambre, *Histoire de l'Astronomie moderne, discours préliminaire*, p. xiv; also Laplace, *Système du Monde*, vol. i, p. 326; and for more careful statements, Kepler's *Opera Omnia*, edit. Frisch, tome ii, p. 464. For Copernicus's prophecy, see Cantu, *Histoire Universelle*, vol. xv, p. 473. (Cantu was an eminent Roman Catholic.)

sors, endeavouring to curry favour with the Church, attacked him with sham science; earnest preachers attacked him with perverted Scripture; theologians, inquisitors, congregations of cardinals, and at last two popes dealt with him, and, as was supposed, silenced his impious doctrine forever.*

I shall present this warfare at some length because, so far as I can find, no careful summary of it has been given in our language, since the whole history was placed in a new light by the revelations of the trial documents in the Vatican Library, honestly published for the first time by L'Épinois in 1867, and since that by Gebler, Berti, Favaro, and others.

The first important attack on Galileo began in 1610, when he announced that his telescope had revealed the moons of the planet Jupiter. The enemy saw that this took the Copernican theory out of the realm of hypothesis, and they gave battle immediately. They denounced both his method and its results as absurd and impious. As to his method, professors bred in the "safe science" favoured by the Church argued that the divinely appointed way of arriving at the truth in astronomy was by theological reasoning on texts of Scripture; and, as to his results, they insisted, first, that Aristotle knew nothing of these new revelations; and, next, that the Bible showed by all applicable types that there could be only seven planets; that this was proved by the seven golden candlesticks of the Apocalypse, by the seven-branched candlestick of the tabernacle, and by the seven churches of Asia; that from Galileo's doctrine consequences must logically result destructive to Christian truth. Bishops and priests therefore warned their flocks, and multitudes of the faithful besought the Inquisition to deal speedily and sharply with the heretic.†

---

* A very curious example of this sham science employed by theologians is seen in the argument, frequently used at that time, that, if the earth really moved, a stone falling from a height would fall back of the point immediately below its point of starting. This is used by Fromundus with great effect. It appears never to have occurred to him to test the matter by dropping a stone from the topmast of a ship. Benzenburg has experimentally demonstrated just such an aberration in falling bodies as is mathematically required by the diurnal motion of the earth. See Jevons, *Principles of Science*, pp. 388, 389, second edition, 1877.

† See Delambre on the discovery of the satellites of Jupiter as the turning-point

In vain did Galileo try to prove the existence of satellites by showing them to the doubters through his telescope: they either declared it impious to look, or, if they did look, denounced the satellites as illusions from the devil. Good Father Clavius declared that "to see satellites of Jupiter, men had to make an instrument which would create them." In vain did Galileo try to save the great truths he had discovered by his letters to the Benedictine Castelli and the Grand-Duchess Christine, in which he argued that literal biblical interpretation should not be applied to science; it was answered that such an argument only made his heresy more detestable; that he was "worse than Luther or Calvin."

The war on the Copernican theory, which up to that time had been carried on quietly, now flamed forth. It was declared that the doctrine was proved false by the standing still of the sun for Joshua, by the declarations that "the foundations of the earth are fixed so firm that they can not be moved," and that the sun "runneth about from one end of the heavens to the other." *

But the little telescope of Galileo still swept the heavens, and another revelation was announced—the mountains and valleys in the moon. This brought on another attack. It was declared that this, and the statement that the moon shines by light reflected from the sun, directly contradict the statement in Genesis that the moon is "a great light." To make the matter worse, a painter, placing the moon in a religious picture in its usual position beneath the feet of the

---

with the heliocentric doctrine. As to its effects on Bacon, see Jevons, p. 638, as above. For argument drawn from the candlestick and the seven churches, see Delambre, p. 20.

* For principal points as given, see Libri, *Histoire des Sciences mathématiques en Italie*, vol. iv, p. 211; De Morgan, *Paradoxes*, p. 26, for account of Father Clavius. It is interesting to know that Clavius, in his last years, acknowledged that "the whole system of the heavens is broken down, and must be mended," Cantu, *Histoire Universelle*, vol. xv, p. 478. See Th. Martin, *Galilée*, pp. 34, 208, and 266; also Heller, *Geschichte der Physik*, Stuttgart, 1882, vol. i, p. 366. For the original documents, see L'Épinois, pp. 34 and 36; or, better, Gebler's careful edition of the trial (*Die Acten des Galileischen Processes*, Stuttgart, 1877), pp. 47 *et seq.* Martin's translation seems somewhat too free. See also Gebler, *Galileo Galilei*, English translation, London, 1879, pp. 76–78; also Reusch, *Der Process Galilei's und die Jesuiten*, Bonn, 1879, chaps. ix, x, xi.

Blessed Virgin, outlined on its surface mountains and valleys; this was denounced as a sacrilege logically resulting from the astronomer's heresy.

Still another struggle was aroused when the hated telescope revealed spots upon the sun, and their motion indicating the sun's rotation. Monsignor Elci, head of the University of Pisa, forbade the astronomer Castelli to mention these spots to his students. Father Busaeus, at the University of Innspruck, forbade the astronomer Scheiner, who had also discovered the spots and proposed a *safe* explanation of them, to allow the new discovery to be known there. At the College of Douay and the University of Louvain this discovery was expressly placed under the ban, and this became the general rule among the Catholic universities and colleges of Europe. The Spanish universities were especially intolerant of this and similar ideas, and up to a recent period their presentation was strictly forbidden in the most important university of all—that of Salamanca.*

Such are the consequences of placing the instruction of men's minds in the hands of those mainly absorbed in saving men's souls. Nothing could be more in accordance with the idea recently put forth by sundry ecclesiastics, Catholic and Protestant, that the Church alone is empowered to promulgate scientific truth or direct university instruction. But science gained a victory here also. Observations of the solar spots were reported not only from Galileo in Italy, but from Fabricius in Holland. Father Scheiner then endeavoured to make the usual compromise between theology and science. He promulgated a pseudo-scientific theory, which only provoked derision.

The war became more and more bitter. The Dominican Father Caccini preached a sermon from the text, "Ye men of Galilee, why stand ye gazing up into heaven?" and this wretched pun upon the great astronomer's name ushered in sharper weapons; for, before Caccini ended, he insisted that "geometry is of the devil," and that "mathematicians should be banished as the authors of all heresies." The Church authorities gave Caccini promotion.

---

* See Ticknor, *History of Spanish Literature*, vol. iii.

Father Lorini proved that Galileo's doctrine was not only heretical but "atheistic," and besought the Inquisition to intervene. The Bishop of Fiesole screamed in rage against the Copernican system, publicly insulted Galileo, and denounced him to the Grand-Duke. The Archbishop of Pisa secretly sought to entrap Galileo and deliver him to the Inquisition at Rome. The Archbishop of Florence solemnly condemned the new doctrines as unscriptural; and Paul V, while petting Galileo, and inviting him as the greatest astronomer of the world to visit Rome, was secretly moving the Archbishop of Pisa to pick up evidence against the astronomer.

But by far the most terrible champion who now appeared was Cardinal Bellarmin, one of the greatest theologians the world has known. He was earnest, sincere, and learned, but insisted on making science conform to Scripture. The weapons which men of Bellarmin's stamp used were purely theological. They held up before the world the dreadful consequences which must result to Christian theology were the heavenly bodies proved to revolve about the sun and not about the earth. Their most tremendous dogmatic engine was the statement that "his pretended discovery vitiates the whole Christian plan of salvation." Father Lecazre declared "it casts suspicion on the doctrine of the incarnation." Others declared, "It upsets the whole basis of theology. If the earth is a planet, and only one among several planets, it can not be that any such great things have been done specially for it as the Christian doctrine teaches. If there are other planets, since God makes nothing in vain, they must be inhabited; but how can their inhabitants be descended from Adam? How can they trace back their origin to Noah's ark? How can they have been redeemed by the Saviour?" Nor was this argument confined to the theologians of the Roman Church; Melanchthon, Protestant as he was, had already used it in his attacks on Copernicus and his school.

In addition to this prodigious theological engine of war there was kept up a fire of smaller artillery in the shape of texts and scriptural extracts.

But the war grew still more bitter, and some weapons

used in it are worth examining. They are very easily examined, for they are to be found on all the battlefields of science; but on that field they were used with more effect than on almost any other. These weapons are the epithets "infidel" and "atheist." They have been used against almost every man who has ever done anything new for his fellow-men. The list of those who have been denounced as "infidel" and "atheist" includes almost all great men of science, general scholars, inventors, and philanthropists. The purest Christian life, the noblest Christian character, have not availed to shield combatants. Christians like Isaac Newton, Pascal, Locke, Milton, and even Fénelon and Howard, have had this weapon hurled against them. Of all proofs of the existence of a God, those of Descartes have been wrought most thoroughly into the minds of modern men; yet the Protestant theologians of Holland sought to bring him to torture and to death by the charge of atheism, and the Roman Catholic theologians of France thwarted him during his life and prevented any due honours to him after his death.*

These epithets can hardly be classed with civilized weapons. They are burning arrows; they set fire to masses of popular prejudice, always obscuring the real question, sometimes destroying the attacking party. They are poisoned weapons. They pierce the hearts of loving women; they alienate dear children; they injure a man after life is ended, for they leave poisoned wounds in the hearts of those who loved him best—fears for his eternal salvation, dread of the Divine wrath upon him. Of course, in these days these weapons, though often effective in vexing good men and in scaring good women, are somewhat blunted; indeed, they not infrequently injure the assailants more than the assailed. So it was not in the days of Galileo; they were then in all their sharpness and venom.†

---

* For various objectors and objections to Galileo by his contemporaries, see Libri, *Histoire des Sciences mathématiques en Italie*, vol. iv, pp. 233, 234; also Martin, *Vie de Galilée*. For Father Lecazre's argument, see Flammarion, *Mondes imaginaires et mondes réels*, 6e éd., pp. 315, 316. For Melanchthon's argument, see his *Initia*, in *Opera*, vol. iii, Halle, 1846.

† For curious exemplification of the way in which these weapons have been

Yet a baser warfare was waged by the Archbishop of Pisa. This man, whose cathedral derives its most enduring fame from Galileo's deduction of a great natural law from the swinging lamp before its altar, was not an archbishop after the noble mould of Borromeo and Fénelon and Cheverus. Sadly enough for the Church and humanity, he was simply a zealot and intriguer: he perfected the plan for entrapping the great astronomer.

Galileo, after his discoveries had been denounced, had written to his friend Castelli and to the Grand-Duchess Christine two letters to show that his discoveries might be reconciled with Scripture. On a hint from the Inquisition at Rome, the archbishop sought to get hold of these letters and exhibit them as proofs that Galileo had uttered heretical views of theology and of Scripture, and thus to bring him into the clutch of the Inquisition. The archbishop begs Castelli, therefore, to let him see the original letter in the handwriting of Galileo. Castelli declines. The archbishop then, while, as is now revealed, writing constantly and bitterly to the Inquisition against Galileo, professes to Castelli the greatest admiration of Galileo's genius and a sincere desire to know more of his discoveries. This not succeeding, the archbishop at last throws off the mask and resorts to open attack.

The whole struggle to crush Galileo and to save him would be amusing were it not so fraught with evil. There were intrigues and counter-intrigues, plots and counter-plots, lying and spying; and in the thickest of this seething, squabbling, screaming mass of priests, bishops, archbishops, and cardinals, appear two popes, Paul V and Urban VIII. It is most suggestive to see in this crisis of the Church, at the tomb of the prince of the apostles, on the eve of the greatest errors in Church policy the world has known, in all the intrigues and deliberations of these consecrated leaders

---

hurled, see lists of persons charged with "infidelity" and "atheism," in the *Dictionnaire des Athées*, Paris, [1800]; also Lecky, *History of Rationalism*, vol. ii, p. 50. For the case of Descartes, see Saisset, *Descartes et ses Précurseurs*, pp. 103, 110. For the facility with which the term "atheist" has been applied from the early Aryans down to believers in evolution, see Tylor, *Primitive Culture*, vol. i, p. 420.

of the Church, no more evidence of the guidance or presence of the Holy Spirit than in a caucus of New York politicians at Tammany Hall.

But the opposing powers were too strong. In 1615 Galileo was summoned before the Inquisition at Rome, and the mine which had been so long preparing was sprung. Sundry theologians of the Inquisition having been ordered to examine two propositions which had been extracted from Galileo's letters on the solar spots, solemnly considered these points during about a month and rendered their unanimous decision as follows: "*The first proposition, that the sun is the centre and does not revolve about the earth, is foolish, absurd, false in theology, and heretical, because expressly contrary to Holy Scripture*"; and "*the second proposition, that the earth is not the centre but revolves about the sun, is absurd, false in philosophy, and, from a theological point of view at least, opposed to the true faith.*"

The Pope himself, Paul V, now intervened again: he ordered that Galileo be brought before the Inquisition. Then the greatest man of science in that age was brought face to face with the greatest theologian—Galileo was confronted by Bellarmin. Bellarmin shows Galileo the error of his opinion and orders him to renounce it. De Lauda, fortified by a letter from the Pope, gives orders that the astronomer be placed in the dungeons of the Inquisition should he refuse to yield. Bellarmin now commands Galileo, "in the name of His Holiness the Pope and the whole Congregation of the Holy Office, to relinquish altogether the opinion that the sun is the centre of the world and immovable, and that the earth moves, nor henceforth to hold, teach, or defend it in any way whatsoever, verbally or in writing." This injunction Galileo acquiesces in and promises to obey.*

This was on the 26th of February, 1616. About a fort-

* I am aware that the theory proposed by Wohlwill and developed by Gebler denies that this promise was ever made by Galileo, and holds that the passage was a forgery devised later by the Church rulers to justify the proceedings of 1632 and 1633. This would make the conduct of the Church worse, but authorities as eminent consider the charge not proved. A careful examination of the documents seems to disprove it.

night later the Congregation of the Index, moved thereto, as the letters and documents now brought to light show, by Pope Paul V, solemnly rendered a decree that "*the doctrine of the double motion of the earth about its axis and about the sun is false, and entirely contrary to Holy Scripture*"; and that this opinion must neither be taught nor advocated. The same decree condemned all writings of Copernicus and "*all writings which affirm the motion of the earth.*" The great work of Copernicus was interdicted until corrected in accordance with the views of the Inquisition; and the works of Galileo and Kepler, though not mentioned by name at that time, were included among those implicitly condemned as "affirming the motion of the earth."

The condemnations were inscribed upon the *Index*; and, finally, the papacy committed itself as an infallible judge and teacher to the world by prefixing to the *Index* the usual papal bull giving its monitions the most solemn papal sanction. To teach or even read the works denounced or passages condemned was to risk persecution in this world and damnation in the next. Science had apparently lost the decisive battle.

For a time after this judgment Galileo remained in Rome, apparently hoping to find some way out of this difficulty; but he soon discovered the hollowness of the protestations made to him by ecclesiastics, and, being recalled to Florence, remained in his hermitage near the city in silence, working steadily, indeed, but not publishing anything save by private letters to friends in various parts of Europe.

But at last a better vista seemed to open for him. Cardinal Barberini, who had seemed liberal and friendly, became pope under the name of Urban VIII. Galileo at this conceived new hopes, and allowed his continued allegiance to the Copernican system to be known. New troubles ensued. Galileo was induced to visit Rome again, and Pope Urban tried to cajole him into silence, personally taking the trouble to show him his errors by argument. Other opponents were less considerate, for works appeared attacking his ideas—works all the more unmanly, since their authors knew that Galileo was restrained by force from defending himself. Then, too, as if to accumulate proofs of the unfitness of the Church to take charge of advanced instruction, his salary as a professor at the University of Pisa was taken from him, and sapping and mining began. Just as the Archbishop of Pisa some years before had tried to betray him with honeyed words to the Inquisition, so now Father Grassi tried it, and, after various attempts to draw him out by flattery, suddenly denounced his scientific ideas as "leading to a denial of the Real Presence in the Eucharist."

For the final assault upon him a park of heavy artillery was at last wheeled into place. It may be seen on all the scientific battlefields. It consists of general denunciation; and in 1631 Father Melchior Inchofer, of the Jesuits, brought his artillery to bear upon Galileo with this declaration: "The opinion of the earth's motion is of all heresies the most abominable, the most pernicious, the most scandalous; the immovability of the earth is thrice sacred; argument against the immortality of the soul, the existence of God, and the incarnation, should be tolerated sooner than an argument to prove that the earth moves."

From the other end of Europe came a powerful echo. From the shadow of the Cathedral of Antwerp, the noted theologian Fromundus gave forth his famous treatise, the *Ant-Aristarchus*. Its very title-page was a contemptuous insult to the memory of Copernicus, since it paraded the assumption that the new truth was only an exploded theory of a pagan astronomer. Fromundus declares that "sacred Scripture fights against the Copernicans." To prove that the sun revolves about the earth, he cites the passage in the Psalms which speaks of the sun "which cometh forth as a bridegroom out of his chamber." To prove that the earth stands still, he quotes a passage from Ecclesiastes, "The earth standeth fast forever." To show the utter futility of the Copernican theory, he declares that, if it were true, "the wind would constantly blow from the east"; and that "buildings and the earth itself would fly off with such a rapid motion that men would have to be provided with claws like cats to enable them to hold fast to the earth's surface." Greatest weapon of all, he works up, by the use of Aristotle and St. Thomas Aquinas, a demonstration from theology and science combined, that the earth *must* stand in the centre, and that the sun *must* revolve about it.* Nor was it merely fanatics who opposed the truth revealed by Copernicus; such strong men as Jean Bodin, in France, and Sir Thomas Browne, in England, declared against it as evidently contrary to Holy Scripture.

IV. VICTORY OF THE CHURCH OVER GALILEO.

While news of triumphant attacks upon him and upon the truth he had established were coming in from all parts of Europe, Galileo prepared a careful treatise in the form of a dialogue, exhibiting the arguments for and against the Copernican and Ptolemaic systems, and offered to submit to any conditions that the Church tribunals might impose, if they would allow it to be printed. At last, after discussions which extended through eight years, they consented, imposing a humiliating condition—a preface written in accordance with the ideas of Father Ricciardi, Master of the Sacred Palace, and signed by Galileo, in which the Copernican theory was virtually exhibited as a play of the imagination, and not at all as opposed to the Ptolemaic doctrine reasserted in 1616 by the Inquisition under the direction of Pope Paul V.

This new work of Galileo—the *Dialogo*—appeared in 1632, and met with prodigious success. It put new weapons into the hands of the supporters of the Copernican theory. The pious preface was laughed at from one end of Europe to the other. This roused the enemy; the Jesuits, Dominicans,

* For Father Inchofer's attack, see his *Tractatus Syllepticus*, cited in Galileo's letter to Deodati, July 28, 1634. For Fromundus's more famous attack, see his *Ant-Aristarchus*, already cited, *passim*, but especially the heading of chapter vi, and the argument in chapters x and xi. A copy of this work may be found in the Astor Library at New York, and another in the White Library at Cornell University. For interesting reference to one of Fromundus's arguments, showing, by a mixture of mathematics and theology, that the earth is the centre of the universe, see Quetelet, *Histoire des Sciences mathématiques et physiques*, Bruxelles, 1864, p. 170; also Mädler, *Geschichte der Astronomie*, vol. i, p. 274. For Bodin's opposition to the Copernican theory, see Hallam, *Literature of Europe*; also Lecky. For Sir Thomas Browne, see his *Vulgar and Common Errors*, book iv, chap. v; and as to the real reason for his disbelief in the Copernican view, see Dr. Johnson's preface to his *Life of Browne*, vol. i, p. xix, of his collected works.

and the great majority of the clergy returned to the attack more violent than ever, and in the midst of them stood Pope Urban VIII, most bitter of all. His whole power was now thrown against Galileo. He was touched in two points: first, in his personal vanity, for Galileo had put the Pope's arguments into the mouth of one of the persons in the dialogue and their refutation into the mouth of another; but, above all, he was touched in his religious feelings. Again and again His Holiness insisted to all comers on the absolute and specific declarations of Holy Scripture, which prove that the sun and heavenly bodies revolve about the earth, and declared that to gainsay them is simply to dispute revelation. Certainly, if one ecclesiastic more than another ever seemed *not* under the care of the Spirit of Truth, it was Urban VIII in all this matter.

Herein was one of the greatest pieces of ill fortune that has ever befallen the older Church. Had Pope Urban been broad-minded and tolerant like Benedict XIV, or had he been taught moderation by adversity like Pius VII, or had he possessed the large scholarly qualities of Leo XIII, now reigning, the vast scandal of the Galileo case would never have burdened the Church: instead of devising endless quibbles and special pleadings to escape responsibility for this colossal blunder, its defenders could have claimed forever for the Church the glory of fearlessly initiating a great epoch in human thought.

But it was not so to be. Urban was not merely Pope; he was also a prince of the house of Barberini, and therefore doubly angry that his arguments had been publicly controverted.

The opening strategy of Galileo's enemies was to forbid the sale of his work; but this was soon seen to be unavailing, for the first edition had already been spread throughout Europe. Urban now became more angry than ever, and both Galileo and his works were placed in the hands of the Inquisition. In vain did the good Benedictine Castelli urge that Galileo was entirely respectful to the Church; in vain did he insist that "nothing that can be done can now hinder the earth from revolving." He was dismissed in disgrace, and Galileo was forced to appear in the presence of the dread tribunal without defender or adviser. There, as was so long concealed, but as is now fully revealed, he was menaced with torture again and again by express order of Pope Urban, and, as is also thoroughly established from the trial documents themselves, forced to abjure under threats, and subjected to imprisonment by command of the Pope; the Inquisition deferring in this whole matter to the papal authority. All the long series of attempts made in the supposed interest of the Church to mystify these transactions have at last failed. The world knows now that Galileo was subjected certainly to indignity, to imprisonment, and to threats equivalent to torture, and was at last forced to pronounce publicly and on his knees his recantation, as follows:

"I, Galileo, being in my seventieth year, being a prisoner and on my knees, and before your Eminences, having before my eyes the Holy Gospel, which I touch with my hands, abjure, curse, and detest the error and the heresy of the movement of the earth." *

He was vanquished indeed, for he had been forced, in the face of all coming ages, to perjure himself. To complete his dishonour, he was obliged to swear that he would denounce to the Inquisition any other man of science whom he should discover to be supporting the "heresy of the motion of the earth."

Many have wondered at this abjuration, and on account of it have denied to Galileo the title of martyr. But let such gainsayers consider the circumstances. Here was an old man—one who had reached the allotted threescore years and ten—broken with disappointments, worn out with labours and cares, dragged from Florence to Rome, with the threat from the Pope himself that if he delayed he should be "brought in chains"; sick in body and mind, given over

---

* For various utterances of Pope Urban against the Copernican theory at this period, see extracts from the original documents given by Gebler. For punishment of those who had shown some favour to Galileo, see various citations, and especially those from the Vatican manuscript, Gebler, p. 216. As to the text of the abjuration, see L'Épinois; also Polacco, *Anticopernicus*, etc., Venice, 1644; and for a discussion regarding its publication, see Favaro, *Miscellanea Galileana*, p. 804. It is not probable that torture in the ordinary sense was administered to Galileo, though it was threatened. See Th. Martin, *Vie de Galilée*, for a fair summing up of the case.

to his oppressors by the Grand-Duke who ought to have protected him, and on his arrival in Rome threatened with torture. What the Inquisition was he knew well. He could remember as but of yesterday the burning of Giordano Bruno in that same city for scientific and philosophic heresy; he could remember, too, that only eight years before this very time De Dominis, Archbishop of Spalatro, having been seized by the Inquisition for scientific and other heresies, had died in a dungeon, and that his body and his writings had been publicly burned.

To the end of his life—nay, after his life was ended—the persecution of Galileo was continued. He was kept in exile from his family, from his friends, from his noble employments, and was held rigidly to his promise not to speak of his theory. When, in the midst of intense bodily sufferings from disease, and mental sufferings from calamities in his family, he besought some little liberty, he was met with threats of committal to a dungeon. When, at last, a special commission had reported to the ecclesiastical authorities that he had become blind and wasted with disease and sorrow, he was allowed a little more liberty, but that little was hampered by close surveillance. He was forced to bear contemptible attacks on himself and on his works in silence; to see the men who had befriended him severely punished; Father Castelli banished; Ricciardi, the Master of the Sacred Palace, and Ciampoli, the papal secretary, thrown out of their positions by Pope Urban, and the Inquisitor at Florence reprimanded for having given permission to print Galileo's work. He lived to see the truths he had established carefully weeded out from all the Church colleges and universities in Europe; and, when in a scientific work he happened to be spoken of as "renowned," the Inquisition ordered the substitution of the word "notorious." *

And now measures were taken to complete the destruction of the Copernican theory, with Galileo's proofs of it. On the 16th of June, 1633, the Holy Congregation, with the permission of the reigning Pope, ordered the sentence upon

---

* For the substitution of the word "notorious" for "renowned" by order of the Inquisition, see Martin, p. 227.

Galileo, and his recantation, to be sent to all the papal nuncios throughout Europe, as well as to all archbishops, bishops, and inquisitors in Italy; and this document gave orders that the sentence and abjuration be made known "to your vicars, that you and all professors of philosophy and mathematics may have knowledge of it, that they may know why we proceeded against the said Galileo, and recognise the gravity of his error, in order that they may avoid it, and thus not incur the penalties which they would have to suffer in case they fell into the same." *

As a consequence, the professors of mathematics and astronomy in various universities of Europe were assembled and these documents were read to them. To the theological authorities this gave great satisfaction. The Rector of the University of Douay, referring to the opinion of Galileo, wrote to the papal nuncio at Brussels: "The professors of our university are so opposed to this fanatical opinion that they have always held that it must be banished from the schools. In our English college at Douay this paradox has never been approved and never will be."

Still another step was taken: the Inquisitors were ordered, especially in Italy, not to permit the publication of a new edition of any of Galileo's works, or of any similar writings. On the other hand, theologians were urged, now that Copernicus and Galileo and Kepler were silenced, to reply to them with tongue and pen. Europe was flooded with these theological refutations of the Copernican system.

To make all complete, there was prefixed to the *Index* of the Church, forbidding "all writings which affirm the motion of the earth," a bull signed by the reigning Pope, which, by virtue of his infallibility as a divinely guided teacher in matters of faith and morals, clinched this condemnation into the consciences of the whole Christian world.

From the mass of books which appeared under the auspices of the Church immediately after the condemnation

---

* For a copy of this document, see Gebler, p. 269. As to the spread of this and similar documents notifying Europe of Galileo's condemnation, see Favaro, pp. 804, 805.

of Galileo, for the purpose of rooting out every vestige of the hated Copernican theory from the mind of the world, two may be taken as typical. The first of these was a work by Scipio Chiaramonti, dedicated to Cardinal Barberini. Among his arguments against the double motion of the earth may be cited the following:

"Animals, which move, have limbs and muscles; the earth has no limbs or muscles, therefore it does not move. It is angels who make Saturn, Jupiter, the sun, etc., turn round. If the earth revolves, it must also have an angel in the centre to set it in motion; but only devils live there; it would therefore be a devil who would impart motion to the earth. . . .

"The planets, the sun, the fixed stars, all belong to one species—namely, that of stars. It seems, therefore, to be a grievous wrong to place the earth, which is a sink of impurity, among these heavenly bodies, which are pure and divine things."

The next, which I select from the mass of similar works, is the *Anticopernicus Catholicus* of Polacco. It was intended to deal a finishing stroke at Galileo's heresy. In this it is declared:

"The Scripture always represents the earth as at rest, and the sun and moon as in motion; or, if these latter bodies are ever represented as at rest, Scripture represents this as the result of a great miracle. . . .

"These writings must be prohibited, because they teach certain principles about the position and motion of the terrestrial globe repugnant to Holy Scripture and to the Catholic interpretation of it, not as hypotheses but as established facts. . . ."

Speaking of Galileo's book, Polacco says that it "smacked of Copernicanism," and that, "when this was shown to the Inquisition, Galileo was thrown into prison and was compelled to utterly abjure the baseness of this erroneous dogma."

As to the authority of the cardinals in their decree, Polacco asserts that, since they are the "Pope's Council" and his "brothers," their work is one, except that the Pope is favoured with special divine enlightenment.

Having shown that the authority of the Scriptures, of popes, and of cardinals is against the new astronomy, he gives a refutation based on physics. He asks: "If we concede the motion of the earth, why is it that an arrow shot into the air falls back to the same spot, while the earth and all things on it have in the meantime moved very rapidly toward the east? Who does not see that great confusion would result from this motion?"

Next he argues from metaphysics, as follows: "The Copernican theory of the earth's motion is against the nature of the earth itself, because the earth is not only cold but contains in itself the principle of cold; but cold is opposed to motion, and even destroys it—as is evident in animals, which become motionless when they become cold."

Finally, he clinches all with a piece of theological reasoning, as follows: "Since it can certainly be gathered from Scripture that the heavens move above the earth, and since a circular motion requires something immovable around which to move, . . . the earth is at the centre of the universe." *

But any sketch of the warfare between theology and science in this field would be incomplete without some reference to the treatment of Galileo after his death. He had begged to be buried in his family tomb in Santa Croce; this request was denied. His friends wished to erect a monument over him; this, too, was refused. Pope Urban said to the ambassador Niccolini that "it would be an evil example for the world if such honours were rendered to a man who had been brought before the Roman Inquisition for an opinion so false and erroneous; who had communicated it to many others, and who had given so great a scandal to Christendom." In accordance, therefore, with the wish of the Pope and the orders of the Inquisition, Galileo was buried ignobly, apart from his family, without fitting ceremony, without monument, without epitaph. Not until forty years after did Pierrozzi dare write an inscription

* For Chiaramonti's book and selections given, see Gebler as above, p. 271. For Polacco, see his work as cited, especially Assertiones i, ii, vii, xi, xiii, lxxiii, clxxxvii, and others. The work is in the White Library at Cornell University. The date of it is 1644.

to be placed above his bones; not until a hundred years after did Nelli dare transfer his remains to a suitable position in Santa Croce, and erect a monument above them. Even then the old conscientious hostility burst forth: the Inquisition was besought to prevent such honours to "a man condemned for notorious errors"; and that tribunal refused to allow any epitaph to be placed above him which had not been submitted to its censorship. Nor has that old conscientious consistency in hatred yet fully relented: hardly a generation since has not seen some ecclesiastic, like Marini or De Bonald or Rallaye or De Gabriac, suppressing evidence, or torturing expressions, or inventing theories to blacken the memory of Galileo and save the reputation of the Church. Nay, more: there are school histories, widely used, which, in the supposed interest of the Church, misrepresent in the grossest manner all these transactions in which Galileo was concerned. *Sancta simplicitas!* The Church has no worse enemies than those who devise and teach these perversions. They are simply rooting out, in the long run, from the minds of the more thoughtful scholars, respect for the great organization which such writings are supposed to serve.*

The Protestant Church was hardly less energetic against this new astronomy than the mother Church. The sacred science of the first Lutheran Reformers was transmitted as a precious legacy, and in the next century was made much of by Calovius. His great learning and determined orthodoxy gave him the Lutheran leadership. Utterly refusing to look at ascertained facts, he cited the turning back of the shadow upon King Hezekiah's dial and the standing still of the sun for Joshua, denied the movement of the earth, and denounced the whole new view as clearly opposed to Scripture. To this day his arguments are repeated by sundry orthodox leaders of American Lutheranism.

* For the persecutions of Galileo's memory after his death, see Gebler and Wohlwill, but especially Th. Martin, p. 243 and chaps. ix and x. For documentary proofs, see L'Épinois. For a collection of the slanderous theories invented against Galileo, see Martin, final chapters and appendix. Both these authors are devoted to the Church, but, unlike Monsignor Marini, are too upright to resort to the pious fraud of suppressing documents or interpolating pretended facts.

As to the other branches of the Reformed Church, we have already seen how Calvinists, Anglicans, and, indeed, Protestant sectarians generally, opposed the new truth.* In England, among the strict churchmen, the great Dr. South denounced the Royal Society as "irreligious," and among the Puritans the eminent John Owen declared that Newton's discoveries were "built on fallible phenomena and advanced by many arbitrary presumptions against evident testimonies of Scripture." Even Milton seems to have hesitated between the two systems. At the beginning of the eighth book of *Paradise Lost* he makes Adam state the difficulties of the Ptolemaic system, and then brings forward an angel to make the usual orthodox answers. Later, Milton seems to lean toward the Copernican theory, for, referring to the earth, he says:

> "Or she from west her silent course advance
> With inoffensive pace, that spinning sleeps
> On her soft axle, while she faces even
> And bears thee soft with the smooth air along."

English orthodoxy continued to assert itself. In 1724 John Hutchinson, professor at Cambridge, published his *Moses' Principia*, a system of philosophy in which he sought to build up a complete physical system of the universe from the Bible. In this he assaulted the Newtonian theory as "atheistic," and led the way for similar attacks by such Church teachers as Horne, Duncan Forbes, and Jones of Nayland. But one far greater than these involved himself in this view. That same limitation of his reason by the simple statements of Scripture which led John Wesley to declare that, "unless witchcraft is true, nothing in the Bible is true," led him, while giving up the Ptolemaic theory and accepting in a general way the Copernican, to suspect the demonstrations of Newton. Happily, his inborn nobility of character lifted him above any bitterness or persecuting spirit, or any imposition of doctrinal tests which could prevent those who came after him from finding their way to the truth.

---

* For Calovius, see Zoeckler, *Geschichte*, vol. i, pp. 684 and 763. For Calvin and Turretin, see Shields, *The Final Philosophy*, pp. 60, 61.

But in the midst of this vast expanse of theologic error signs of right reason began to appear, both in England and America. Noteworthy is it that Cotton Mather, bitter as was his orthodoxy regarding witchcraft, accepted, in 1721, the modern astronomy fully, with all its consequences.

In the following year came an even more striking evidence that the new scientific ideas were making their way in England. In 1722 Thomas Burnet published the sixth edition of his *Sacred Theory of the Earth*. In this he argues, as usual, to establish the scriptural doctrine of the earth's stability; but in his preface he sounds a remarkable warning. He mentions the great mistake into which St. Augustine led the Church regarding the doctrine of the antipodes, and says, "If within a few years or in the next generation it should prove as certain and demonstrable that the earth is moved, as it is now that there are antipodes, those that have been zealous against it, and engaged the Scripture in the controversy, would have the same reason to repent of their forwardness that St. Augustine would now, if he were still alive."

Fortunately, too, Protestantism had no such power to oppose the development of the Copernican ideas as the older Church had enjoyed. Yet there were some things in its warfare against science even more indefensible. In 1772 the famous English expedition for scientific discovery sailed from England under Captain Cook. Greatest by far of all the scientific authorities chosen to accompany it was Dr. Priestley. Sir Joseph Banks had especially invited him. But the clergy of Oxford and Cambridge interfered. Priestley was considered unsound in his views of the Trinity; it was evidently suspected that this might vitiate his astronomical observations; he was rejected, and the expedition crippled.

The orthodox view of astronomy lingered on in other branches of the Protestant Church. In Germany even Leibnitz attacked the Newtonian theory of gravitation on theological grounds, though he found some little consolation in thinking that it might be used to support the Lutheran doctrine of consubstantiation.

In Holland the Calvinistic Church was at first strenuous

against the whole new system, but we possess a comical proof that Calvinism even in its strongholds was powerless against it; for in 1642 Blaer published at Amsterdam his book on the use of globes, and, in order to be on the safe side, devoted one part of his work to the Ptolemaic and the other to the Copernican scheme, leaving the benevolent reader to take his choice.*

Nor have efforts to renew the battle in the Protestant Church been wanting in these latter days. The attempt in the Church of England, in 1864, to fetter science, which was brought to ridicule by Herschel, Bowring, and De Morgan; the assemblage of Lutheran clergy at Berlin, in 1868, to protest against "science falsely so called," are examples of these. Fortunately, to the latter came Pastor Knak, and his denunciations of the Copernican theory as absolutely incompatible with a belief in the Bible, dissolved the whole assemblage in ridicule.

In its recent dealings with modern astronomy the wisdom of the Catholic Church in the more civilized countries has prevented its yielding to some astounding errors into which one part of the Protestant Church has fallen heedlessly.

Though various leaders in the older Church have committed the absurd error of allowing a text-book and sundry review articles to appear which grossly misstate the Galileo episode, with the certainty of ultimately undermining confidence in her teachings among her more thoughtful young men, she has kept clear of the folly of continuing to tie her instruction, and the acceptance of our sacred books, to an adoption of the Ptolemaic theory.

Not so with American Lutheranism. In 1873 was published in St. Louis, at the publishing house of the Lutheran Synod of Missouri, a work entitled *Astronomische Unterre-*

* For the attitude of Leibnitz, Hutchinson, and the others named toward the Newtonian theory, see Lecky, *History of England in the Eighteenth Century*, chap. ix. For John Wesley, see his *Compendium of Natural Philosophy, being a Survey of the Wisdom of God in the Creation*, London, 1784. See also Leslie Stephen, *Eighteenth Century*, vol. ii, p. 413. For Owen, see his *Works*, vol. xix, p. 310. For Cotton Mather's view, see *The Christian Philosopher*, London, 1721, especially pp. 16 and 17. For the case of Priestley, see Weld, *History of the Royal Society*, vol. ii, p. 56, for the facts and the admirable letter of Priestley upon this rejection. For Blaer, see his *L'Usage des Globes*, Amsterdam, 1642.

dung, the author being well known as a late president of a Lutheran Teachers' Seminary.

No attack on the whole modern system of astronomy could be more bitter. On the first page of the introduction the author, after stating the two theories, asks, "Which is right?" and says: "It would be very simple to me which is right, if it were only a question of human import. But the wise and truthful God has expressed himself on this matter in the Bible. The entire Holy Scripture settles the question that the earth is the principal body (*Hauptkörper*) of the universe, that it stands fixed, and that sun and moon only serve to light it."

The author then goes on to show from Scripture the folly, not only of Copernicus and Newton, but of a long line of great astronomers in more recent times. He declares: "Let no one understand me as inquiring first where truth is to be found—in the Bible or with the astronomers. No; I know that beforehand—that my God never lies, never makes a mistake; out of his mouth comes only truth, when he speaks of the structure of the universe, of the earth, sun, moon, and stars. . . .

"Because the truth of the Holy Scripture is involved in this, therefore the above question is of the highest importance to me. . . . Scientists and others lean upon the miserable reed (*Rohrstab*) that God teaches only the order of salvation, but not the order of the universe."

Very noteworthy is the fact that this late survival of an ancient belief based upon text-worship is found, not in the teachings of any zealous priest of the mother Church, but in those of an eminent professor in that branch of Protestantism which claims special enlightenment.*

Nor has the warfare against the dead champions of science been carried on by the older Church alone.

On the 10th of May, 1859, Alexander von Humboldt was

---

* For the amusing details of the attempt in the English Church to repress science, and of the way in which it was met, see De Morgan, *Paradoxes*, p. 42. For Pastor Knak and his associates, see the *Revue des Deux Mondes*, 1868. Of the recent Lutheran works against the Copernican astronomy, see especially the *Astronomische Unterredung zwischen einem Liebhaber der Astronomie und mehreren berühmten Astronomer der Neuzeit*, by J. C. W. L., St. Louis, 1873.

buried. His labours had been among the glories of the century, and his funeral was one of the most imposing that Berlin had ever seen. Among those who honoured themselves by their presence was the prince regent, afterward the Emperor William I; but of the clergy it was observed that none were present save the officiating clergyman and a few regarded as unorthodox.*

V. RESULTS OF THE VICTORY OVER GALILEO.

We return now to the sequel of the Galileo case.

Having gained their victory over Galileo, living and dead, having used it to scare into submission the professors of astronomy throughout Europe, conscientious churchmen exulted. Loud was their rejoicing that the "heresy," the "infidelity," the "atheism" involved in believing that the earth revolves about its axis and moves around the sun had been crushed by the great tribunal of the Church, acting in strict obedience to the expressed will of one Pope and the written order of another. As we have seen, all books teaching this hated belief were put upon the *Index* of books forbidden to Christians, and that *Index* was prefaced by a bull enforcing this condemnation upon the consciences of the faithful throughout the world, and signed by the reigning Pope.

The losses to the world during this complete triumph of theology were even more serious than at first appears: one must especially be mentioned. There was then in Europe one of the greatest thinkers ever given to mankind—René Descartes. Mistaken though many of his reasonings were, they bore a rich fruitage of truth. He had already done a vast work. His theory of vortices—assuming a uniform material regulated by physical laws—as the beginning of the visible universe, though it was but a provisional hypothesis, had ended the whole old theory of the heavens with the vaulted firmament and the direction of the planetary movements by angels, which even Kepler had allowed. The

* See Bruhns and Lassell, *Life of Humboldt*, London, 1873, vol. ii, p. 411.

scientific warriors had stirred new life in him, and he was working over and summing up in his mighty mind all the researches of his time. The result would have made an epoch in history. His aim was to combine all knowledge and thought into a *Treatise on the World*, and in view of this he gave eleven years to the study of anatomy alone. But the fate of Galileo robbed him of all hope, of all courage; the battle seemed lost; he gave up his great plan forever.*

But ere long it was seen that this triumph of the Church was in reality a prodigious defeat. From all sides came proofs that Copernicus and Galileo were right; and although Pope Urban and the Inquisition held Galileo in strict seclusion, forbidding him even to *speak* regarding the double motion of the earth; and although this condemnation of "all books which affirm the motion of the earth" was kept on the *Index*; and although the papal bull still bound the *Index* and the condemnations in it on the consciences of the faithful; and although colleges and universities under Church control were compelled to teach the old doctrine—it was seen by clear-sighted men everywhere that this victory of the Church was a disaster to the victors.

New champions pressed on. Campanella, full of vagaries as he was, wrote his *Apology for Galileo*, though for that and other heresies, religious and political, he seven times underwent torture.

And Kepler comes: he leads science on to greater victories. Copernicus, great as he was, could not disentangle scientific reasoning entirely from the theological bias: the doctrines of Aristotle and Thomas Aquinas as to the necessary superiority of the circle had vitiated the minor features of his system, and left breaches in it through which the enemy was not slow to enter; but Kepler sees these errors, and by wonderful genius and vigour he gives to the world the three laws which bear his name, and this fortress of sci-

---

* For Descartes's discouragement, see Humboldt, *Cosmos*, London, 1851, vol. iii, p. 21; also Lange, *Geschichte des Materialismus*, English translation, vol. i, pp. 248, 249, where the letters of Descartes are given, showing his despair, and the relinquishment of his best thoughts and works in order to preserve peace with the Church; also Saisset, *Descartes et ses Précurseurs*, pp. 100 *et seq.*; also Jolly, *Histoire du Mouvement intellectuel au XVIe Siècle*, vol. i, p. 390.

ence is complete. He thinks and speaks as one inspired. His battle is severe. He is solemnly warned by the Protestant Consistory of Stuttgart "not to throw Christ's kingdom into confusion with his silly fancies," and as solemnly ordered to "bring his theory of the world into harmony with Scripture": he is sometimes abused, sometimes ridiculed, sometimes imprisoned. Protestants in Styria and Würtemberg, Catholics in Austria and Bohemia, press upon him; but Newton, Halley, Bradley, and other great astronomers follow, and to science remains the victory.*

Yet this did not end the war. During the seventeenth century, in France, after all the splendid proofs added by Kepler, no one dared openly teach the Copernican theory, and Cassini, the great astronomer, never declared for it. In 1672 the Jesuit Father Riccioli declared that there were precisely forty-nine arguments for the Copernican theory and seventy-seven against it. Even after the beginning of the eighteenth century—long after the demonstrations of Sir Isaac Newton—Bossuet, the great Bishop of Meaux, the foremost theologian that France has ever produced, declared it contrary to Scripture.

Nor did matters seem to improve rapidly during that century. In England, John Hutchinson, as we have seen, published in 1724 his *Moses' Principia* maintaining that the Hebrew Scriptures are a perfect system of natural philosophy, and are opposed to the Newtonian system of gravitation; and, as we have also seen, he was followed by a long list of noted men in the Church. In France, two eminent mathematicians published in 1748 an edition of Newton's

---

* For Campanella, see Amabile, *Fra Tommaso Campanella*, Naples, 1882, especially vol. iii; also Libri, vol. iv, pp. 149 *et seq.* Fromundus, speaking of Kepler's explanation, says, "*Vix teneo ebullientem risum.*" This is almost equal to the *New York Church Journal*, speaking of John Stuart Mill as "that small sciolist," and of the preface to Dr. Draper's great work as "chippering." How a journal, generally so fair in its treatment of such subjects, can condescend to such weapons, is one of the wonders of modern journalism. For the persecution of Kepler, see Heller, *Geschichte der Physik*, vol. i, pp. 281 *et seq.*; also Reuschle, *Kepler und die Astronomie*, Frankfurt a. M., 1871, pp. 87 *et seq.*; also Prof. Sigwart, *Kleine Schriften*, pp. 211 *et seq.* There is poetic justice in the fact that these two last-named books come from Würtemberg professors. See also *The New-Englander* for March, 1884, p. 178.

Principia; but, in order to avert ecclesiastical censure, they felt obliged to prefix to it a statement absolutely false. Three years later, Boscovich, the great mathematician of the Jesuits, used these words: "As for me, full of respect for the Holy Scriptures and the decree of the Holy Inquisition, I regard the earth as immovable; nevertheless, for simplicity in explanation I will argue as if the earth moves; for it is proved that of the two hypotheses the appearances favour this idea."

In Germany, especially in the Protestant part of it, the war was even more bitter, and it lasted through the first half of the eighteenth century. Eminent Lutheran doctors of divinity flooded the country with treatises to prove that the Copernican theory could not be reconciled with Scripture. In the theological seminaries and in many of the universities where clerical influence was strong they seemed to sweep all before them; and yet at the middle of the century we find some of the clearest-headed of them aware of the fact that their cause was lost.*

In 1757 the most enlightened perhaps in the whole line of the popes, Benedict XIV, took up the matter, and the Congregation of the *Index* secretly allowed the ideas of Copernicus to be tolerated. Yet in 1765 Lalande, the great French astronomer, tried in vain at Rome to induce the authorities to remove Galileo's works from the *Index*. Even at a date far within our own nineteenth century the authorities of many universities in Catholic Europe, and especially those in Spain, excluded the Newtonian system. In 1771 the greatest of them all, the University of Salamanca, being urged to teach physical science, refused, making answer as follows: "Newton teaches nothing that would make a good

---

* For Cassini's position, see Henri Martin, *Histoire de France*, vol. xiii, p. 175. For Riccioli, see Daunou, *Études Historiques*, vol. ii, p. 439. For Bossuet, see Bertrand, p. 41. For Hutchinson, see Lyell, *Principles of Geology*, p. 48. For Wesley, see his work, already cited. As to Boscovich, his declaration, mentioned in the text, was in 1746, but in 1785 he seemed to feel his position in view of history, and apologized abjectly: Bertrand, pp. 60, 61. See also Whewell's notice of Le Sueur and Jacquier's introduction to their edition of Newton's *Principia*. For the struggle in Germany, see Zoeckler, *Geschichte der Beziehungen zwischen Theologie und Naturwissenschaft*, vol. ii, pp. 45 *et seq.*

logician or metaphysician; and Gassendi and Descartes do not agree so well with revealed truth as Aristotle does."

Vengeance upon the dead also has continued far into our own century. On the 5th of May, 1829, a great multitude assembled at Warsaw to honour the memory of Copernicus and to unveil Thorwaldsen's statue of him.

Copernicus had lived a pious, Christian life; he had been beloved for unostentatious Christian charity; with his religious belief no fault had ever been found; he was a canon of the Church at Frauenberg, and over his grave had been written the most touching of Christian epitaphs. Naturally, then, the people expected a religious service; all was understood to be arranged for it; the procession marched to the church and waited. The hour passed, and no priest appeared; none could be induced to appear. Copernicus, gentle, charitable, pious, one of the noblest gifts of God to religion as well as to science, was evidently still under the ban. Five years after that, his book was still standing on the *Index* of books prohibited to Christians.

The edition of the *Index* published in 1819 was as inexorable toward the works of Copernicus and Galileo as its predecessors had been; but in the year 1820 came a crisis. Canon Settele, Professor of Astronomy at Rome, had written an elementary book in which the Copernican system was taken for granted. The Master of the Sacred Palace, Anfossi, as censor of the press, refused to allow the book to be printed unless Settele revised his work and treated the Copernican theory as merely a hypothesis. On this Settele appealed to Pope Pius VII, and the Pope referred the matter to the Congregation of the Holy Office. At last, on the 16th of August, 1820, it was decided that Settele might teach the Copernican system as established, and this decision was approved by the Pope. This aroused considerable discussion, but finally, on the 11th of September, 1822, the cardinals of the Holy Inquisition graciously agreed that "the printing and publication of works treating of the motion of the earth and the stability of the sun, in accordance with the general opinion of modern astronomers, is permitted at Rome." This decree was ratified by Pius VII, but it was not until thirteen years later, in 1835, that there was issued an edition of the *Index* from which the condemnation of works defending the double motion of the earth was left out.

This was not a moment too soon, for, as if the previous proofs had not been sufficient, each of the motions of the earth was now absolutely demonstrated anew, so as to be recognised by the ordinary observer. The parallax of fixed stars, shown by Bessel as well as other noted astronomers in 1838, clinched forever the doctrine of the revolution of the earth around the sun, and in 1851 the great experiment of Foucault with the pendulum showed to the human eye the earth in motion around its own axis. To make the matter complete, this experiment was publicly made in one of the churches at Rome by the eminent astronomer, Father Secchi, of the Jesuits, in 1852—just two hundred and twenty years after the Jesuits had done so much to secure Galileo's condemnation.*

---

* For good statements of the final action of the Church in the matter, see Gebler; also Zoeckler, ii, 352. See also Bertrand, *Fondateurs de l'Astronomie moderne*, p. 61; Flammarion, *Vie de Copernic*, chap. ix. As to the time when the decree of condemnation was repealed, there have been various pious attempts to make it earlier than the reality. Artaud, p. 307, cited in an apologetic article in the *Dublin Review*, September, 1865, says that Galileo's famous dialogue was published in 1714, at Padua, entire, and with the usual approbations. The same article also declares that in 1818 the ecclesiastical decrees were repealed by Pius VII in full Consistory. Whewell accepts this; but Cantu, an authority favourable to the Church, acknowledges that Copernicus's work remained on the *Index* as late as 1835 (Cantu, *Histoire universelle*, vol. xv, p. 483); and with this Th. Martin, not less favourable to the Church, but exceedingly careful as to the facts, agrees; and the most eminent authority of all, Prof. Reusch, of Bonn, in his *Der Index der verbotenen Bücher*, Bonn, 1885, vol. ii, p. 396, confirms the above statement in the text. For a clear statement of Bradley's exquisite demonstration of the Copernican theory by reasonings upon the rapidity of light, etc., and Foucault's exhibition of the rotation of the earth by the pendulum experiment, see Hoefer, *Histoire de l'Astronomie*, pp. 492 *et seq.* For more recent proofs of the Copernican theory, by the discoveries of Bunsen, Bischoff, Benzenburg, and others, see Jevons, *Principles of Science.*

VI. THE RETREAT OF THE CHURCH AFTER ITS VICTORY OVER GALILEO.

Any history of the victory of astronomical science over dogmatic theology would be incomplete without some account of the retreat made by the Church from all its former positions in the Galileo case.

The retreat of the Protestant theologians was not difficult. A little skilful warping of Scripture, a little skilful use of that time-honoured phrase, attributed to Cardinal Baronius, that the Bible is given to teach us, not how the heavens go, but how men go to heaven, and a free use of explosive rhetoric against the pursuing army of scientists, sufficed.

But in the older Church it was far less easy. The retreat of the sacro-scientific army of Church apologists lasted through two centuries.

In spite of all that has been said by these apologists, there no longer remains the shadow of a doubt that the papal infallibility was committed fully and irrevocably against the double revolution of the earth. As the documents of Galileo's trial now published show, Paul V, in 1616, pushed on with all his might the condemnation of Galileo and of the works of Copernicus and of all others teaching the motion of the earth around its own axis and around the sun. So, too, in the condemnation of Galileo in 1633, and in all the proceedings which led up to it and which followed it, Urban VIII was the central figure. Without his sanction no action could have been taken.

True, the Pope did not formally sign the decree against the Copernican theory *then*; but this came later. In 1664 Alexander VII prefixed to the *Index* containing the condemnations of the works of Copernicus and Galileo and "all books which affirm the motion of the earth" a papal bull signed by himself, binding the contents of the *Index* upon the consciences of the faithful. This bull confirmed and approved in express terms, finally, decisively, and infallibly, the condemnation of "all books teaching the movement of the earth and the stability of the sun." *

* See Rev. William W. Roberts, *The Pontifical Decrees against the Doctrine*

The position of the mother Church had been thus made especially difficult; and the first important move in retreat by the apologists was the statement that Galileo was condemned, not because he affirmed the motion of the earth, but because he supported it from Scripture. There was a slight appearance of truth in this. Undoubtedly, Galileo's letters to Castelli and the grand duchess, in which he attempted to show that his astronomical doctrines were not opposed to Scripture, gave a new stir to religious bigotry. For a considerable time, then, this quibble served its purpose; even a hundred and fifty years after Galileo's condemnation it was renewed by the Protestant Mallet du Pan, in his wish to gain favour from the older Church.

But nothing can be more absurd, in the light of the original documents recently brought out of the Vatican archives, than to make this contention now. The letters of Galileo to Castelli and the Grand-Duchess were not published until after the condemnation; and, although the Archbishop of Pisa had endeavoured to use them against him, they were but casually mentioned in 1616, and entirely left out of view in 1633. What was condemned in 1616 by the Sacred Congregation held in the presence of Pope Paul V, as "*absurd, false in theology, and heretical, because absolutely contrary to Holy Scripture*," was the proposition that "*the sun is the centre about which the earth revolves*"; and what was condemned as "*absurd, false in philosophy, and from a theologic point of view, at least, opposed to the true faith*," was the proposition that "*the earth is not the centre of the universe and immovable, but has a diurnal motion.*"

And again, what Galileo was made, by express order of Pope Urban, and by the action of the Inquisition under threat of torture, to abjure in 1633, was "*the error and heresy of the movement of the earth.*"

What the *Index* condemned under sanction of the bull

---

of the Earth's Movement, London, 1885, p. 94; and for the text of the papal bull, *Speculatores domus Israel*, pp. 132, 133, see also St. George Mivart's article in the *Nineteenth Century* for July, 1885. For the authentic publication of the bull, see preface to the *Index* of 1664, where the bull appears, signed by the Pope. The Rev. Mr. Roberts and Mr. St. George Mivart are Roman Catholics, and both acknowledge that the papal sanction was fully given.

issued by Alexander VII in 1664 was, "*all books teaching the movement of the earth and the stability of the sun.*"

What the *Index*, prefaced by papal bulls, infallibly binding its contents upon the consciences of the faithful, for nearly two hundred years steadily condemned was, "*all books which affirm the motion of the earth.*"

Not one of these condemnations was directed against Galileo "for reconciling his ideas with Scripture."*

Having been dislodged from this point, the Church apologists sought cover under the statement that Galileo was condemned not for heresy, but for contumacy and want of respect toward the Pope.

There was a slight chance, also, for this quibble: no doubt Urban VIII, one of the haughtiest of pontiffs, was induced by Galileo's enemies to think that he had been treated with some lack of proper etiquette: first, by Galileo's adhesion to his own doctrines after his condemnation in 1616; and, next, by his supposed reference in the *Dialogue* of 1632 to the arguments which the Pope had used against him.

But it would seem to be a very poor service rendered to the doctrine of papal infallibility to claim that a decision so immense in its consequences could be influenced by the personal resentment of the reigning pontiff.

Again, as to the first point, the very language of the various sentences shows the folly of this assertion; for these sentences speak always of "heresy," and never of "contumacy." As to the last point, the display of the original documents settled that forever. They show Galileo from first to last as most submissive toward the Pope, and patient under the papal arguments and exactions. He had, indeed, expressed his anger at times against his traducers; but to hold this the cause of the judgment against him is to degrade the whole proceedings, and to convict Paul V, Urban

* For the original trial documents, copied carefully from the Vatican manuscripts, see the Roman Catholic authority, L'Épinois, especially p. 35, where the principal document is given in its original Latin; see also Gebler, *Die Acten des Galilei'schen Processes*, for still more complete copies of the same documents. For minute information regarding these documents and their publication, see Favaro, *Miscellanea Galileana Inedita*, forming vol. xxii, part iii, of the *Memoirs of the Venetian Institute* for 1887, and especially pp. 891 and following.

VIII, Bellarmin, the other theologians, and the Inquisition, of direct falsehood, since they assigned entirely different reasons for their conduct. From this position, therefore, the assailants retreated.*

The next rally was made about the statement that the persecution of Galileo was the result of a quarrel between Aristotelian professors on one side and professors favouring the experimental method on the other. But this position was attacked and carried by a very simple statement. If the divine guidance of the Church is such that it can be dragged into a professorial squabble, and made the tool of a faction in bringing about a most disastrous condemnation of a proved truth, how did the Church at that time differ from any human organization sunk into decrepitude, managed nominally by simpletons, but really by schemers? If that argument be true, the condition of the Church was even worse than its enemies have declared it; and amid the jeers of an unfeeling world the apologists sought new shelter.

The next point at which a stand was made was the assertion that the condemnation of Galileo was "provisory"; but this proved a more treacherous shelter than the others. The wording of the decree of condemnation itself is a sufficient answer to this claim. When doctrines have been solemnly declared, as those of Galileo were solemnly declared under sanction of the highest authority in the Church, "contrary to the sacred Scriptures," "opposed to the true faith," and "false and absurd in theology and philosophy"—to say that such declarations are "provisory" is to say that the truth held by the Church is not immutable; from this, then, the apologists retreated.†

Still another contention was made, in some respects more curious than any other: it was, mainly, that Galileo "was no more a victim of Catholics than of Protestants; for they

---

* The invention of the "contumacy" quibble seems due to Monsignor Marini, who appears also to have manipulated the original documents to prove it. Even Whewell was evidently somewhat misled by him, but Whewell wrote before L'Épinois had shown all the documents, and under the supposition that Marini was an honest man.

† This argument also seems to have been foisted upon the world by the wily Monsignor Marini.

more than the Catholic theologians impelled the Pope to the action taken." *

But if Protestantism could force the papal hand in a matter of this magnitude, involving vast questions of belief and far-reaching questions of policy, what becomes of "inerrancy"—of special protection and guidance of the papal authority in matters of faith?

While this retreat from position to position was going on, there was a constant discharge of small-arms, in the shape of innuendoes, hints, and sophistries: every effort was made to blacken Galileo's private character: the irregularities of his early life were dragged forth, and stress was even laid upon breaches of etiquette; but this succeeded so poorly that even as far back as 1850 it was thought necessary to cover the retreat by some more careful strategy.

This new strategy is instructive. The original documents of the Galileo trial had been brought during the Napoleonic conquests to Paris; but in 1846 they were returned to Rome by the French Government, on the express pledge by the papal authorities that they should be published. In 1850, after many delays on various pretexts, the long-expected publication appeared. The personage charged with presenting them to the world was Monsignor Marini. This ecclesiastic was of a kind which has too often afflicted both the Church and the world at large. Despite the solemn promise of the papal court, the wily Marini became the instrument of the Roman authorities in evading the promise. By suppressing a document here, and interpolating a statement there, he managed to give plausible standing-ground for nearly every important sophistry ever broached to save the infallibility of the Church and destroy the reputation of Galileo. He it was who supported the idea that Galileo was "condemned not for heresy, but for contumacy."

The first effect of Monsignor Marini's book seemed useful in covering the retreat of the Church apologists. Aided by him, such vigorous writers as Ward were able to throw

---

* See the Rev. A. M. Kirsch on *Professor Huxley and Evolution*, in *The American Catholic Quarterly*, October, 1877. The article is, as a whole, remarkably fair-minded, and in the main just, as to the Protestant attitude, and as to the causes underlying the whole action against Galileo.

up temporary intrenchments between the Roman authorities and the indignation of the world.

But some time later came an investigator very different from Monsignor Marini. This was a Frenchman, M. L'Épinois. Like Marini, L'Epinois was devoted to the Church; but, unlike Marini, he could not lie. Having obtained access in 1867 to the Galileo documents at the Vatican, he published several of the most important, without suppression or pious-fraudulent manipulation. This made all the intrenchments based upon Marini's statements untenable. Another retreat had to be made.

And now came the most desperate effort of all. The apologetic army, reviving an idea which the popes and the Church had spurned for centuries, declared that the popes *as popes* had never condemned the doctrines of Copernicus and Galileo; that they had condemned them as men simply; that therefore the Church had never been committed to them; that the condemnation was made by the cardinals of the Inquisition and Index; and that the Pope had evidently been restrained by interposition of Providence from signing their condemnation. Nothing could show the desperation of the retreating party better than jugglery like this. The fact is, that in the official account of the condemnation by Bellarmin, in 1616, he declares distinctly that he makes this condemnation "in the name of His Holiness the Pope."*

Again, from Pope Urban downward, among the Church authorities of the seventeenth century the decision was always acknowledged to be made by the Pope and the Church. Urban VIII spoke of that of 1616 as made by Pope Paul V and the Church, and of that of 1633 as made by himself and the Church. Pope Alexander VII in 1664, in his bull *Speculatores*, solemnly sanctioned the condemnation of all books affirming the earth's movement.†

When Gassendi attempted to raise the point that the de-

---

* See the citation from the Vatican manuscript given in Gebler, p. 78.

† For references by Urban VIII to the condemnation as made by Pope Paul V see pp. 136, 144, and elsewhere in Martin, who much against his will is forced to allow this. See also Roberts, *Pontifical Decrees against the Earth's Movement*, and St. George Mivart's article, as above quoted; also Reusch, *Index der verbotenen Bücher*, Bonn, 1885, vol. ii, pp. 29 *et seq.*

cision against Copernicus and Galileo was not sanctioned by the Church as such, an eminent theological authority, Father Lecazre, rector of the College of Dijon, publicly contradicted him, and declared that it "was not certain cardinals, but the supreme authority of the Church," that had condemned Galileo; and to this statement the Pope and other Church authorities gave consent either openly or by silence. When Descartes and others attempted to raise the same point, they were treated with contempt. Father Castelli, who had devoted himself to Galileo, and knew to his cost just what the condemnation meant and who made it, takes it for granted, in his letter to the papal authorities, that it was made by the Church. Cardinal Querenghi, in his letters; the ambassador Guicciardini, in his dispatches; Polacco, in his refutation; the historian Viviani, in his biography of Galileo—all writing under Church inspection and approval at the time, took the view that the Pope and the Church condemned Galileo, and this was never denied at Rome. The Inquisition itself, backed by the greatest theologian of the time (Bellarmin), took the same view. Not only does he declare that he makes the condemnation "in the name of His Holiness the Pope," but we have the Roman *Index*, containing the condemnation for nearly two hundred years, prefaced by a solemn bull of the reigning Pope binding this condemnation on the consciences of the whole Church, and declaring year after year that "all books which affirm the motion of the earth" are damnable. To attempt to face all this, added to the fact that Galileo was required to abjure "the heresy of the movement of the earth" by written order of the Pope, was soon seen to be impossible. Against the assertion that the Pope was not responsible we have all this mass of testimony, and the bull of Alexander VII in 1664.*

---

* For Lecazre's answer to Gassendi, see Martin, pp. 146, 147. For the attempt to make the crime of Galileo a breach of etiquette, see *Dublin Review*, as above. Whewell, vol. i, p. 283. Citation from Marini: "Galileo was punished for trifling with the authorities, to which he refused to submit, and was punished for obstinate contumacy, not heresy." The sufficient answer to all this is that the words of the inflexible sentence designating the condemned books are "*Libri omnes qui affirmant telluris motum.*" See Bertrand, p. 59. As to the idea that "Galileo was pun-

This contention, then, was at last utterly given up by honest Catholics themselves. In 1870 a Roman Catholic clergyman in England, the Rev. Mr. Roberts, evidently thinking that the time had come to tell the truth, published a book entitled *The Pontifical Decrees against the Earth's Movement*, and in this exhibited the incontrovertible evidences that the papacy had committed itself and its infallibility fully against the movement of the earth. This Catholic clergyman showed from the original record that Pope Paul V, in 1616, had presided over the tribunal condemning the doctrine of the earth's movement, and ordering Galileo to give up the opinion. He showed that Pope Urban VIII, in 1633, pressed on, directed, and promulgated the final condemnation, making himself in all these ways responsible for it. And, finally, he showed that Pope Alexander VII, in 1664, by his bull—*Speculatores domus Israel*—attached to the *Index*, condemning "all books which affirm the motion of the earth," had absolutely pledged the papal infallibility against the earth's movement. He also confessed that under the rules laid down by the highest authorities in the Church, and especially by Sixtus V and Pius IX, there was no escape from this conclusion.

Various theologians attempted to evade the force of the argument. Some, like Dr. Ward and Bouix, took refuge in verbal niceties; some, like Dr. Jeremiah Murphy, comforted themselves with declamation. The only result was, that in 1885 came another edition of the Rev. Mr. Roberts's work, even more cogent than the first; and, besides this, an essay by that eminent Catholic, St. George Mivart, acknowledging the Rev. Mr. Roberts's position to be impregnable, and

---

ished not for his opinion, but for basing it on Scripture," the answer may be found in the Roman *Index* of 1704, in which are noted for condemnation "*Libri omnes docentes mobilitatem terræ et immobilitatem solis.*" For the way in which, when it was found convenient in argument, Church apologists insisted that it *was* "the Supreme Chief of the Church by a pontifical decree, and not certain cardinals," who condemned Galileo and his doctrine, see Father Lecazre's letter to Gassendi, in Flammarion, *Pluralité des Mondes*, p. 427, and Urban VIII's own declarations as given by Martin. For the way in which, when necessary, Church apologists asserted the very contrary of this, declaring that "it was issued in a doctrinal decree of the Congregation of the Index, and *not* as the Holy Father's teaching," see *Dublin Review*, September, 1865.

declaring virtually that the Almighty allowed Pope and Church to fall into complete error regarding the Copernican theory, in order to teach them that science lies outside their province, and that the true priesthood of scientific truth rests with scientific investigators alone.*

In spite, then, of all casuistry and special pleading, this sturdy honesty ended the controversy among Catholics themselves, so far as fair-minded men are concerned.

In recalling it at this day there stand out from its later phases two efforts at compromise especially instructive, as showing the embarrassment of militant theology in the nineteenth century.

The first of these was made by John Henry Newman in the days when he was hovering between the Anglican and Roman Churches. In one of his sermons before the University of Oxford he spoke as follows:

"Scripture says that the sun moves and the earth is stationary, and science that the earth moves and the sun is comparatively at rest. How can we determine which of these opposite statements is the very truth till we know what motion is? If our idea of motion is but an accidental result of our present senses, neither proposition is true and both are true: neither true philosophically; both true for certain practical purposes in the system in which they are respectively found."

In all anti-theological literature there is no utterance more hopelessly skeptical. And for what were the youth of Oxford led into such bottomless depths of disbelief as to any real existence of truth or any real foundation for it? Simply to save an outworn system of interpretation into which the gifted preacher happened to be born.

The other utterance was suggested by De Bonald and developed in the *Dublin Review*, as is understood, by one of Newman's associates. This argument was nothing less than an attempt to retreat under the charge of deception against the Almighty himself. It is as follows: "But it may well

* For this crushing answer by two eminent Roman Catholics to the sophistries cited—an answer which does infinitely more credit to the older Church than all the perverted ingenuity used in concealing the truth or breaking the force of it—see Roberts and St. George Mivart, as already cited.

be doubted whether the Church did retard the progress of scientific truth. What retarded it was the circumstance that God has thought fit to express many texts of Scripture in words which have every appearance of denying the earth's motion. But it is God who did this, not the Church; and, moreover, since he saw fit so to act as to retard the progress of scientific truth, it would be little to her discredit, even if it were true, that she had followed his example."

This argument, like Mr. Gosse's famous attempt to reconcile geology to Genesis—by supposing that for some inscrutable purpose God deliberately deceived the thinking world by giving to the earth all the appearances of development through long periods of time, while really creating it in six days, each of an evening and a morning—seems only to have awakened the amazed pity of thinking men. This, like the argument of Newman, was a last desperate effort of Anglican and Roman divines to save something from the wreckage of dogmatic theology.*

All these well-meaning defenders of the faith but wrought into the hearts of great numbers of thinking men the idea that there is a necessary antagonism between science and religion. Like the landsman who lashes himself to the anchor of the sinking ship, they simply attached Christianity by the strongest cords of logic which they could spin

---

* For the quotation from Newman, see his *Sermons on the Theory of Religious Belief*, sermon xiv, cited by Bishop Goodwin in *Contemporary Review* for January, 1892. For the attempt to take the blame off the shoulders of both Pope and cardinals and place it upon the Almighty, see the article above cited, in the *Dublin Review*, September, 1865, p. 419, and July, 1871, pp. 157 *et seq.* For a good summary of the various attempts, and for replies to them in a spirit of judicial fairness, see Th. Martin, *Vie de Galilée*, though there is some special pleading to save the infallibility of Pope and Church. The bibliography at the close is very valuable. For details of Mr. Gosse's theory, as developed in his *Omphalos*, see the chapter on Geology in this work. As to a still later attempt, see Wegg-Prosser, *Galileo and his Judges*, London, 1889, the main thing in it being an attempt to establish, against the honest and honourable concessions of Catholics like Roberts and Mivart, sundry far-fetched and wire-drawn distinctions between dogmatic and disciplinary bulls—an attempt which will only deepen the distrust of straightforward reasoners. The author's point of view is stated in the words, "I have maintained that the Church has a right to lay her restraining hand on the speculations of natural science" (p. 167).

to these mistaken ideas in science, and, could they have had their way, the advance of knowledge would have ingulfed both together.

On the other hand, what had science done for religion? Simply this: Copernicus, escaping persecution only by death; Giordano Bruno, burned alive as a monster of impiety; Galileo, imprisoned and humiliated as the worst of misbelievers; Kepler, accused of "throwing Christ's kingdom into confusion with his silly fancies"; Newton, bitterly attacked for "dethroning Providence," gave to religion stronger foundations and more ennobling conceptions.

Under the old system, that princely astronomer, Alphonso of Castile, seeing the inadequacy of the Ptolemaic theory, yet knowing no other, startled Europe with the blasphemy that, if he had been present at creation, he could have suggested a better order of the heavenly bodies. Under the new system, Kepler, filled with a religious spirit, exclaimed, "I do think the thoughts of God." The difference in religious spirit between these two men marks the conquest made in this long struggle by Science for Religion.*

Nothing is more unjust than to cast especial blame for all this resistance to science upon the Roman Church. The Protestant Church, though rarely able to be so severe, has been more blameworthy. The persecution of Galileo and his compeers by the older Church was mainly at the beginning of the seventeenth century; the persecution of Robertson Smith, and Winchell, and Woodrow, and Toy, and the young professors at Beyrout, by various Protestant authorities, was near the end of the nineteenth century. Those earlier persecutions by Catholicism were strictly in accordance with principles held at that time by all religionists, Catholic and Protestant, throughout the world; these later persecutions by Protestants were in defiance of principles which all Protestants to-day hold or pretend to hold, and none make louder claim to hold them than the very sects

* As a pendant to this ejaculation of Kepler may be cited the words of Linnæus: "*Deum omnipotentem a tergo transeuntem vidi et obstupui.*"

which persecuted these eminent Christian men of our day, men whose crime was that they were intelligent enough to accept the science of their time, and honest enough to acknowledge it.

Most unjustly, then, would Protestantism taunt Catholicism for excluding knowledge of astronomical truths from European Catholic universities in the seventeenth and eighteenth centuries, while real knowledge of geological and biological and anthropological truth is denied or pitifully diluted in so many American Protestant colleges and universities in the nineteenth century.

Nor has Protestantism the right to point with scorn to the Catholic *Index*, and to lay stress on the fact that nearly every really important book in the last three centuries has been forbidden by it, so long as young men in so many American Protestant universities and colleges are nursed with "ecclesiastical pap" rather than with real thought, and directed to the works of "solemnly constituted impostors," or to sundry "approved courses of reading," while they are studiously kept aloof from such leaders in modern thought as Darwin, Spencer, Huxley, Draper, and Lecky.

It may indeed be justly claimed by Protestantism that some of the former strongholds of her bigotry have become liberalized; but, on the other hand, Catholicism can point to the fact that Pope Leo XIII, now happily reigning, has made a noble change as regards open dealing with documents. The days of Monsignor Marini, it may be hoped, are gone. The Vatican Library, with its masses of historical material, has been thrown open to Protestant and Catholic scholars alike, and this privilege has been freely used by men representing all shades of religious thought.

As to the older errors, the whole civilized world was at fault, Protestant as well as Catholic. It was not the fault of religion; it was the fault of that short-sighted linking of theological dogmas to scriptural texts which, in utter defiance of the words and works of the Blessed Founder of Christianity, narrow-minded, loud-voiced men are ever prone to substitute for religion. Justly is it said by one of the most eminent among contemporary Anglican divines, that "it is because they have mistaken the dawn for a conflagration that theologians have so often been foes of light." *

* For an exceedingly striking statement, by a Roman Catholic historian of genius, as to the *popular* demand for persecution and the pressure of the lower strata in ecclesiastical organizations for cruel measures, see Balmès's *Le Protestantisme comparé au Catholicisme*, etc., fourth edition, Paris, 1855, vol. ii. Archbishop Spaulding has something of the same sort in his *Miscellanies*. L'Épinois, *Galilée*, pp. 22 *et seq.*, stretches this as far as possible to save the reputation of the Church in the Galileo matter. As to the various branches of the Protestant Church in England and the United States, it is a matter of notoriety that the smug, well-to-do laymen, whether elders, deacons, or vestrymen, are, as a rule, far more prone to heresy-hunting than are their better educated pastors. As to the cases of Messrs. Winchell, Woodrow, Toy, and the professors at Beyrout, with details, see the chapter in this series on *The Fall of Man and Anthropology*. Among Protestant historians who have been recently allowed full and free examination of the treasures in the Vatican Library, and even those involving questions between Catholicism and Protestantism, are Von Sybel, of Berlin, and Philip Schaff, of New York. It should be added that the latter went with commendatory letters from eminent prelates of the Catholic Church in Europe and America. For the closing citation, see Canon Farrar, *History of Interpretation*, p. 432.

CHAPTER IV.

FROM "SIGNS AND WONDERS" TO LAW IN THE HEAVENS.

I. THE THEOLOGICAL VIEW.

FEW things in the evolution of astronomy are more suggestive than the struggle between the theological and the scientific doctrine regarding comets—the passage from the conception of them as fire-balls flung by an angry God for the purpose of scaring a wicked world, to a recognition of them as natural in origin and obedient to law in movement. Hardly anything throws a more vivid light upon the danger of wresting texts of Scripture to preserve ideas which observation and thought have superseded, and upon the folly of arraying ecclesiastical power against scientific discovery.*

Out of the ancient world had come a mass of beliefs regarding comets, meteors, and eclipses; all these were held to be signs displayed from heaven for the warning of mankind. Stars and meteors were generally thought to presage happy events, especially the births of gods, heroes, and great men. So firmly rooted was this idea that we constantly find among the ancient nations traditions of lights in the heavens preceding the birth of persons of note. The sacred books of India show that the births of Crishna and of Buddha were announced by such heavenly lights.† The

---

* The present study, after its appearance in the *Popular Science Monthly* as a "new chapter in the Warfare of Science," was revised and enlarged to nearly its present form, and read before the American Historical Association, among whose papers it was published, in 1887, under the title of *A History of the Doctrine of Comets.*

† For Crishna, see Cox, *Aryan Mythology*, vol. ii, p. 133; the *Vishnu Purana* (Wilson's translation), book v, chap. iv. As to lights at the birth, or rather at the

sacred books of China tell of similar appearances at the births of Yu, the founder of the first dynasty, and of the inspired sage, Lao-tse. According to the Jewish legends, a star appeared at the birth of Moses, and was seen by the Magi of Egpyt, who informed the king; and when Abraham was born an unusual star appeared in the east. The Greeks and Romans cherished similar traditions. A heavenly light accompanied the birth of Æsculapius, and the births of various Cæsars were heralded in like manner.*

The same conception entered into our Christian sacred books. Of all the legends which grew in such luxuriance and beauty about the cradle of Jesus of Nazareth, none appeals more directly to the highest poetic feeling than that given by one of the evangelists, in which a star, rising in the east, conducted the wise men to the manger where the Galilean peasant-child—the Hope of Mankind, the Light of the World—was lying in poverty and helplessness.

Among the Mohammedans we have a curious example of the same tendency toward a kindly interpretation of stars and meteors, in the belief of certain Mohammedan teachers that meteoric showers are caused by good angels hurling missiles to drive evil angels out of the sky.

Eclipses were regarded in a very different light, being supposed to express the distress of Nature at earthly calamities. The Greeks believed that darkness overshadowed the earth at the deaths of Prometheus, Atreus, Hercules, Æsculapius, and Alexander the Great. The Roman legends held

---

conception, of Buddha, see Bunsen, *Angel Messiah*, pp. 22, 23; Alabaster, *Wheel of the Law* (illustrations of Buddhism), p. 102; Edwin Arnold, *Light of Asia*; Bp. Bigandet, *Life of Gaudama*, the Burmese Buddha, p. 30; Oldenberg, *Buddha* (English translation), part i, chap. ii.

* For Chinese legends regarding stars at the birth of Yu and Lao-tse, see Thornton, *History of China*, vol. i, p. 137; also Pingré, *Cométographie*, p. 245. Regarding stars at the births of Moses and Abraham, see Calmet, *Fragments*, part viii; Baring-Gould, *Legends of Old Testament Characters*, chap. xxiv; Farrar, *Life of Christ*, chap. iii. As to the Magi, see Higgins, *Anacalypsis*; Hooykaas, Ort, and Kuenen, *Bible for Learners*, vol. iii. For Greek and Roman traditions, see Bell, *Pantheon*, s. v. *Æsculapius* and *Atreus*; Gibbon, *Decline and Fall*, vol. i, pp. 151, 590; Farrar, *Life of Christ* (Amer. ed.), p. 52; Cox, *Tales of Ancient Greece*, pp. 41, 61, 62; Higgins, *Anacalypsis*, vol. i, p. 322; also Suetonius, *Caes.*, Julius, p. 88, Claud., p. 463; Seneca, *Nat. Quaest.*, vol. i, p. 1; Virgil, *Ecl.*, vol. ix, p. 47; as well as Ovid, Pliny, and others.

that at the death of Romulus there was darkness for six hours. In the history of the Cæsars occur portents of all three kinds; for at the death of Julius the earth was shrouded in darkness, the birth of Augustus was heralded by a star, and the downfall of Nero by a comet. So, too, in one of the Christian legends clustering about the crucifixion, darkness overspread the earth from the sixth to the ninth hour. Neither the silence regarding it of the only evangelist who claims to have been present, nor the fact that observers like Seneca and Pliny, who, though they carefully described much less striking occurrences of the same sort and in more remote regions, failed to note any such darkness even in Judea, have availed to shake faith in an account so true to the highest poetic instincts of humanity.

This view of the relations between Nature and man continued among both Jews and Christians. According to Jewish tradition, darkness overspread the earth for three days when the books of the Law were profaned by translation into Greek. Tertullian thought an eclipse an evidence of God's wrath against unbelievers. Nor has this mode of thinking ceased in modern times. A similar claim was made at the execution of Charles I; and Increase Mather thought an eclipse in Massachusetts an evidence of the grief of Nature at the death of President Chauncey, of Harvard College. Archbishop Sandys expected eclipses to be the final tokens of woe at the destruction of the world, and traces of this feeling have come down to our own time. The quaint story of the Connecticut statesman who, when his associates in the General Assembly were alarmed by an eclipse of the sun, and thought it the beginning of the Day of Judgment, quietly ordered in candles, that he might in any case be found doing his duty, marks probably the last noteworthy appearance of the old belief in any civilized nation.*

---

* For Hindu theories, see Alabaster, *Wheel of the Law*, 11. For Greek and Roman legends, see Higgins, *Anacalypsis*, vol. i, pp. 616, 617; also Suetonius, *Caes.*, Julius, p. 88, Claud., p. 46; Seneca, *Quaest. Nat.*, vol. i, p. 1, vol. vii, p. 17; Pliny, *Hist. Nat.*, vol. ii, p. 25; Tacitus, *Ann.*, vol. xiv, p. 22; Josephus, *Antiq.*, vol. xiv, p. 12; and the authorities above cited. For the tradition of the Jews regarding the darkness of three days, see citation in Renan, *Histoire du Peuple Israël*, vol. iv, chap. iv. For Tertullian's belief regarding the significance of an eclipse, see the *Ad*

In these beliefs regarding meteors and eclipses there was little calculated to do harm by arousing that superstitious terror which is the worst breeding-bed of cruelty. Far otherwise was it with the belief regarding comets. During many centuries it gave rise to the direst superstition and fanaticism. The Chaldeans alone among the ancient peoples generally regarded comets without fear, and thought them bodies wandering as harmless as fishes in the sea; the Pythagoreans alone among philosophers seem to have had a vague idea of them as bodies returning at fixed periods of time; and in all antiquity, so far as is known, one man alone, Seneca, had the scientific instinct and prophetic inspiration to give this idea definite shape, and to declare that the time would come when comets would be found to move in accordance with natural law. Here and there a few strong men rose above the prevailing superstition. The Emperor Vespasian tried to laugh it down, and insisted that a certain comet in his time could not betoken his death, because it was hairy, and he bald; but such scoffing produced little permanent effect, and the prophecy of Seneca was soon forgotten. These and similar isolated utterances could not stand against the mass of opinion which upheld the doctrine that comets are "signs and wonders." *

The belief that every comet is a ball of fire flung from the right hand of an angry God to warn the grovelling dwellers of earth was received into the early Church, transmitted through the Middle Ages to the Reformation period, and in its transmission was made all the more precious by

---

Scapulam, chap. iii, in Migne, *Patrolog. Lat.*, vol. i, p. 701. For the claim regarding Charles I, see a sermon preached before Charles II, cited by Lecky, *England in the Eighteenth Century*, vol. i, p. 65. Mather thought, too, that it might have something to do with the death of sundry civil functionaries of the colonies: see his *Discourse concerning Comets*, 1682. For Archbishop Sandys's belief, see his eighteenth sermon (in *Parker Soc. Publications*). The story of Abraham Davenport has been made familiar by the poem of Whittier.

* For terror caused in Rome by comets, see Pingré, *Cométographie*, pp. 165, 166. For the Chaldeans, see Wolf, *Geschichte der Astronomie*, p. 10 *et seq.*, and p. 181 *et seq.*; also Pingré, chap. ii. For the Pythagorean notions, see citation from Plutarch in Costard, *History of Astronomy*, p. 283. For Seneca's prediction, see Guillemin, *World of Comets* (translated by Glaisher), pp. 4, 5; also Watson, *On Comets*, p. 126. For this feeling in antiquity generally, see the preliminary chapters of the two works last cited.

supposed textual proofs from Scripture. The great fathers of the Church committed themselves unreservedly to it. In the third century Origen, perhaps the most influential of the earlier fathers of the universal Church in all questions between science and faith, insisted that comets indicate catastrophes and the downfall of empires and worlds. Bede, so justly revered by the English Church, declared in the eighth century that "comets portend revolutions of kingdoms, pestilence, war, winds, or heat"; and John of Damascus, his eminent contemporary in the Eastern Church, took the same view. Rabanus Maurus, the great teacher of Europe in the ninth century, an authority throughout the Middle Ages, adopted Bede's opinion fully. St. Thomas Aquinas, the great light of the universal Church in the thirteenth century, whose works the Pope now reigning commends as the centre and source of all university instruction, accepted and handed down the same opinion. The sainted Albert the Great, the most noted genius of the mediæval Church in natural science, received and developed this theory. These men and those who followed them founded upon scriptural texts and theological reasonings a system that for seventeen centuries defied every advance of thought.*

The main evils thence arising were three: the paralysis of self-help, the arousing of fanaticism, and the strengthening of ecclesiastical and political tyranny. The first two of these evils—the paralysis of self-help and the arousing of fanaticism—are evident throughout all these ages. At the appearance of a comet we constantly see all Christendom, from pope to peasant, instead of striving to avert war by wise statesmanship, instead of striving to avert pestilence by observation and reason, instead of striving to avert famine by skilful economy, whining before fetiches, trying to bribe them to remove these signs of God's wrath, and planning to wreak this supposed wrath of God upon misbelievers.

As to the third of these evils—the strengthening of eccle-

---

* For Origen, see his *De Princip.*, vol. i, p. 7; also Maury, *Lég. Pieuses*, p. 203, note. For Bede and others, see *De Nat.*, vol. xxiv; Joh. Dam., *De Fid. Or.*, vol. ii, p. 7; Maury, *La Magie et l'Astronomie*, pp. 181, 182. For Albertus Magnus, see his *Opera*, vol. i, tr. iii, chaps. x, xi. Among the texts of Scripture on which this belief rested was especially Joel ii, 30, 31.

siastical and civil despotism—examples appear on every side. It was natural that hierarchs and monarchs whose births were announced by stars, or whose deaths were announced by comets, should regard themselves as far above the common herd, and should be so regarded by mankind; passive obedience was thus strengthened, and the most monstrous assumptions of authority were considered simply as manifestations of the Divine will. Shakespeare makes Calphurnia say to Cæsar:

> "When beggars die, there are no comets seen;
> The heavens themselves blaze forth the death of princes."

Galeazzo, the tyrant of Milan, expressing satisfaction on his deathbed that his approaching end was of such importance as to be heralded by a comet, is but a type of many thus encouraged to prey upon mankind; and Charles V, one of the most powerful monarchs the world has known, abdicating under fear of the comet of 1556, taking refuge in the monastery of San Yuste, and giving up the best of his vast realms to such a scribbling bigot as Philip II, furnishes an example even more striking.*

But for the retention of this belief there was a moral cause. Myriads of good men in the Christian Church down to a recent period saw in the appearance of comets not merely an exhibition of "signs in the heavens" foretold in Scripture, but also Divine warnings of vast value to humanity as incentives to repentance and improvement of life—warnings, indeed, so precious that they could not be spared without danger to the moral government of the world. And this belief in the portentous character of comets as an essential part of the Divine government, being, as it was thought, in full accord with Scripture, was made for centuries a source of terror to humanity. To say nothing of examples in the earlier periods, comets in the tenth century especially increased the distress of all Europe. In the middle of the eleventh century a comet was thought to accompany the death of Edward the Confessor and to presage the Norman

* For Cæsar, see Shakespeare, *Julius Cæsar*, act ii, sc. 2. For Galeazzo, see Guillemin, *World of Comets*, p. 19. For Charles V, see Prof. Wolf's essay in the *Monatschrift des wissenschaftlichen Vereins*, Zürich, 1857, p. 228.

conquest; the traveller in France to-day may see this belief as it was then wrought into the Bayeux tapestry.*

Nearly every decade of years throughout the Middle Ages saw Europe plunged into alarm by appearances of this sort, but the culmination seems to have been reached in 1456. At that time the Turks, after a long effort, had made good their footing in Europe. A large statesmanship or generalship might have kept them out; but, while different religious factions were disputing over petty shades of dogma, they had advanced, had taken Constantinople, and were evidently securing their foothold. Now came the full bloom of this superstition. A comet appeared. The Pope of that period, Calixtus III, though a man of more than ordinary ability, was saturated with the ideas of his time. Alarmed at this monster, if we are to believe the contemporary historian, this infallible head of the Church solemnly "decreed several days of prayer for the averting of the wrath of God, that whatever calamity impended might be turned from the Christians and against the Turks." And, that all might join daily in this petition, there was then established that midday Angelus which has ever since called good Catholics to prayer against the powers of evil. Then, too, was incorporated into a litany the plea, "From the Turk and the comet, good Lord, deliver us." Never was papal intercession less effective; for the Turk has held Constantinople from that day to this, while the obstinate comet, being that now known under the name of Halley, has returned imperturbably at short periods ever since.†

* For evidences of this widespread terror, see chronicles of Raoul Glaber, Guillaume de Nangis, William of Malmesbury, Florence of Worcester, Ordericus Vitalis, *et al.*, *passim*, and the *Anglo-Saxon Chronicle* (in the *Rolls Series*). For very thrilling pictures of this horror in England, see Freeman, *Norman Conquest*, vol. iii, pp. 640–644, and *William Rufus*, vol. ii, p. 118. For the Bayeux tapestry, see Bruce, *Bayeux Tapestry Elucidated*, plate vii and p. 86; also Guillemin, *World of Comets*, p. 24. There is a large photographic copy, in the South Kensington Museum at London, of the original, wrought, as is generally believed, by the wife of William the Conqueror and her ladies, and still preserved in the town museum at Bayeux.

† The usual statement is, that Calixtus excommunicated the comet by a bull, and this is accepted by Arago, Grant, Hoefer, Guillemin, Watson, and many historians of astronomy. Hence the parallel made on a noted occasion by President Lincoln. No such bull, however, is to be found in the published *Bullaria*, and

But the superstition went still further. It became more and more incorporated into what was considered "scriptural science" and "sound learning." The encyclopedic summaries, in which the science of the Middle Ages and the Reformation period took form, furnish abundant proofs of this.

Yet scientific observation was slowly undermining this structure. The inspired prophecy of Seneca had not been forgotten. Even as far back as the ninth century, in the midst of the sacred learning so abundant at the court of Charlemagne and his successors, we find a scholar protesting against the accepted doctrine. In the thirteenth century we have a mild question by Albert the Great as to the supposed influence of comets upon individuals; but the prevailing theological current was too strong, and he finally yielded to it in this as in so many other things.

So, too, in the sixteenth century, we have Copernicus refusing to accept the usual theory, Paracelsus writing to Zwingli against it, and Julius Cæsar Scaliger denouncing it as "ridiculous folly." *

At first this scepticism only aroused the horror of theologians and increased the vigour of ecclesiastics; both asserted the theological theory of comets all the more strenuously as based on scriptural truth. During the sixteenth century France felt the influence of one of her greatest men on the side of this superstition. Jean Bodin, so far before his time in political theories, was only thoroughly abreast of it in religious theories: the same reverence for

---

that establishing the *Angelus* (as given by Raynaldus in the *Annales Eccl.*) contains no mention of the comet. But the authority of Platina (in his *Vitæ Pontificum*, Venice, 1479, *sub* Calistus III), who was not only in Rome at the time, but, when he wrote his history, archivist of the Vatican, is final as to the Pope's attitude. Platina's authority was never questioned until modern science had changed the ideas of the world. The recent attempt of Pastor (in his *Geschichte der Päpste*) to pooh-pooh down the whole matter is too evident an evasion to carry weight with those who know how even the most careful histories have to be modified to suit the views of the censorship at Rome.

* As to encyclopedic summaries, see Vincent of Beauvais, *Speculum Naturale*, and the various editions of Reisch's *Margarita Philosophica*. For Charlemagne's time, see Champion, *La Fin du Monde*, p. 156; Leopardi, *Errori Popolari*, p. 165. As to Albert the Great's question, see Heller, *Geschichte der Physik*, vol. i, p. 188. As to scepticism in the sixteenth century, see Champion, *La Fin du Monde*, pp. 155, 156; and for Scaliger, Dudith's book, cited below.

the mere letter of Scripture which made him so fatally powerful in supporting the witchcraft delusion, led him to support this theological theory of comets—but with a difference: he thought them the souls of men, wandering in space, bringing famine, pestilence, and war.

Not less strong was the same superstition in England. Based upon mediæval theology, it outlived the revival of learning. From a multitude of examples a few may be selected as typical. Early in the sixteenth century Polydore Virgil, an ecclesiastic of the unreformed Church, alludes, in his *English History*, to the presage of the death of the Emperor Constantine by a comet as to a simple matter of fact; and in his work on prodigies he pushes this superstition to its most extreme point, exhibiting comets as preceding almost every form of calamity.

In 1532, just at the transition period from the old Church to the new, Cranmer, paving the way to his archbishopric, writes from Germany to Henry VIII, and says of the comet then visible: "What strange things these tokens do signify to come hereafter, God knoweth; for they do not lightly appear but against some great matter."

Twenty years later Bishop Latimer, in an Advent sermon, speaks of eclipses, rings about the sun, and the like, as signs of the approaching end of the world.*

In 1580, under Queen Elizabeth, there was set forth an "order of prayer to avert God's wrath from us, threatened by the late terrible earthquake, to be used in all parish churches." In connection with this there was also commended to the faithful "a godly admonition for the time present"; and among the things referred to as evidence of God's wrath are comets, eclipses, and falls of snow.

This view held sway in the Church of England during Elizabeth's whole reign and far into the Stuart period: Strype, the ecclesiastical annalist, gives ample evidence of this, and among the more curious examples is the surmise

---

* For Bodin, see *Theatr.*, lib. ii, cited by Pingré, vol. i, p. 45; also a vague citation in Baudrillart, *Bodin et son Temps*, p. 360. For Polydore Virgil, see *English History*, p. 97 (in *Camden Society Publications*). For Cranmer, see *Remains*, vol. ii, p. 535 (in *Parker Society Publications*). For Latimer, see *Sermons*, second Sunday in Advent, 1552.

that the comet of 1572 was a token of Divine wrath provoked by the St. Bartholomew massacre.

As to the Stuart period, Archbishop Spottiswoode seems to have been active in carrying the superstition from the sixteenth century to the seventeenth, and Archbishop Bramhall cites Scripture in support of it. Rather curiously, while the diary of Archbishop Laud shows so much superstition regarding dreams as portents, it shows little or none regarding comets; but Bishop Jeremy Taylor, strong as he was, evidently favoured the usual view. John Howe, the eminent Nonconformist divine in the latter part of the century, seems to have regarded the comet superstition as almost a fundamental article of belief; he laments the total neglect of comets and portents generally, declaring that this neglect betokens want of reverence for the Ruler of the world; he expresses contempt for scientific inquiry regarding comets, insists that they may be natural bodies and yet supernatural portents, and ends by saying, "I conceive it very safe to suppose that some very considerable thing, either in the way of judgment or mercy, may ensue, according as the cry of persevering wickedness or of penitential prayer is more or less loud at that time." *

The Reformed Church of Scotland supported the superstition just as strongly. John Knox saw in comets tokens of the wrath of Heaven; other authorities considered them "a warning to the king to extirpate the Papists"; and as late as 1680, after Halley had won his victory, comets were announced on high authority in the Scottish Church to be "prodigies of great judgment on these lands for our sins, for never was the Lord more provoked by a people."

While such was the view of the clergy during the sixteenth and seventeenth centuries, the laity generally ac-

---

* For *Liturgical Services of the Reign of Queen Elizabeth*, see *Parker Society Publications*, pp. 569, 570. For Strype, see his *Ecclesiastical Memorials*, vol. iii, part i, p. 472; also his *Annals of the Reformation*, vol. ii, part ii, p. 151; and his *Life of Sir Thomas Smith*, pp. 161, 162. For Spottiswoode, see *History of the Church of Scotland* (Edinburgh reprint, 1851), vol. i, pp. 185, 186. For Bramhall, see his *Works*, Oxford, 1844, vol. iv, pp. 60, 307, etc. For Jeremy Taylor, see his *Sermons on the Life of Christ*. For John Howe, see his *Works*, London, 1862, vol. iv, pp. 140, 141.

cepted it as a matter of course. Among the great leaders in literature there was at least general acquiescence in it. Both Shakespeare and Milton recognise it, whether they fully accept it or not. Shakespeare makes the Duke of Bedford, lamenting at the bier of Henry V, say:

> "Comets, importing change of time and states,
> Brandish your crystal tresses in the sky;
> And with them scourge the bad revolting stars,
> That have consented unto Henry's death."

Milton, speaking of Satan preparing for combat, says:

> "On the other side,
> Incensed with indignation, Satan stood
> Unterrified, and like a comet burned,
> That fires the length of Ophiuchus huge
> In the arctic sky, and from its horrid hair
> Shakes pestilence and war."

We do indeed find that in some minds the discoveries of Tycho Brahe and Kepler begin to take effect, for, in 1621, Burton in his *Anatomy of Melancholy* alludes to them as changing public opinion somewhat regarding comets; and, just before the middle of the century, Sir Thomas Browne expresses a doubt whether comets produce such terrible effects, "since it is found that many of them are above the moon." * Yet even as late as the last years of the seventeenth century we have English authors of much power battling for this supposed scriptural view; and among the natural and typical results we find, in 1682, Ralph Thoresby, a Fellow of the Royal Society, terrified at the comet of that year, and writing in his diary the following passage: "Lord, fit us for whatever changes it may portend; for, though I am not ignorant that such meteors proceed from natural causes, yet are they frequently also the presages of imminent calamities." Interesting is it to note here that this was Halley's comet, and that Halley was at this very moment making those scientific studies upon it which were to free

---

* For John Knox, see his *Historie of the Reformation of Religion within the Realm of Scotland* (Edinburgh, 1732), lib. iv; also Chambers, *Domestic Annals of Scotland*, vol. ii, pp. 410–412. For Burton, see his *Anatomy of Melancholy*, part ii, sect. 2 For Browne, see the *Vulgar and Common Errors*, book vi, chap. xiv.

the civilized world forever from such terrors as distressed Thoresby.

The belief in comets as warnings against sin was especially one of those held "always, everywhere, and by all," and by Eastern Christians as well as by Western. One of the most striking scenes in the history of the Eastern Church is that which took place at the condemnation of Nikon, the great Patriarch of Moscow. Turning toward his judges, he pointed to a comet then blazing in the sky, and said, "God's besom shall sweep you all away!"

Of all countries in western Europe, it was in Germany and German Switzerland that this superstition took strongest hold. That same depth of religious feeling which produced in those countries the most terrible growth of witchcraft persecution, brought superstition to its highest development regarding comets. No country suffered more from it in the Middle Ages. At the Reformation Luther declared strongly in favour of it. In one of his Advent sermons he said, "The heathen write that the comet may arise from natural causes, but God creates not one that does not foretoken a sure calamity." Again he said, "Whatever moves in the heaven in an unusual way is certainly a sign of God's wrath." And sometimes, yielding to another phase of his belief, he declared them works of the devil, and declaimed against them as "harlot stars." *

Melanchthon, too, in various letters refers to comets as heralds of Heaven's wrath, classing them, with evil conjunctions of the planets and abortive births, among the "signs" referred to in Scripture. Zwingli, boldest of the greater Reformers in shaking off traditional beliefs, could not shake off this, and insisted that the comet of 1531 betokened calamity. Arietus, a leading Protestant theologian, declared, "The heavens are given us not merely for our pleasure, but also

* For Thoresby, see his *Diary* (London, 1830), vol. i, p. 132. Halley's great service is described further on in this chapter. For Nikon's speech, see Dean Stanley's *History of the Eastern Church*, p. 485. For very striking examples of this mediæval terror in Germany, see Von Raumer, *Geschichte der Hohenstaufen*, vol. vi, p. 538. For the Reformation period, see Wolf, *Gesch. d. Astronomie*; also Prætorius, *Ueber d. Cometstern* (Erfurt, 1580), in which the above sentences of Luther are printed on the title-page as epigraphs. For "Huren-Sternen," see the sermon of Celichius, described later.

as a warning of the wrath of God for the correction of our lives." Lavater insisted that comets are signs of death or calamity, and cited proofs from Scripture.

Catholic and Protestant strove together for the glory of this doctrine. It was maintained with especial vigour by Fromundus, the eminent professor and Doctor of Theology at the Catholic University of Louvain, who so strongly opposed the Copernican system; at the beginning of the seventeenth century, even so gifted an astronomer as Kepler yielded somewhat to the belief; and near the end of that century Voigt declared that the comet of 1618 clearly presaged the downfall of the Turkish Empire, and he stigmatized as "atheists and Epicureans" all who did not believe comets to be God's warnings.*

II. THEOLOGICAL EFFORTS TO CRUSH THE SCIENTIFIC VIEW.

Out of this belief was developed a great series of efforts to maintain the theological view of comets, and to put down forever the scientific view. These efforts may be divided into two classes: those directed toward learned men and scholars, through the universities, and those directed toward the people at large, through the pulpits. As to the first of these, that learned men and scholars might be kept in the paths of "sacred science" and "sound learning," especial pains was taken to keep all knowledge of the scientific view of comets as far as possible from students in the universities. Even to the end of the seventeenth century the oath generally required of professors of astronomy over a large part of Europe prevented their teaching that comets are heavenly bodies obedient to law. Efforts just as earnest were made to fasten into students' minds the theological theory. Two or three examples out of many may serve as

* For Melanchthon, see Wolf, *ubi supra*. For Zwingli, see Wolf, p. 235. For Arietus, see Mädler, *Geschichte der Himmelskunde*, vol. ii. For Kepler's superstition, see Wolf, p 281. For Voigt, see *Himmels-Magnaten Reichstage*, Hamburg, 1676. For both Fromundus and Voigt, see also Mädler, vol. ii, p. 399, and Lecky, *Rationalism in Europe*, vol. i, p. 28.

types. First of these may be named the teaching of Jacob Heerbrand, professor at the University of Tübingen, who in 1577 illustrated the moral value of comets by comparing the Almighty sending a comet, to the judge laying the executioner's sword on the table between himself and the criminal in a court of justice; and, again, to the father or schoolmaster displaying the rod before naughty children. A little later we have another churchman of great importance in that region, Schickhart, head pastor and superintendent at Göppingen, preaching and publishing a comet sermon, in which he denounces those who stare at such warnings of God without heeding them, and compares them to "calves gaping at a new barn door." Still later, at the end of the seventeenth century, we find Conrad Dieterich, director of studies at the University of Marburg, denouncing all scientific investigation of comets as impious, and insisting that they are only to be regarded as "signs and wonders." *

The results of this ecclesiastical pressure upon science in the universities were painfully shown during generation after generation, as regards both professors and students; and examples may be given typical of its effects upon each of these two classes.

The first of these is the case of Michael Maestlin. He was by birth a Swabian Protestant, was educated at Tübingen as a pupil of Apian, and, after a period of travel, was settled as deacon in the little parish of Backnang, when the comet of 1577 gave him an occasion to apply his astronomical studies. His minute and accurate observation of it is to this day one of the wonders of science. It seems almost impossible that so much could be accomplished by the naked eye. His observations agreed with those of Tycho Brahe, and won for Maestlin the professorship of astronomy in the University of Heidelberg. No man had so clearly proved the supralunar position of a comet, or shown so conclusively that its motion was not erratic, but regular. The young astronomer, though Apian's pupil, was an avowed Copernican

* For the effect of the anti-Pythagorean oath, see Prowe, *Copernicus*; also Mädler and Wolf. For Heerbrand, see his *Von dem erschrockenlichen Wunderzeichen*, Tübingen, 1577. For Schickhart, see his *Predigt vom Wunderzeichen*, Stuttgart, 1621. For Dieterich, see his sermon, described more fully below.

and the destined master and friend of Kepler. Yet, in the treatise embodying his observations, he felt it necessary to save his reputation for orthodoxy by calling the comet a "new and horrible prodigy," and by giving a chapter of "conjectures on the signification of the present comet," in which he proves from history that this variety of comet betokens peace, but peace purchased by a bloody victory. That he really believed in this theological theory seems impossible; the very fact that his observations had settled the supralunar character and regular motion of comets proves this. It was a humiliation only to be compared to that of Osiander when he wrote his grovelling preface to the great book of Copernicus. Maestlin had his reward: when, a few years later, his old teacher, Apian, was driven from his chair at Tübingen for refusing to sign the Lutheran *Concord-Book*, Maestlin was elected to his place.

Not less striking was the effect of this theological pressure upon the minds of students. Noteworthy as an example of this is the book of the Leipsic lawyer, Büttner. From no less than eighty-six biblical texts he proves the Almighty's purpose of using the heavenly bodies for the instruction of men as to future events, and then proceeds to frame exhaustive tables, from which, the time and place of the comet's first appearance being known, its signification can be deduced. This manual he gave forth as a triumph of religious science, under the name of the *Comet Hour-Book*.*

The same devotion to the portent theory is found in the universities of Protestant Holland. Striking is it to see in the sixteenth century, after Tycho Brahe's discovery, the Dutch theologian, Gerard Vossius, Professor of Theology and Eloquence at Leyden, lending his great weight to the superstition. "The history of all times," he says, "shows comets to be the messengers of misfortune. It does not follow that they are endowed with intelligence, but that there is a deity who makes use of them to call the human race to repentance." Though familiar with the works of Tycho Brahe, he finds it "hard to believe" that all comets are

* For Maestlin, see his *Observatio et Demonstratio Cometæ*, Tübingen, 1578. For Büttner, see his *Cometen Stundbüchlein*, Leipsic, 1605.

ethereal, and adduces several historical examples of sublunary ones.

Nor was this attempt to hold back university teaching to the old view of comets confined to Protestants. The Roman Church was, if possible, more strenuous in the same effort. A few examples will serve as types, representing the orthodox teaching at the great centres of Catholic theology.

One of these is seen in Spain. The eminent jurist Torreblanca was recognised as a controlling authority in all the universities of Spain, and from these he swayed in the seventeenth century the thought of Catholic Europe, especially as to witchcraft and the occult powers in Nature. He lays down the old cometary superstition as one of the foundations of orthodox teaching. Begging the question, after the fashion of his time, he argues that comets can not be stars, because new stars always betoken good, while comets betoken evil.

The same teaching was given in the Catholic universities of the Netherlands. Fromundus, at Louvain, the enemy of Galileo, steadily continued his crusade against all cometary heresy.*

But a still more striking case is seen in Italy. The reverend Father Augustin de Angelis, rector of the Clementine College at Rome, as late as 1673, after the new cometary theory had been placed beyond reasonable doubt, and even while Newton was working out its final demonstration, published a third edition of his *Lectures on Meteorology*. It was dedicated to the Cardinal of Hesse, and bore the express sanction of the Master of the Sacred Palace at Rome and of the head of the religious order to which De Angelis belonged. This work deserves careful analysis, not only as representing the highest and most approved university teaching of the time at the centre of Roman Catholic Christendom, but still more because it represents that attempt to make a compromise between theology and science, or rather the attempt to confiscate science to the uses of theology,

* For Vossius, see the *De Idololatria* (in his *Opera*, vol. v, pp. 283–285). For Torreblanca, see his *De Magia*, Seville, 1618, and often reprinted. For Fromundus, see his *Meteorologica*.

which we so constantly find whenever the triumph of science in any field has become inevitable.

As to the scientific element in this compromise, De Angelis holds, in his general introduction regarding meteorology, that the main material cause of comets is "exhalation," and says, "If this exhalation is thick and sticky, it blazes into a comet." And again he returns to the same view, saying that "one form of exhalation is dense, hence easily inflammable and long retentive of fire, from which sort are especially generated comets." But it is in his third lecture that he takes up comets specially, and his discussion of them is extended through the fourth, fifth, and sixth lectures. Having given in detail the opinions of various theologians and philosophers, he declares his own in the form of two conclusions. The first of these is that "comets are not heavenly bodies, but originate in the earth's atmosphere below the moon; for everything heavenly is eternal and incorruptible, but comets have a beginning and ending—*ergo*, comets can not be heavenly bodies." This, we may observe, is levelled at the observations and reasonings of Tycho Brahe and Kepler, and is a very good illustration of the scholastic and mediæval method—the method which blots out an ascertained fact by means of a metaphysical formula. His second conclusion is that "comets are of elemental and sublunary nature; for they are an exhalation hot and dry, fatty and well condensed, inflammable and kindled in the uppermost regions of the air." He then goes on to answer sundry objections to this mixture of metaphysics and science, and among other things declares that "the fatty, sticky material of a comet may be kindled from sparks falling from fiery heavenly bodies or from a thunderbolt"; and, again, that the thick, fatty, sticky quality of the comet holds its tail in shape, and that, so far are comets from having their paths beyond the moon's orbit, as Tycho Brahe and Kepler thought, he himself in 1618 saw "a bearded comet so near the summit of Vesuvius that it almost seemed to touch it." As to sorts and qualities of comets, he accepts Aristotle's view, and divides them into bearded and tailed.* He goes on into

* *Barbata et caudata.*

long disquisitions upon their colours, forms, and motions. Under this latter head he again plunges deep into a sea of metaphysical considerations, and does not reappear until he brings up his compromise in the opinion that their movement is as yet uncertain and not understood, but that, if we must account definitely for it, we must say that it is effected by angels especially assigned to this service by Divine Providence. But, while proposing this compromise between science and theology as to the origin and movement of comets, he will hear to none as regards their mission as "signs and wonders" and presages of evil. He draws up a careful table of these evils, arranging them in the following order: Drought, wind, earthquake, tempest, famine, pestilence, war, and, to clinch the matter, declares that the comet observed by him in 1618 brought not only war, famine, pestilence, and earthquake, but also a general volcanic eruption, "which would have destroyed Naples, had not the blood of the invincible martyr Januarius withstood it."

It will be observed, even from this sketch, that, while the learned Father Augustin thus comes infallibly to the mediæval conclusion, he does so very largely by scientific and essentially modern processes, giving unwonted prominence to observation, and at times twisting scientific observation into the strand with his metaphysics. The observations and methods of his science are sometimes shrewd, sometimes comical. Good examples of the latter sort are such as his observing that the comet stood very near the summit of Vesuvius, and his reasoning that its tail was kept in place by its stickiness. But observations and reasonings of this sort are always the first homage paid by theology to science as the end of their struggle approaches.*

Equally striking is an example seen a little later in another part of Europe; and it is the more noteworthy because Halley and Newton had already fully established the modern scientific theory. Just at the close of the seventeenth century the Jesuit Reinzer, professor at Linz, put forth his *Meteorologia Philosophico-Politica*, in which all natural phenomena received both a physical and a moral interpretation.

---

* See De Angelis, *Lectiones Meteorologicæ*, Rome, 1669.

It was profusely and elaborately illustrated, and on account of its instructive contents was in 1712 translated into German for the unlearned reader. The comet receives, of course, great attention. "It appears," says Reinzer, "only then in the heavens when the latter punish the earth, and through it [the comet] not only predict but bring to pass all sorts of calamity. . . . And, to that end, its tail serves for a rod, its hair for weapons and arrows, its light for a threat, and its heat for a sign of anger and vengeance." Its warnings are threefold: (1) "Comets, generated in the air, betoken *naturally* drought, wind, earthquake, famine, and pestilence." (2) "Comets can indirectly, in view of their material, betoken wars, tumults, and the death of princes; for, being hot and dry, they bring the moistnesses [*Feuchtigkeiten*] in the human body to an extraordinary heat and dryness, increasing the gall; and, since the emotions depend on the temperament and condition of the body, men are through this change driven to violent deeds, quarrels, disputes, and finally to arms: especially is this the result with princes, who are more delicate and also more arrogant than other men, and whose moistnesses are more liable to inflammation of this sort, inasmuch as they live in luxury and seldom restrain themselves from those things which in such a dry state of the heavens are especially injurious." (3) "All comets, whatever prophetic significance they may have naturally in and of themselves, are yet principally, according to the Divine pleasure, heralds of the death of great princes, of war, and of other such great calamities; and this is known and proved, first of all, from the words of Christ himself: 'Nation shall rise against nation, and kingdom against kingdom; and great earthquakes shall be in divers places, and famines, and pestilences; and fearful sights and great signs shall there be from heaven.'" *

While such pains was taken to keep the more highly educated classes in the "paths of scriptural science and sound learning" at the universities, equal efforts were made to preserve the cometary orthodoxy of the people at large

---

* See Reinzer, *Meteorologia Philosophico-Politica* (edition of Augsburg, 1712), pp. 101–103.

by means of the pulpits. Out of the mass of sermons for this purpose which were widely circulated I will select just two as typical, and they are worthy of careful study as showing some special dangers of applying theological methods to scientific facts. In the second half of the sixteenth century the recognised capital of orthodox Lutheranism was Magdeburg, and in the region tributary to this metropolis no Church official held a more prominent station than the "Superintendent," or Lutheran bishop, of the neighbouring Altmark. It was this dignitary, Andreas Celichius by name, who at Magdeburg, in 1578, gave to the press his *Theological Reminder of the New Comet*. After deprecating as blasphemous the attempt of Aristotle to explain the phenomenon otherwise than as a supernatural warning from God to sinful man, he assures his hearers that "whoever would know the comet's real source and nature must not merely gape and stare at the scientific theory that it is an earthy, greasy, tough, and sticky vapour and mist, rising into the upper air and set ablaze by the celestial heat." Far more important for them is it to know what this vapour is. It is really, in the opinion of Celichius, nothing more or less than "the thick smoke of human sins, rising every day, every hour, every moment, full of stench and horror, before the face of God, and becoming gradually so thick as to form a comet, with curled and plaited tresses, which at last is kindled by the hot and fiery anger of the Supreme Heavenly Judge." He adds that it is probably only through the prayers and tears of Christ that this blazing monument of human depravity becomes visible to mortals. In support of this theory, he urges the "coming up before God" of the wickedness of Sodom and Gomorrah and of Nineveh, and especially the words of the prophet regarding Babylon, "Her stench and rottenness is come up before me." That the anger of God can produce the conflagration without any intervention of Nature is proved from the Psalms, "He sendeth out his word and melteth them." From the position of the comet, its course, and the direction of its tail he augurs especially the near approach of the judgment day, though it may also betoken, as usual, famine, pestilence, and war. "Yet even in these days," he mourns, "there are people reckless and giddy enough to pay no heed to such celestial warnings, and these even cite in their own defence the injunction of Jeremiah not to fear signs in the heavens." This idea he explodes, and shows that good and orthodox Christians, while not superstitious like the heathen, know well "that God is not bound to his creation and the ordinary course of Nature, but must often, especially in these last dregs of the world, resort to irregular means to display his anger at human guilt." *

The other typical case occurred in the following century and in another part of Germany. Conrad Dieterich was, during the first half of the seventeenth century, a Lutheran ecclesiastic of the highest authority. His ability as a theologian had made him Archdeacon of Marburg, Professor of Philosophy and Director of Studies at the University of Giessen, and "Superintendent," or Lutheran bishop, in southwestern Germany. In the year 1620, on the second Sunday in Advent, in the great Cathedral of Ulm, he developed the orthodox doctrine of comets in a sermon, taking up the questions: 1. What are comets? 2. What do they indicate? 3. What have we to do with their significance? This sermon marks an epoch. Delivered in that stronghold of German Protestantism and by a prelate of the highest standing, it was immediately printed, prefaced by three laudatory poems from different men of note, and sent forth to drive back the scientific, or, as it was called, the "godless," view of comets. The preface shows that Dieterich was sincerely alarmed by the tendency to regard comets as natural appearances. His text was taken from the twenty-fifth verse of the twenty-first chapter of St. Luke: "And there shall be signs in the sun, and in the moon, and in the stars; and upon the earth distress of nations, with perplexity; the sea and the waves roaring." As to what comets are, he cites a multitude of philosophers, and, finding that they differ among themselves, he uses a form of argument not uncommon from that day to this, declaring that this difference of opinion proves that there is no solution of the problem save in revelation, and insisting that comets are "signs especially sent by the Al-

* For Celichius, or Celich, see his own treatise, as above.

mighty to warn the earth." An additional proof of this he finds in the forms of comets. One, he says, took the form of a trumpet; another, of a spear; another, of a goat; another, of a torch; another, of a sword; another, of an arrow; another, of a sabre; still another, of a bare arm. From these forms of comets he infers that we may divine their purpose. As to their creation, he quotes John of Damascus and other early Church authorities in behalf of the idea that each comet is a star newly created at the Divine command, out of nothing, and that it indicates the wrath of God. As to their purpose, having quoted largely from the Bible and from Luther, he winds up by insisting that, as God can make nothing in vain, comets must have some distinct object; then, from Isaiah and Joel among the prophets, from Matthew, Mark, and Luke among the evangelists, from Origen and John Chrysostom among the fathers, from Luther and Melanchthon among the Reformers, he draws various texts more or less conclusive to prove that comets indicate evil and only evil; and he cites Luther's Advent sermon to the effect that, though comets may arise in the course of Nature, they are still signs of evil to mankind. In answer to the theory of sundry naturalists that comets are made up of "a certain fiery, warm, sulphurous, saltpetery, sticky fog," he declaims: "Our sins, our sins: they are the fiery heated vapours, the thick, sticky, sulphurous clouds which rise from the earth toward heaven before God." Throughout the sermon Dieterich pours contempt over all men who simply investigate comets as natural objects, calls special attention to a comet then in the heavens resembling a long broom or bundle of rods, and declares that he and his hearers can only consider it rightly "when we see standing before us our Lord God in heaven as an angry father with a rod for his children." In answer to the question what comets signify, he commits himself entirely to the idea that they indicate the wrath of God, and therefore calamities of every sort. Page after page is filled with the records of evils following comets. Beginning with the creation of the world, he insists that the first comet brought on the deluge of Noah, and cites a mass of authorities, ranging from Moses and Isaiah to Albert the Great and Melanchthon, in support of the view that comets precede earthquakes, famines, wars, pestilences, and every form of evil. He makes some parade of astronomical knowledge as to the greatness of the sun and moon, but relapses soon into his old line of argument. Imploring his audience not to be led away from the well-established belief of Christendom and the principles of their fathers, he comes back to his old assertion, insists that "our sins are the inflammable material of which comets are made," and winds up with a most earnest appeal to the Almighty to spare his people.*

Similar efforts from the pulpit were provoked by the great comet of 1680. Typical among these was the effort in Switzerland of Pastor Heinrich Erni, who, from the Cathedral of Zürich, sent a circular letter to the clergy of that region showing the connection of the eleventh and twelfth verses of the first chapter of Jeremiah with the comet, giving notice that at his suggestion the authorities had proclaimed a solemn fast, and exhorting the clergy to preach earnestly on the subject of this warning.

Nor were the interpreters of the comet's message content with simple prose. At the appearance of the comet of 1618, Grasser and Gross, pastors and doctors of theology at Basle, put forth a collection of doggerel rhymes to fasten the orthodox theory into the minds of school-children and peasants. One of these may be translated:

> "I am a Rod in God's right hand
> Threatening the German and foreign land."

Others for a similar purpose taught:

> "Eight things there be a Comet brings,
> When it on high doth horrid range:
> Wind, Famine, Plague, and Death to Kings,
> War, Earthquakes, Floods, and Direful Change."

Great ingenuity was shown in meeting the advance of science, in the universities and schools, with new texts of

* For Dieterich, see *Ulmische Cometen-Predigt, von dem Cometen, so nechst abgewischen 1618 Jahrs im Wintermonat erstenmahls in Schwaben sehen lassen, . . . gehalten zu Ulm . . . durch Conrad Dieterich*, Ulm, 1620. For a life of the author, see article *Dieterich* in the *Allgemeine Deutsche Biographie*. See also Wolf.

Scripture; and Stephen Spleiss, Rector of the Gymnasium at Schaffhausen, got great credit by teaching that in the vision of Jeremiah the "almond rod" was a tailed comet, and the "seething pot" a bearded one.*

It can be easily understood that such authoritative utterances as that of Dieterich must have produced a great effect throughout Protestant Christendom; and in due time we see their working in New England. That same tendency to provincialism, which, save at rare intervals, has been the bane of Massachusetts thought from that day to this, appeared; and in 1664 we find Samuel Danforth arguing from the Bible that "comets are portentous signals of great and notable changes," and arguing from history that they "have been many times heralds of wrath to a secure and impenitent world." He cites especially the comet of 1652, which appeared just before Mr. Cotton's sickness and disappeared after his death. Morton also, in his *Memorial* recording the death of John Putnam, alludes to the comet of 1662 as "a very signal testimony that God had then removed a bright star and a shining light out of the heaven of his Church here into celestial glory above." Again he speaks of another comet, insisting that "it was no fiery meteor caused by exhalation, but it was sent immediately by God to awaken the secure world," and goes on to show how in that year "it pleased God to smite the fruits of the earth—namely, the wheat in special—with blasting and mildew, whereby much of it was spoiled and became profitable for nothing, and much of it worth little, being light and empty. This was looked upon by the judicious and conscientious of the land as a speaking providence against the unthankfulness of many, . . . as also against voluptuousness and abuse of the good creatures of God by licentiousness in drinking and fashions in apparel, for the obtaining whereof a great part of the principal grain was oftentimes unnecessarily expended."

But in 1680 a stronger than either of these seized upon the doctrine and wielded it with power. Increase Mather,

* For Erni, see Wolf, *Gesch. d. Astronomie*, p. 239. For Grasser and Gross, see their *Christenliches Bedencken . . . von dem erschrockenlichen Cometen*, etc., Zürich, 1664. For Spleiss, see *Beiläuftiger Bericht von dem jetzigen Cometsternen*, etc., Schaffhausen, 1664.

so open always to ideas from Europe, and always so powerful for good or evil in the colonies, preached his sermon on "Heaven's Alarm to the World, . . . wherein is shown that fearful sights and signs in the heavens are the presages of great calamities at hand." The texts were taken from the book of Revelation: "And the third angel sounded, and there fell a great star from heaven, burning, as it were a lamp," and "Behold, the third woe cometh quickly." In this, as in various other sermons, he supports the theological cometary theory fully. He insists that "we are fallen into the dregs of time," and that the day of judgment is evidently approaching. He explains away the words of Jeremiah—"Be not dismayed at signs in the heavens"—and shows that comets have been forerunners of nearly every form of evil. Having done full justice to evils thus presaged in scriptural times, he begins a similar display in modern history by citing blazing stars which foretold the invasions of Goths, Huns, Saracens, and Turks, and warns gainsayers by citing the example of Vespasian, who, after ridiculing a comet, soon died. The general shape and appearance of comets, he thinks, betoken their purpose, and he cites Tertullian to prove them "God's sharp razors on mankind, whereby he doth poll, and his scythe whereby he doth shear down multitudes of sinful creatures." At last, rising to a fearful height, he declares: "For the Lord hath fired his beacon in the heavens among the stars of God there; the fearful sight is not yet out of sight. The warning piece of heaven is going off. Now, then, if the Lord discharge his murdering pieces from on high, and men be found in their sins unfit for death, their blood shall be upon them." And again, in an agony of supplication, he cries out: "Do we see the sword blazing over us? Let it put us upon crying to God, that the judgment be diverted and not return upon us again so speedily. . . . Doth God threaten our very heavens? O pray unto him, that he would not take away stars and send comets to succeed them." *

---

* For Danforth, see his *Astronomical Description of the Late Comet or Blazing Star, Together with a Brief Theological Application Thereof*, 1664. For Morton, see his *Memorial*, pp. 251, 252; also 309, 310. Texts cited by Mather were Rev. viii, 10, and xi, 14.

Two years later, in August, 1682, he followed this with another sermon on "The Latter Sign," "wherein is showed that the voice of God in signal providences, especially when repeated and iterated, ought to be hearkened unto." Here, too, of course, the comet comes in for a large share of attention. But his tone is less sure: even in the midst of all his arguments appears an evident misgiving. The thoughts of Newton in science and Bayle in philosophy were evidently tending to accomplish the prophecy of Seneca. Mather's alarm at this is clear. His natural tendency is to uphold the idea that a comet is simply a fire-ball flung from the hand of an avenging God at a guilty world, but he evidently feels obliged to yield something to the scientific spirit; hence, in the *Discourse concerning Comets*, published in 1683, he declares: "There are those who think that, inasmuch as comets may be supposed to proceed from natural causes, there is no speaking voice of Heaven in them beyond what is to be said of all other works of God. But certain it is that many things which may happen according to the course of Nature are portentous signs of Divine anger and prognostics of great evils hastening upon the world." He then notices the eclipse of August, 1672, and adds: "That year the college was eclipsed by the death of the learned president there, worthy Mr. Chauncey; and two colonies—namely, Massachusetts and Plymouth—by the death of two governors, who died within a twelvemonth after. . . . Shall, then, such mighty works of God as comets are be insignificant things?"*

III. THE INVASION OF SCEPTICISM.

Vigorous as Mather's argument is, we see scepticism regarding "signs" continuing to invade the public mind; and, in spite of his threatenings, about twenty years after we find a remarkable evidence of this progress in the fact that this

* Increase Mather's *Heaven's Alarm to the World* was first printed at Boston in 1681, but was reprinted in 1682, and was appended, with the sermon on *The Latter Sign*, to the *Discourse on Comets* (Boston, 1683).

scepticism has seized upon no less a personage than that colossus of orthodoxy, his thrice illustrious son, Cotton Mather himself; and him we find, in 1726, despite the arguments of his father, declaring in his *Manuductio*: "Perhaps there may be some need for me to caution you against being dismayed at the signs of the heavens, or having any superstitious fancies upon eclipses and the like. . . . I am willing that you be apprehensive of nothing portentous in blazing stars. For my part, I know not whether all our worlds, and even the sun itself, may not fare the better for them." *

Curiously enough, for this scientific scepticism in Cotton Mather there was a cause identical with that which had developed superstition in the mind of his father. The same provincial tendency to receive implicitly any new European fashion in thinking or speech wrought upon both, plunging one into superstition and drawing the other out of it.

European thought, which New England followed, had at last broken away in great measure from the theological view of comets as signs and wonders. The germ of this emancipating influence was mainly in the great utterance of Seneca; and we find in nearly every century some evidence that this germ was still alive. This life became more and more evident after the Reformation period, even though theologians in every Church did their best to destroy it. The first series of attacks on the old theological doctrine were mainly founded in philosophic reasoning. As early as the first half of the sixteenth century we hear Julius Cæsar Scaliger protesting against the cometary superstition as "ridiculous folly." † Of more real importance was the treatise of Blaise de Vigenère, published at Paris in 1578. In this little book various statements regarding comets as signs of wrath or causes of evils are given, and then followed by a very gentle and quiet discussion, usually tending to develop that healthful scepticism which is the parent of investigation. A fair example of his mode of treating the subject is seen in his

* For Cotton Mather, see the *Manuductio*, pp. 54, 55.

† For Scaliger, see p. 20 of Dudith's book, cited below.

dealing with a bit of "sacred science." This was simply that "comets menace princes and kings with death because they live more delicately than other people; and, therefore, the air thickened and corrupted by a comet would be naturally more injurious to them than to common folk who live on coarser food." To this De Vigenère answers that there are very many persons who live on food as delicate as that enjoyed by princes and kings, and yet receive no harm from comets. He then goes on to show that many of the greatest monarchs in history have met death without any comet to herald it.

In the same year thoughtful scepticism of a similar sort found an advocate in another part of Europe. Thomas Erastus, the learned and devout professor of medicine at Heidelberg, put forth a letter dealing in the plainest terms with the superstition. He argued especially that there could be no natural connection between the comet and pestilence, since the burning of an exhalation must tend to purify rather than to infect the air. In the following year the eloquent Hungarian divine Dudith published a letter in which the theological theory was handled even more shrewdly; for he argued that, if comets were caused by the sins of mortals, they would never be absent from the sky. But these utterances were for the time brushed aside by the theological leaders of thought as shallow or impious.

In the seventeenth century able arguments against the superstition, on general grounds, began to be multiplied. In Holland, Balthasar Bekker opposed this, as he opposed the witchcraft delusion, on general philosophic grounds; and Lubienitzky wrote in a compromising spirit to prove that comets were as often followed by good as by evil events. In France, Pierre Petit, formerly geographer of Louis XIII, and an intimate friend of Descartes, addressed to the young Louis XIV a vehement protest against the superstition, basing his arguments not on astronomy, but on common sense. A very effective part of the little treatise was devoted to answering the authority of the fathers of the early Church. To do this, he simply reminded his readers that St. Augustine and St. John Damascenus had also opposed the doctrine of the antipodes. The book did good service in France, and was translated in Germany a few years later.*

All these were denounced as infidels and heretics, yet none the less did they set men at thinking, and prepare the way for a far greater genius; for toward the end of the same century the philosophic attack was taken up by Pierre Bayle, and in the whole series of philosophic champions he is chief. While professor at the University of Sedan he had observed the alarm caused by the comet of 1680, and he now brought all his reasoning powers to bear upon it. Thoughts deep and witty he poured out in volume after volume. Catholics and Protestants were alike scandalized. Catholic France spurned him, and Jurieu, the great Reformed divine, called his cometary views "atheism," and tried hard to have Protestant Holland condemn him. Though Bayle did not touch immediately the mass of mankind, he wrought with power upon men who gave themselves the trouble of thinking. It was indeed unfortunate for the Church that theologians, instead of taking the initiative in this matter, left it to Bayle; for, in tearing down the pretended scriptural doctrine of comets, he tore down much else: of all men in his time, no one so thoroughly prepared the way for Voltaire.

Bayle's whole argument is rooted in the prophecy of Seneca. He declares: "Comets are bodies subject to the ordinary law of Nature, and not prodigies amenable to no law." He shows historically that there is no reason to regard comets as portents of earthly evils. As to the fact that such evils occur after the passage of comets across the sky, he compares the person believing that comets cause these evils to a woman looking out of a window into a Paris street and believing that the carriages pass because she looks out. As to the accomplishment of some predictions, he cites the shrewd saying of Henry IV, to the effect that "the public

---

* For Blaise de Vigenère, see his *Traité des Comètes*, Paris, 1578. For Dudith, see his *De Cometarum Significatione*, Basle, 1579, to which the letter of Erastus is appended. Bekker's views may be found in his *Onderzoek van de Betekening der Cometen*, Leeuwarden, 1683. For Lubienitzky's, see his *Theatrum Cometicum*, Amsterdam, 1667, in part ii: *Historia Cometarum*, preface "to the reader." For Petit, see his *Dissertation sur la Nature des Comètes*, Paris, 1665 (German translation, Dresden and Zittau, 1681).

will remember one prediction that comes true better than all the rest that have proved false." Finally, he sums up by saying: "The more we study man, the more does it appear that pride is his ruling passion, and that he affects grandeur even in his misery. Mean and perishable creature that he is, he has been able to persuade men that he can not die without disturbing the whole course of Nature and obliging the heavens to put themselves to fresh expense in order to light his funeral pomp. Foolish and ridiculous vanity! If we had a just idea of the universe, we should soon comprehend that the death or birth of a prince is too insignificant a matter to stir the heavens." *

This great philosophic champion of right reason was followed by a literary champion hardly less famous; for Fontenelle now gave to the French theatre his play of *The Comet*, and a point of capital importance in France was made by rendering the army of ignorance ridiculous.†

Such was the line of philosophic and literary attack, as developed from Scaliger to Fontenelle. But beneath and in the midst of all of it, from first to last, giving firmness, strength, and new sources of vitality to it, was the steady development of scientific effort; and to the series of great men who patiently wrought and thought out the truth by scientific methods through all these centuries belong the honours of the victory.

For generations men in various parts of the world had been making careful observations on these strange bodies. As far back as the time when Luther and Melanchthon and Zwingli were plunged into alarm by various comets from 1531 to 1539, Peter Apian kept his head sufficiently cool to make scientific notes of their paths through the heavens. A little later, when the great comet of 1556 scared popes, emperors, and reformers alike, such men as Fabricius at Vienna and Heller at Nuremberg quietly observed its path.

* Regarding Bayle, see Mädler, *Himmelskunde*, vol. i, p. 327. For special points of interest in Bayle's argument, see his *Pensées Diverses sur les Comètes*, Amsterdam, 1749, pp. 79, 102, 134, 206. For the response to Jurieu, see the *Continuation des Pensées*, Rotterdam, 1705; also Champion, p. 164, Lecky, *ubi supra*, and Guillemin, pp. 29, 30.

† See Fontenelle, cited by Champion, p. 167.

In vain did men like Dieterich and Heerbrand and Celich from various parts of Germany denounce such observations and investigations as impious; they were steadily continued, and in 1577 came the first which led to the distinct foundation of the modern doctrine. In that year appeared a comet which again plunged Europe into alarm. In every European country this alarm was strong, but in Germany strongest of all. The churches were filled with terror-stricken multitudes. Celich preaching at Magdeburg was echoed by Heerbrand preaching at Tübingen, and both these from thousands of other pulpits, Catholic and Protestant, throughout Europe. In the midst of all this din and outcry a few men quietly but steadily observed the monster; and Tycho Brahe announced, as the result, that its path lay farther from the earth than the orbit of the moon. Another great astronomical genius, Kepler, confirmed this. This distinct beginning of the new doctrine was bitterly opposed by theologians; they denounced it as one of the evil results of that scientific meddling with the designs of Providence against which they had so long declaimed in pulpits and professors' chairs; they even brought forward some astronomers ambitious or wrong-headed enough to testify that Tycho and Kepler were in error.*

Nothing could be more natural than such opposition; for this simple announcement by Tycho Brahe began a new era. It shook the very foundation of cometary superstition. The Aristotelian view, developed by the theologians, was that what lies within the moon's orbit appertains to the earth and is essentially transitory and evil, while what lies beyond it belongs to the heavens and is permanent, regular, and pure. Tycho Brahe and Kepler, therefore, having by means of scientific observation and thought taken comets out of the category of meteors and appearances in the neighbourhood of the earth, and placed them among the heavenly bodies, dealt a blow at the very foundations of the theological argument, and gave a great impulse to the idea that comets are

---

* See Mädler, *Himmelskunde*, vol. i, pp. 181, 197; also Wolf, *Gesch. d. Astronomie*, and Janssen, *Gesch. d. deutschen Volkes*, vol. v, p. 350. Heerbrand's sermon, cited above, is a good specimen of the theologic attitude. See Pingré, vol. ii, p. 81.

themselves heavenly bodies moving regularly and in obedience to law.

IV. THEOLOGICAL EFFORTS AT COMPROMISE.—THE FINAL VICTORY OF SCIENCE.

Attempts were now made to compromise. It was declared that, while some comets were doubtless supralunar, some must be sublunar. But this admission was no less fatal on another account. During many centuries the theory favoured by the Church had been, as we have seen, that the earth was surrounded by hollow spheres, concentric and transparent, forming a number of glassy strata incasing one another "like the different coatings of an onion," and that each of these in its movement about the earth carries one or more of the heavenly bodies. Some maintained that these spheres were crystal; but Lactantius, and with him various fathers of the Church, spoke of the heavenly vault as made of ice. Now, the admission that comets could move beyond the moon was fatal to this theory, for it sent them crashing through these spheres of ice or crystal, and therefore through the whole sacred fabric of the Ptolemaic theory.*

Here we may pause for a moment to note one of the chief differences between scientific and theological reasoning considered in themselves. Kepler's main reasoning as to the existence of a law for cometary movement was right; but his secondary reasoning, that comets move nearly in straight lines, was wrong. His right reasoning was developed by Gassendi in France, by Borelli in Italy, by Hevel and Doerfel in Germany, by Eysat and Bernouilli in Switzerland, by Percy and—most important of all, as regards mathematical demonstration—by Newton in England. The general theory, which was true, they accepted and developed; the secondary theory, which was found untrue, they rejected; and, as a result, both of what they thus accepted

* For these features in cometary theory, see Pingre, vol. i, p. 89; also Humboldt, *Cosmos* (English translation, London, 1868), vol. iii p. 169.

and of what they rejected, was evolved the basis of the whole modern cometary theory.

Very different was this from the theological method. As a rule, when there arises a thinker as great in theology as Kepler in science, the whole mass of his conclusions ripens into a dogma. His disciples labour not to test it, but to establish it; and while, in the Catholic Church, it becomes a dogma to be believed or disbelieved under the penalty of damnation, it becomes in the Protestant Church the basis for one more sect.

Various astronomers laboured to develop the truth discovered by Tycho and strengthened by Kepler. Cassini seemed likely to win for Italy the glory of completing the great structure; but he was sadly fettered by Church influences, and was obliged to leave most of the work to others. Early among these was Hevel. He gave reasons for believing that comets move in parabolic curves toward the sun. Then came a man who developed this truth further—Samuel Doerfel; and it is a pleasure to note that he was a clergyman. The comet of 1680, which set Erni in Switzerland, Mather in New England, and so many others in all parts of the world at declaiming, set Doerfel at thinking. Undismayed by the authority of Origen and St. John Chrysostom, the arguments of Luther, Melanchthon, and Zwingli, the outcries of Celich, Heerbrand, and Dieterich, he pondered over the problem in his little Saxon parsonage, until in 1681 he set forth his proofs that comets are heavenly bodies moving in parabolas of which the sun is the focus. Bernouilli arrived at the same conclusion; and, finally, this great series of men and works was closed by the greatest of all, when Newton, in 1686, having taken the data furnished by the comet of 1680, demonstrated that comets are guided in their movements by the same principle that controls the planets in their orbits. Thus was completed the evolution of this new truth in science.

Yet we are not to suppose that these two great series of philosophical and scientific victories cleared the field of all opponents. Declamation and pretended demonstration of the old theologic view were still heard; but the day of complete victory dawned when Halley, after most thorough observation and calculation, recognised the comet of 1682 as one which had already appeared at stated periods, and foretold its return in about seventy-five years; and the battle was fully won when Clairaut, seconded by Lalande and Mme. Lepaute, predicted distinctly the time when the comet would arrive at its perihelion, and this prediction was verified.* Then it was that a Roman heathen philosopher was proved more infallible and more directly under Divine inspiration than a Roman Christian pontiff; for the very comet which the traveller finds to-day depicted on the Bayeux tapestry as portending destruction to Harold and the Saxons at the Norman invasion of England, and which was regarded by Pope Calixtus as portending evil to Christendom, was found six centuries later to be, as Seneca had prophesied, a heavenly body obeying the great laws of the universe, and coming at regular periods. Thenceforth the whole ponderous enginery of this superstition, with its proof-texts regarding "signs in the heavens," its theological reasoning to show the moral necessity of cometary warnings, and its ecclesiastical fulminations against the "atheism, godlessness, and infidelity" of scientific investigation, was seen by all thinking men to be as weak against the scientific method as Indian arrows against needle guns. Copernicus, Galileo, Cassini, Doerfel, Newton, Halley, and Clairaut had gained the victory.†

It is instructive to note, even after the main battle was lost, a renewal of the attempt, always seen under like circumstances, to effect a compromise, to establish a "safe science" on grounds pseudo-scientific and pseudo-theologic. Luther, with his strong common sense, had foreshadowed this; Kepler had expressed a willingness to accept it. It was insisted that comets might be heavenly bodies moving in regular

---

* See Pingré, vol. i, p. 53; Grant, *History of Physical Astronomy*, p. 305, etc., etc. For a curious partial anticipation by Hooke, in 1664, of the great truth announced by Halley in 1682, see *Pepys's Diary* for March 1, 1664. For excellent summaries of the whole work of Halley and Clairaut and their forerunners and associates, see Pingré, Mädler, Wolf, Arago, *et al.*

† In accordance with Halley's prophecy, the comet of 1682 has returned in 1759 and 1835. See Mädler, Guillemin, Watson, Grant, Delambre, Proctor, article *Astronomy* in *Encycl. Brit.*, and especially, for details, Wolf, pp. 407–412 and 701–722. For clear statement regarding Doerfel, see Wolf, p. 411.

orbits, and even obedient to law, and yet be sent as "signs in the heavens." Many good men clung longingly to this phase of the old belief, and in 1770 Semler, professor at Halle, tried to satisfy both sides. He insisted that, while from a scientific point of view comets could not exercise any physical influence upon the world, yet from a religious point of view they could exercise a moral influence as reminders of the Just Judge of the Universe.

So hard was it for good men to give up the doctrine of "signs in the heavens," seemingly based upon Scripture and exercising such a healthful moral tendency! As is always the case after such a defeat, these votaries of "sacred science" exerted the greatest ingenuity in devising statements and arguments to avert the new doctrine. Within our own century the great Catholic champion, Joseph de Maistre, echoed these in declaring his belief that comets are special warnings of evil. So, too, in Protestant England, in 1818, the *Gentleman's Magazine* stated that under the malign influence of a recent comet "flies became blind and died early in the season," and "the wife of a London shoemaker had four children at a birth." And even as late as 1829 Mr. Forster, an English physician, published a work to prove that comets produce hot summers, cold winters, epidemics, earthquakes, clouds of midges and locusts, and nearly every calamity conceivable. He bore especially upon the fact that the comet of 1665 was coincident with the plague in London, apparently forgetting that the other great cities of England and the Continent were not thus visited; and, in a climax, announces the fact that the comet of 1663 "made all the cats in Westphalia sick."

There still lingered one little cloud-patch of superstition, arising mainly from the supposed fact that comets had really been followed by a marked rise in temperature. Even this poor basis for the belief that they might, after all, affect earthly affairs was swept away, and science won here another victory; for Arago, by thermometric records carefully kept at Paris from 1735 to 1781, proved that comets had produced no effect upon temperature. Among multitudes of similar examples he showed that, in some years when several comets appeared, the temperature was lower than in other years when few or none appeared. In 1737 there were two comets, and the weather was cool; in 1785 there was no comet, and the weather was hot; through the whole fifty years it was shown that comets were sometimes followed by hot weather, sometimes by cool, and that no rule was deducible. The victory of science was complete at every point.*

But in this history there was one little exhibition so curious as to be worthy of notice, though its permanent effect upon thought was small. Whiston and Burnet, so devoted to what they considered sacred science, had determined that in some way comets must be instruments of Divine wrath. One of them maintained that the deluge was caused by the tail of a comet striking the earth; the other put forth the theory that comets are places of punishment for the damned—in fact, "flying hells." The theories of Whiston and Burnet found wide acceptance also in Germany, mainly through the all-powerful mediation of Gottsched, so long, from his professor's chair at Leipsic, the dictator of orthodox thought, who not only wrote a brief tractate of his own upon the subject, but furnished a voluminous historical introduction to the more elaborate treatise of Heyn. In this book, which appeared at Leipsic in 1742, the agency of comets in the creation, the flood, and the final destruction of the world is fully proved. Both these theories were, however, soon discredited.

Perhaps the more interesting of them can best be met by another, which, if not fully established, appears much better based—namely, that in 1868 the earth passed directly through the tail of a comet, with no deluge, no sound of any wailings of the damned, with but slight appearances here and there, only to be detected by the keen sight of the meteorological or astronomical observer.

In our own country superstitious ideas regarding comets continued to have some little currency; but their life was

---

* For Forster, see his *Illustrations of the Atmospherical Origin of Epidemic Diseases*, Chelmsford, 1829, cited by Arago; also in *Quarterly Review* for April, 1835. For the writings of several on both sides, and especially of those who sought to save, as far as possible, the sacred theory of comets, see Mädler, vol. ii, p. 384 *et seq.*, and Wolf, p. 186.

short. The tendency shown by Cotton Mather, at the beginning of the eighteenth century, toward acknowledging the victory of science, was completed by the utterances of Winthrop, professor at Harvard, who in 1759 published two lectures on comets, in which he simply and clearly revealed the truth, never scoffing, but reasoning quietly and reverently. In one passage he says: "To be thrown into a panic whenever a comet appears, on account of the ill effects which some few of them might possibly produce, if they were not under proper direction, betrays a weakness unbecoming a reasonable being."

A happy influence in this respect was exercised on both continents by John Wesley. Tenaciously as he had held to the supposed scriptural view in so many other matters of science, in this he allowed his reason to prevail, accepted the demonstrations of Halley, and gloried in them.*

The victory was indeed complete. Happily, none of the fears expressed by Conrad Dieterich and Increase Mather were realized. No catastrophe has ensued either to religion or to morals. In the realm of religion the Psalms of David remain no less beautiful, the great utterances of the Hebrew prophets no less powerful; the Sermon on the Mount, "the first commandment, and the second, which is like unto it," the definition of "pure religion and undefiled" by St. James, appeal no less to the deepest things in the human heart. In the realm of morals, too, serviceable as the idea of firebrands thrown by the right hand of an avenging God to scare a naughty world might seem, any competent historian must find that the destruction of the old theological cometary theory was followed by moral improvement rather than by deterioration. We have but to compare the general moral tone of society to-day, wretchedly imperfect as it is, with that existing in the time when this superstition had its

---

* For Heyn, see his *Versuch einer Betrachtung über die Cometen, die Sündfluth und das Vorspiel des jüngsten Gerichts*, Leipsic, 1742. A Latin version, of the same year, bears the title, *Specimen Cometologiæ Sacræ*. For the theory that the earth encountered the tail of a comet, see Guillemin and Watson. For survival of the old idea in America, see a *Sermon* of Israel Loring, of Sudbury, published in 1722. For Prof. J. Winthrop, see his *Comets*. For Wesley, see his *Natural Philosophy*, London, 1784, vol. iii, p. 303.

strongest hold. We have only to compare the court of Henry VIII with the court of Victoria, the reign of the later Valois and earlier Bourbon princes with the present French Republic, the period of the Medici and Sforzas and Borgias with the period of Leo XIII and Humbert, the monstrous wickedness of the Thirty Years' War with the ennobling patriotism of the Franco-Prussian struggle, and the despotism of the miserable German princelings of the sixteenth and seventeenth centuries with the reign of the Emperor William.

The gain is not simply that mankind has arrived at a clearer conception of law in the universe; not merely that thinking men see more clearly that we are part of a system not requiring constant patching and arbitrary interference; but perhaps best of all is the fact that science has cleared away one more series of those dogmas which tend to debase rather than to develop man's whole moral and religious nature. In this emancipation from terror and fanaticism, as in so many other results of scientific thinking, we have a proof of the inspiration of those great words, "THE TRUTH SHALL MAKE YOU FREE."

CHAPTER V

FROM GENESIS TO GEOLOGY.

I. GROWTH OF THEOLOGICAL EXPLANATIONS.

AMONG the philosophers of Greece we find, even at an early period, germs of geological truth, and, what is of vast importance, an atmosphere in which such germs could grow. These germs were transmitted to Roman thought; an atmosphere of tolerance continued; there was nothing which forbade unfettered reasoning regarding either the earth's strata or the remains of former life found in them, and under the Roman Empire a period of fruitful observation seemed sure to begin.

But, as Christianity took control of the world, there came a great change. The earliest attitude of the Church toward geology and its kindred sciences was indifferent, and even contemptuous. According to the prevailing belief, the earth was a "fallen world," and was soon to be destroyed. Why, then, should it be studied? Why, indeed, give a thought to it? The scorn which Lactantius and St. Augustine had cast upon the study of astronomy was extended largely to other sciences.*

But the germs of scientific knowledge and thought developed in the ancient world could be entirely smothered neither by eloquence nor by logic; some little scientific ob-

* For a compact and admirable statement as to the dawn of geological conceptions in Greece and Rome, see Mr. Lester Ward's essay on paleobotany in the *Fifth Annual Report of the United States Geological Survey*, for 1883–'84. As to the reasons why Greek philosophers did comparatively so little for geology, see D'Archiac, *Géologie*, p. 18. For the contempt felt by Lactantius and St. Augustine toward astronomical science, see foregoing chapters on Astronomy and Geography.

servation must be allowed, though all close reasoning upon it was fettered by theology. Thus it was that St. Jerome insisted that the broken and twisted crust of the earth exhibits the wrath of God against sin, and Tertullian asserted that fossils resulted from the flood of Noah.

To keep all such observation and reasoning within orthodox limits, St. Augustine, about the beginning of the fifth century, began an effort to develop from these germs a growth in science which should be sacred and safe. With this intent he prepared his great commentary on the work of creation, as depicted in Genesis, besides dwelling upon the subject in other writings. Once engaged in this work, he gave himself to it more earnestly than any other of the earlier fathers ever did; but his vast powers of research and thought were not directed to actual observation or reasoning upon observation. The keynote of his whole method is seen in his famous phrase, "Nothing is to be accepted save on the authority of Scripture, since greater is that authority than all the powers of the human mind." All his thought was given to studying the letter of the sacred text, and to making it explain natural phenomena by methods purely theological.*

Among the many questions he then raised and discussed may be mentioned such as these: "What caused the creation of the stars on the fourth day?" "Were beasts of prey and venomous animals created before, or after, the fall of Adam? If before, how can their creation be reconciled with God's goodness; if afterward, how can their creation be reconciled to the letter of God's Word?" "Why were only beasts and birds brought before Adam to be named, and not fishes and marine animals?" "Why did the Creator not say, 'Be fruitful and multiply,' to plants as well as to animals?"†

Sundry answers to these and similar questions formed the main contributions of the greatest of the Latin fathers to

* For citations and authorities on these points, see the chapter on Meteorology.

† See Augustine, *De Genesi*, ii, 13; iii, 13, 15 *et seq.*; ix, 12 *et seq.* For the reference to St. Jerome, see Shields, *Final Philosophy*, p. 119; also Lyell, *Introduction to Geology*, vol. i, chap. ii.

the scientific knowledge of the world, after a most thorough study of the biblical text and a most profound application of theological reasoning. The results of these contributions were most important. In this, as in so many other fields, Augustine gave direction to the main current of thought in western Europe, Catholic and Protestant, for nearly thirteen centuries.

In the ages that succeeded, the vast majority of prominent scholars followed him implicitly. Even so strong a man as Pope Gregory the Great yielded to his influence, and such leaders of thought as St. Isidore, in the seventh century, and the Venerable Bede, in the eighth, planting themselves upon Augustine's premises, only ventured timidly to extend their conclusions upon lines he had laid down.

In his great work on *Etymologies*, Isidore took up Augustine's attempt to bring the creation into satisfactory relations with the book of Genesis, and, as to fossil remains, he, like Tertullian, thought that they resulted from the Flood of Noah. In the following century Bede developed the same orthodox traditions.*

The best guess, in a geological sense, among the followers of St. Augustine was made by an Irish monkish scholar, who, in order to diminish the difficulty arising from the distribution of animals, especially in view of the fact that the same animals are found in Ireland as in England, held that various lands now separated were once connected. But, alas! the exigencies of theology forced him to place their separation later than the Flood. Happily for him, such facts were not yet known as that the kangaroo is found only on an island in the South Pacific, and must therefore, according to his theory, have migrated thither with all his progeny, and along a causeway so curiously constructed that none of the beasts of prey, who were his fellow-voyagers in the ark, could follow him.

These general lines of thought upon geology and its kindred science of zoölogy were followed by St. Thomas Aqui-

---

* For Isidore, see the *Etymologiæ*, xi, 4, xiii, 22. For Bede, see the *Hexæmeron*, i, ii, in Migne, tome xci.

nas and by the whole body of mediæval theologians, so far as they gave any attention to such subjects.

The next development of geology, mainly under Church guidance, was by means of the scholastic theology. Phrase-making was substituted for investigation. Without the Church and within it wonderful contributions were thus made. In the eleventh century Avicenna accounted for the fossils by suggesting a "stone-making force";* in the thirteenth, Albert the Great attributed them to a "formative quality;"† in the following centuries some philosophers ventured the idea that they grew from seed; and the Aristotelian doctrine of spontaneous generation was constantly used to prove that these stony fossils possessed powers of reproduction like plants and animals.‡

Still, at various times and places, germs implanted by Greek and Roman thought were warmed into life. The Arabian schools seem to have been less fettered by the letter of the Koran than the contemporary Christian scholars by the letter of the Bible; and to Avicenna belongs the credit of first announcing substantially the modern geological theory of changes in the earth's surface.‖

The direct influence of the Reformation was at first unfavourable to scientific progress, for nothing could be more at variance with any scientific theory of the development of the universe than the ideas of the Protestant leaders. That strict adherence to the text of Scripture which made Luther and Melanchthon denounce the idea that the planets revolve about the sun, was naturally extended to every other scientific statement at variance with the sacred text. There is much reason to believe that the fetters upon scientific thought were closer under the strict interpretation of Scripture by the early Protestants than they had been under the older Church. The dominant spirit among the Reformers is shown by the declaration of Peter Martyr to the effect that, if a wrong opinion should obtain regarding the creation as described in Genesis, "all the promises of Christ

* *Vis lapidifica.*

† *Virtus formativa.*

‡ See authorities given in Mr. Ward's essay, as above.

‖ For Avicenna, see Lyell and D'Archiac.

fall into nothing, and all the life of our religion would be lost." *

In the times immediately succeeding the Reformation matters went from bad to worse. Under Luther and Melanchthon there was some little freedom of speculation, but under their successors there was none; to question any interpretation of Luther came to be thought almost as wicked as to question the literal interpretation of the Scriptures themselves. Examples of this are seen in the struggles between those who held that birds were created entirely from water and those who held that they were created out of water and mud. In the city of Lübeck, the ancient centre of the Hanseatic League, close at the beginning of the seventeenth century, Pfeiffer, "General Superintendent" or bishop in those parts, published his *Pansophia Mosaica*, calculated, as he believed, to beat back science forever. In a long series of declamations he insisted that in the strict text of Genesis alone is safety; that it contains all wisdom and knowledge, human and divine. This being the case, who could care to waste time on the study of material things and give thought to the structure of the world? Above all who, after such a proclamation by such a ruler in the Lutheran Israel, would dare to talk of the "days" mentioned in Genesis as "periods of time"; or of the "firmament" as not meaning a solid vault over the universe; or of the "waters above the heavens" as not contained in a vast cistern supported by the heavenly vault; or of the "windows of heaven" as a figure of speech? †

In England the same spirit was shown even as late as the time of Sir Matthew Hale. We find in his book on the *Origination of Mankind*, published in 1685, the strictest devotion to a theory of creation based upon the mere letter of Scripture, and a complete inability to draw knowledge regarding the earth's origin and structure from any other source.

While the Lutheran, Calvinistic, and Anglican Reformers clung to literal interpretations of the sacred books, and

---

* See his *Commentary on Genesis*, cited by Zoeckler, *Geschichte der Beziehungen zwischen Theologie und Naturwissenschaft*, vol. i, p. 690.

† For Pfeiffer, see Zoeckler, vol. i, pp. 688, 689.

turned their faces away from scientific investigation, it was among their contemporaries at the revival of learning that there began to arise fruitful thought in this field. Then it was, about the beginning of the sixteenth century, that Leonardo da Vinci, as great a genius in science as in art, broached the true idea as to the origin of fossil remains; and his compatriot, Fracastoro, developed this on the modern lines of thought. Others in other parts of Europe took up the idea, and, while mixing with it many crudities, drew from it more and more truth. Toward the end of the sixteenth century Bernard Palissy, in France, took hold of it with the same genius which he showed in artistic creation; but, remarkable as were his assertions of scientific realities, they could gain little hearing. Theologians, philosophers, and even some scientific men of value, under the sway of scholastic phrases, continued to insist upon such explanations as that fossils were the product of "fatty matter set into a fermentation by heat"; or of a "lapidific juice";* or of a "seminal air";† or of a "tumultuous movement of terrestrial exhalations"; and there was a prevailing belief that fossil remains, in general, might be brought under the head of "sports of Nature," a pious turn being given to this phrase by the suggestion that these "sports" indicated some inscrutable purpose of the Almighty.

This remained a leading orthodox mode of explanation in the Church, Catholic and Protestant, for centuries.

II. EFFORTS TO SUPPRESS THE SCIENTIFIC VIEW.

But the scientific method could not be entirely hidden; and, near the beginning of the seventeenth century, De Clave, Bitaud, and De Villon revived it in France. Straightway the theological faculty of Paris protested against the scientific doctrine as unscriptural, destroyed the offending treatises, banished their authors from Paris, and forbade them to live in towns or enter places of public resort.‡

---

* *Succus lapidificus.* † *Aura seminalis.*

‡ See Morley, *Life of Palissy the Potter*, vol. ii, p. 315 *et seq.*

The champions of science, though repressed for a time, quietly laboured on, especially in Italy. Half a century later, Steno, a Dane, and Scilla, an Italian, went still further in the right direction; and, though they and their disciples took great pains to throw a tub to the whale, in the shape of sundry vague concessions to the Genesis legends, they developed geological truth more and more.

In France, the old theological spirit remained exceedingly powerful. About the middle of the eighteenth century Buffon made another attempt to state simple geological truths; but the theological faculty of the Sorbonne dragged him at once from his high position, forced him to recant ignominiously, and to print his recantation. It runs as follows: "I declare that I had no intention to contradict the text of Scripture; that I believe most firmly all therein related about the creation, both as to order of time and matter of fact. I abandon everything in my book respecting the formation of the earth, and generally all which may be contrary to the narrative of Moses." This humiliating document reminds us painfully of that forced upon Galileo a hundred years before.

It has been well observed by one of the greatest of modern authorities that the doctrine which Buffon thus "abandoned" is as firmly established as that of the earth's rotation upon its axis.* Yet one hundred and fifty years were required to secure for it even a fair hearing; the prevailing doctrine of the Church continued to be that "all things were made at the beginning of the world," and that to say that stones and fossils were made before or since "the beginning" is contrary to Scripture. Again we find theological substitutes for scientific explanation ripening into phrases more and more hollow—making fossils "sports of Nature," or "mineral concretions," or "creations of plastic force," or "models" made by the Creator before he had fully decided upon the best manner of creating various beings.

Of this period, when theological substitutes for science were carrying all before them, there still exists a monument

---

* See citation and remark in Lyell's *Principles of Geology*, chap. iii, p. 57; also Huxley, *Essays on Controverted Questions*, p. 62.

commemorating at the same time a farce and a tragedy. This is the work of Johann Beringer, professor in the University of Würzburg and private physician to the Prince-Bishop—the treatise bearing the title *Lithographiæ Wirceburgensis Specimen Primum*, "illustrated with the marvellous likenesses of two hundred figured or rather insectiform stones." Beringer, for the greater glory of God, had previously committed himself so completely to the theory that fossils are simply "stones of a peculiar sort, hidden by the Author of Nature for his own pleasure," * that some of his students determined to give his faith in that pious doctrine a thorough trial. They therefore prepared a collection of sham fossils in baked clay, imitating not only plants, reptiles, and fishes of every sort that their knowledge or imagination could suggest, but even Hebrew and Syriac inscriptions, one of them the name of the Almighty; and these they buried in a place where the professor was wont to search for specimens. The joy of Beringer on unearthing these proofs of the immediate agency of the finger of God in creating fossils knew no bounds. At great cost he prepared this book, whose twenty-two elaborate plates of facsimiles were forever to settle the question in favour of theology and against science, and prefixed to the work an allegorical title page, wherein not only the glory of his own sovereign, but that of heaven itself, was pictured as based upon a pyramid of these miraculous fossils. So robust was his faith that not even a premature exposure of the fraud could dissuade him from the publication of his book. Dismissing in one contemptuous chapter this exposure as a slander by his rivals, he appealed to the learned world. But the shout of laughter that welcomed the work soon convinced even its author. In vain did he try to suppress it; and, according to tradition, having wasted his fortune in vain attempts to buy up all the copies of it, and being taunted by the rivals whom he had thought to overwhelm, he died of chagrin. Even death did not end his misfortunes. The copies of the first edition having been sold by a graceless descendant to a Leipsic bookseller, a second edition was brought out under a new title,

* See Beringer's *Lithographiæ*, etc , p. 91.

and this, too, is now much sought as a precious memorial of human credulity.*

But even this discomfiture did not end the idea which had caused it, for, although some latitude was allowed among the various theologico-scientific explanations, it was still held meritorious to believe that all fossils were placed in the strata on one of the creative days by the hand of the Almighty, and that this was done for some mysterious purpose, probably for the trial of human faith.

Strange as it may at first seem, the theological war against a scientific method in geology was waged more fiercely in Protestant countries than in Catholic. The older Church had learned by her costly mistakes, especially in the cases of Copernicus and Galileo, what dangers to her claim of infallibility lay in meddling with a growing science. In Italy, therefore, comparatively little opposition was made, while England furnished the most bitter opponents to geology so long as the controversy could be maintained, and the most active negotiators in patching up a truce on the basis of a sham science afterward. The Church of England did, indeed, produce some noble men, like Bishop Clayton and John Mitchell, who stood firmly by the scientific method; but these appear generally to have been overwhelmed by a chorus of churchmen and dissenters, whose mixtures of theology and science, sometimes tragic in their results and sometimes comic, are among the most instructive things in modern history.†

---

* See Carus, *Geschichte der Zoologie*, Munich, 1872, p. 467, note, and Reusch, *Bibel und Natur*, p. 197. A list of the authorities upon this episode, with the text of one of the epigrams circulated at poor Beringer's expense, is given by Dr. Reuss in the *Serapeum* for 1852, p. 203. The book itself (the original impression) is in the White Library at Cornell University. For Beringer himself, see especially the encyclopædia of Ersch and Gruber, and the *Allgemeine deutsche Biographie*.

† For a comparison between the conduct of Italian and English ecclesiastics as regards geology, see Lyell, *Principles of Geology*, tenth English edition, vol. i, p. 33. For a philosophical statement of reasons why the struggle was more bitter and the attempt at deceptive compromises more absurd in England than elsewhere, see Maury, *L'Ancienne Académie des Sciences*, second edition, p. 152. For very frank confessions of the reasons why the Roman Catholic Church has become more careful in her dealings with science, see Roberts, *The Pontifical Decrees against the Earth's Movement*, London, 1885, especially pp. 94 and 132, 133, and St. George Mivart's article in the *Nineteenth Century* for July, 1885. The first of

We have already noted that there are generally three periods or phases in a theological attack upon any science. The first of these is marked by the general use of scriptural texts and statements against the new scientific doctrine; the third by attempts at compromise by means of far-fetched reconciliations of textual statements with ascertained fact; but the second or intermediate period between these two is frequently marked by the pitting against science of some great doctrine in theology. We saw this in astronomy, when Bellarmin and his followers insisted that the scientific doctrine of the earth revolving about the sun is contrary to the theological doctrine of the incarnation. So now against geology it was urged that the scientific doctrine that fossils represent animals which died before Adam contradicts the theological doctrine of Adam's fall and the statement that "death entered the world by sin."

In this second stage of the theological struggle with geology, England was especially fruitful in champions of orthodoxy, first among whom may be named Thomas Burnet. In the last quarter of the seventeenth century, just at the time when Newton's great discovery was given to the world, Burnet issued his *Sacred Theory of the Earth.* His position was commanding; he was a royal chaplain and a cabinet officer. Planting himself upon the famous text in the second epistle of Peter,* he declares that the flood had destroyed the old and created a new world. The Newtonian theory he refuses to accept. In his theory of the deluge he lays less stress upon the "opening of the windows of heaven" than upon the "breaking up of the fountains of the great deep." On this latter point he comes forth with great strength. His theory is that the earth is hollow, and filled with fluid like an egg. Mixing together sundry texts from Genesis and from the second epistle of Peter, the theological

---

these gentlemen, it must not be forgotten, is a Roman Catholic clergyman, and the second an eminent layman of the same Church, and both admit that it was the Pope, speaking *ex cathedra*, who erred in the Galileo case; but their explanation is that God allowed the Pope and Church to fall into this grievous error, which has cost so dear, in order to show once and for all that the Church has no right to decide questions in science.

* See II Peter iii, 6.

doctrine of the "Fall," an astronomical theory regarding the ecliptic, and various notions adapted from Descartes, he insisted that, before sin brought on the Deluge, the earth was of perfect mathematical form, smooth and beautiful, "like an egg," with neither seas nor islands nor valleys nor rocks, "with not a wrinkle, scar, or fracture," and that all creation was equally perfect.

In the second book of his great work Burnet went still further. As in his first book he had mixed his texts of Genesis and St. Peter with Descartes, he now mixed the account of the Garden of Eden in Genesis with heathen legends of the golden age, and concluded that before the flood there was over the whole earth perpetual spring, disturbed by no rain more severe than the falling of the dew.

In addition to his other grounds for denying the earlier existence of the sea, he assigned the reason that, if there had been a sea before the Deluge, sinners would have learned to build ships, and so, when the Deluge set in, could have saved themselves.

The work was written with much power, and attracted universal attention. It was translated into various languages, and called forth a multitude of supporters and opponents in all parts of Europe. Strong men rose against it, especially in England, and among them a few dignitaries of the Church; but the Church generally hailed the work with joy. Addison praised it in a Latin ode, and for nearly a century it exercised a strong influence upon European feeling, and aided to plant more deeply than ever the theological opinion that the earth as now existing is merely a ruin; whereas, before sin brought on the Flood, it was beautiful in its "egg-shaped form," and free from every imperfection.

A few years later came another writer of the highest standing—William Whiston, professor at Cambridge, who in 1696 published his *New Theory of the Earth.* Unlike Burnet, he endeavoured to avail himself of the Newtonian idea, and brought in, to aid the geological catastrophe caused by human sin, a comet, which broke open "the fountains of the great deep."

But, far more important than either of these champions,

there arose in the eighteenth century, to aid in the subjection of science to theology, three men of extraordinary power —John Wesley, Adam Clarke, and Richard Watson. All three were men of striking intellectual gifts, lofty character, and noble purpose, and the first-named one of the greatest men in English history; yet we find them in geology hopelessly fettered by the mere letter of Scripture, and by a temporary phase in theology. As in regard to witchcraft and the doctrine of comets, so in regard to geology, this theological view drew Wesley into enormous error.* The great doctrine which Wesley, Watson, Clarke, and their compeers, following St. Augustine, Bede, Peter Lombard, and a long line of the greatest minds in the universal Church, thought it especially necessary to uphold against geologists was, that death entered the world by sin—by the first transgression of Adam and Eve. The extent to which the supposed necessity of upholding this doctrine carried Wesley seems now almost beyond belief. Basing his theology on the declaration that the Almighty after creation found the earth and all created things "very good," he declares, in his sermon on the *Cause and Cure of Earthquakes*, that no one who believes the Scriptures can deny that "sin is the moral cause of earthquakes, whatever their natural cause may be." Again, he declares that earthquakes are the "effect of that curse which was brought upon the earth by the original transgression." Bringing into connection with Genesis the declaration of St. Paul that "the whole creation groaneth and travaileth together in pain until now," he finds additional scriptural proof that the earthquakes were the result of Adam's fall. He declares, in his sermon on *God's Approbation of His Works*, that "before the sin of Adam there were no agitations within the bowels of the earth, no violent convulsions, no concussions of the earth, no earthquakes, but all was unmoved as the pillars of heaven. There were then no such things as eruptions of fires; no volcanoes or burning mountains." Of course, a science which showed that earthquakes had been in operation for ages before the appearance of man on the

---

* For his statement that "the giving up of witchcraft is in effect the giving up of the Bible," see Wesley's *Journal*, 1766–'68.

planet, and which showed, also, that those very earthquakes which he considered as curses resultant upon the Fall were really blessings, producing the fissures in which we find to-day those mineral veins so essential to modern civilization, was entirely beyond his comprehension. He insists that earthquakes are "God's strange works of judgment, the proper effect and punishment of sin."

So, too, as to death and pain. In his sermon on the *Fall of Man* he took the ground that death and pain entered the world by Adam's transgression, insisting that the carnage now going on among animals is the result of Adam's sin. Speaking of the birds, beasts, and insects, he says that, before sin entered the world by Adam's fall, "none of these attempted to devour or in any way hurt one another"; that "the spider was then as harmless as the fly and did not then lie in wait for blood." Here, again, Wesley arrayed his early followers against geology, which reveals, in the fossil remains of carnivorous animals, pain and death countless ages before the appearance of man. The half-digested fragments of weaker animals within the fossilized bodies of the stronger have destroyed all Wesley's arguments in behalf of his great theory.*

Dr. Adam Clarke held similar views. He insisted that thorns and thistles were given as a curse to human labour, on account of Adam's sin, and appeared upon the earth for the first time after Adam's fall. So, too, Richard Watson, the most prolific writer of the great evangelical reform period, and the author of the *Institutes*, the standard theological treatise on the evangelical side, says, in a chapter treating of the Fall, and especially of the serpent which tempted Eve: "We have no reason at all to believe that the animal had a serpentine form in any mode or degree until his transformation. That he was then degraded to a reptile, to go upon his belly, imports, on the contrary, an entire alteration and loss of the original form." All that admirable adjustment of the serpent to its environment which delights naturalists was to the Wesleyan divine simply an evil result of the sin of Adam and Eve. Yet here again geology was

* See Wesley's sermon on *God's Approbation of His Works*, parts xi and xii.

obliged to confront theology in revealing the *python* in the Eocene, ages before man appeared.*

The immediate results of such teaching by such men was to throw many who would otherwise have resorted to observation and investigation back upon scholastic methods. Again reappears the old system of solving the riddle by phrases. In 1733, Dr. Theodore Arnold urged the theory of "models," and insisted that fossils result from "infinitesimal particles brought together in the creation to form the outline of all the creatures and objects upon and within the earth"; and Arnold's work gained wide acceptance.†

Such was the influence of this succession of great men that toward the close of the last century the English opponents of geology on biblical grounds seemed likely to sweep all before them. Cramping our whole inheritance of sacred literature within the rules of a historical compend, they showed the terrible dangers arising from the revelations of geology, which make the earth older than the six thousand years required by Archbishop Usher's interpretation of the Old Testament. Nor was this feeling confined to ecclesiastics. Williams, a thoughtful layman, declared that such researches led to infidelity and atheism, and are "nothing less than to depose the Almighty Creator of the universe from his office." The poet Cowper, one of the mildest of men, was also roused by these dangers, and in his most elaborate poem wrote:

> "Some drill and bore
> The solid earth, and from the strata there
> Extract a register, by which we learn
> That He who made it, and revealed its date
> To Moses, was mistaken in its age!"

John Howard summoned England to oppose "those scientific systems which are calculated to tear up in the public mind every remaining attachment to Christianity."

With this special attack upon geological science by means of the dogma of Adam's fall, the more general attack by the lit-

---

* See *Westminster Review*, October, 1870, article on *John Wesley's Cosmogony*, with citations from Wesley's *Sermons*, Watson's *Institutes of Theology*, Adam Clarke's *Commentary on the Holy Scriptures*, etc.

† See citation in Mr. Ward's article, as above, p. 390.

eral interpretation of the text was continued. The legendary husks and rinds of our sacred books were insisted upon as equally precious and nutritious with the great moral and religious truths which they envelop. Especially precious were the six days—each "the evening and the morning"—and the exact statements as to the time when each part of creation came into being. To save these, the struggle became more and more desperate.

Difficult as it is to realize it now, within the memory of many now living the battle was still raging most fiercely in England, and both kinds of artillery usually brought against a new science were in full play, and filling the civilized world with their roar.

About half a century since, the Rev. J. Mellor Brown, the Rev. Henry Cole, and others were hurling at all geologists alike, and especially at such Christian scholars as Dr. Buckland and Dean Conybeare and Pye Smith and Prof. Sedgwick, the epithets of "infidel," "impugner of the sacred record," and "assailant of the volume of God." *

The favourite weapon of the orthodox party was the charge that the geologists were "attacking the truth of God." They declared geology "not a subject of lawful inquiry," denouncing it as "a dark art," as "dangerous and disreputable," as "a forbidden province," as "infernal artillery," and as "an awful evasion of the testimony of revelation." †

This attempt to scare men from the science having failed, various other means were taken. To say nothing about England, it is humiliating to human nature to remember the annoyances, and even trials, to which the pettiest and narrowest of men subjected such Christian scholars in our own country as Benjamin Silliman and Edward Hitchcock and Louis Agassiz.

But it is a duty and a pleasure to state here that one great Christian scholar did honour to religion and to himself by quietly accepting the claims of science and making the best of them, despite all these clamours. This man was

---

* For these citations, see Lyell, *Principles of Geology*, introduction.

† See Pye Smith, D. D., *Geology and Scripture*, pp. 156, 157, 168, 169.

Nicholas Wiseman, better known afterward as Cardinal Wiseman. The conduct of this pillar of the Roman Catholic Church contrasts admirably with that of timid Protestants, who were filling England with shrieks and denunciations.*

And here let it be noted that one of the most interesting skirmishes in this war occurred in New England. Prof. Stuart, of Andover, justly honoured as a Hebrew scholar, declared that to speak of six periods of time for the creation was flying in the face of Scripture; that Genesis expressly speaks of six days, each made up of "the evening and the morning," and not six periods of time.

To him replied a professor in Yale College, James Kingsley. In an article admirable for keen wit and kindly temper, he showed that Genesis speaks just as clearly of a solid firmament as of six ordinary days, and that, if Prof. Stuart had surmounted one difficulty and accepted the Copernican theory, he might as well get over another and accept the revelations of geology. The encounter was quick and decisive, and the victory was with science and the broader scholarship of Yale.†

Perhaps the most singular attempt against geology was made by a fine survival of the eighteenth century Don—Dean Cockburn, of York—to *scold* its champions off the field. Having no adequate knowledge of the new science, he opened a battery of abuse, giving it to the world at large from the pulpit and through the press, and even through private letters. From his pulpit in York Minster he denounced Mary Somerville by name for those studies in physical geography which have made her name honoured throughout the world.

But the special object of his antipathy was the British Association for the Advancement of Science. He issued a pamphlet against it which went through five editions in two years, sent solemn warnings to its president, and in various

---

* Wiseman, *Twelve Lectures on the Connection between Science and Revealed Religion*, first American edition, New York, 1837. As to the comparative severity of the struggle regarding astronomy, geology, etc., in Catholic and Protestant countries, see Lecky, *England in the Eighteenth Century*, chap. ix, p. 525.

† See *Silliman's Journal*, vol. xxx, p. 114.

ways made life a burden to Sedgwick, Buckland, and other eminent investigators who ventured to state geological facts as they found them.

These weapons were soon seen to be ineffective; they were like Chinese gongs and dragon lanterns against rifled cannon; the work of science went steadily on.*

III. THE FIRST GREAT EFFORT AT COMPROMISE, BASED ON THE FLOOD OF NOAH.

Long before the end of the struggle already described, even at a very early period, the futility of the usual scholastic weapons had been seen by the more keen-sighted champions of orthodoxy; and, as the difficulties of the ordinary attack upon science became more and more evident, many of these champions endeavoured to patch up a truce. So began the third stage in the war—the period of attempts at compromise.

The position which the compromise party took was that the fossils were produced by the Deluge of Noah.

This position was strong, for it was apparently based upon Scripture. Moreover, it had high ecclesiastical sanction, some of the fathers having held that fossil remains, even on the highest mountains, represented animals destroyed at the Deluge. Tertullian was especially firm on this point, and St. Augustine thought that a fossil tooth discovered in North Africa must have belonged to one of the giants mentioned in Scripture.†

---

* Prof. Goldwin Smith informs me that the papers of Sir Robert Peel, yet unpublished, contain very curious specimens of the epistles of Dean Cockburn. See also *Personal Recollections of Mary Somerville*, Boston, 1874, pp. 139 and 375. Compare with any statement of his religious views that Dean Cockburn was able to make, the following from Mrs. Somerville: "Nothing has afforded me so convincing a proof of the Deity as these purely mental conceptions of numerical and mathematical science which have been, by slow degrees, vouchsafed to man—and are still granted in these latter times by the differential calculus, now superseded by the higher algebra—all of which must have existed in that sublimely omniscient mind from eternity." See also *The Life and Letters of Adam Sedgwick*, Cambridge, 1890, vol ii, pp. 76 and following.

† For Tertullian, see his *De Pallio*, c. ii (Migne, *Patr. Lat.*, vol. ii, p. 1033). For Augustine's view, see Cuvier, *Recherches sur les Ossements fossiles*, fourth edition, vol. ii, p. 143.

In the sixteenth century especially, weight began to be attached to this idea by those who felt the worthlessness of various scholastic explanations. Strong men in both the Catholic and the Protestant camps accepted it; but the man who did most to give it an impulse into modern theology was Martin Luther. He easily saw that scholastic phrase-making could not meet the difficulties raised by fossils, and he naturally urged the doctrine of their origin at Noah's Flood.*

With such support, it soon became the dominant theory in Christendom: nothing seemed able to stand against it; but before the end of the same sixteenth century it met some serious obstacles. Bernard Palissy, one of the most keen-sighted of scientific thinkers in France, as well as one of the most devoted of Christians, showed that it was utterly untenable. Conscientious investigators in other parts of Europe, and especially in Italy, showed the same thing; all in vain.† In vain did good men protest against the injury sure to be brought upon religion by tying it to a scientific theory sure to be exploded; the doctrine that fossils are the remains of animals drowned at the Flood continued to be upheld by the great majority of theological leaders for nearly three centuries as "sound doctrine," and as a blessed means of reconciling science with Scripture. To sustain this scriptural view, efforts energetic and persistent were put forth both by Catholics and Protestants.

In France, the learned Benedictine, Calmet, in his great works on the Bible, accepted it as late as the beginning of the eighteenth century, believing the mastodon's bones exhibited by Mazurier to be those of King Teutobocus, and holding them valuable testimony to the existence of the giants mentioned in Scripture and of the early inhabitants of the earth overwhelmed by the Flood.‡

---

* For Luther's opinion, see his *Commentary on Genesis.*

† For a very full statement of the honourable record of Italy in this respect, and for the enlightened views of some Italian churchmen, see Stoppani, *Il Dogma e le Scienze Positive*, Milan, 1886, pp. 203 *et seq.*

‡ For the steady adherence to this sacred theory, see Audiat, *Vie de Palissy*, p. 412, and Cantu, *Histoire Universelle*, vol. xv, p. 492. For Calmet, see

But the greatest champion appeared in England. We have already seen how, near the close of the seventeenth century, Thomas Burnet prepared the way in his *Sacred Theory of the Earth* by rejecting the discoveries of Newton, and showing how sin led to the breaking up of the "foundations of the great deep"; and we have also seen how Whiston, in his *New Theory of the Earth*, while yielding a little and accepting the discoveries of Newton, brought in a comet to aid in producing the Deluge; but far more important than these in permanent influence was John Woodward, professor at Gresham College, a leader in scientific thought at the University of Cambridge, and, as a patient collector of fossils and an earnest investigator of their meaning, deserving of the highest respect. In 1695 he published his *Natural History of the Earth*, and rendered one great service to science, for he yielded another point, and thus destroyed the foundations for the old theory of fossils. He showed that they were not "sports of Nature," or "models inserted by the Creator in the strata for some inscrutable purpose," but that they were really remains of living beings, as Xenophanes had asserted two thousand years before him. So far, he rendered a great service both to science and religion; but, this done, the text of the Old Testament narrative and the famous passage in St. Peter's Epistle were too strong for him, and he, too, insisted that the fossils were produced by the Deluge. Aided by his great authority, the assault on the true scientific position was vigorous: Mazurier exhibited certain fossil remains of a mammoth discovered in France as bones of the giants mentioned in Scripture; Father Torrubia did the same thing in Spain; Increase Mather sent to England similar remains discovered in America, with a like statement.

For the edification of the faithful, such "bones of the giants mentioned in Scripture" were hung up in public places. Jurieu saw some of them thus suspended in one of the churches of Valence; and Henrion, apparently under the stimulus thus given, drew up tables showing the size of

---

his *Dissertation sur les Géants*, cited in Berger de Xivrey, *Traditions Tératologiques*, p. 191.

our antediluvian ancestors, giving the height of Adam as 123 feet 9 inches and that of Eve as 118 feet 9 inches and 9 lines.*

But the most brilliant service rendered to the theological theory came from another quarter; for, in 1726, Scheuchzer, having discovered a large fossil lizard, exhibited it to the world as the "human witness of the Deluge": † this great discovery was hailed everywhere with joy, for it seemed to prove not only that human beings were drowned at the Deluge, but that "there were giants in those days." Cheered by the applause thus gained, he determined to make the theological position impregnable. Mixing together various texts of Scripture with notions derived from the philosophy of Descartes and the speculations of Whiston, he developed the theory that "the fountains of the great deep" were broken up by the direct physical action of the hand of God, which, being literally applied to the axis of the earth, suddenly stopped the earth's rotation, broke up "the fountains of the great deep," spilled the water therein contained, and produced the Deluge. But his service to sacred science did not end here, for he prepared an edition of the Bible, in which magnificent engravings in great number illustrated his view and enforced it upon all readers. Of these engravings no less than thirty-four were devoted to the Deluge alone. ‡

---

* See Cuvier, *Recherches sur les Ossements fossiles*, fourth edition, vol. ii, p. 56; also Geoffroy St.-Hilaire, cited by Berger de Xivrey, *Traditions Tératologiques*, p. 190.

† *Homo diluvii testis.*

‡ See Zoeckler, vol. ii, p. 172; also Scheuchzer, *Physica Sacra*, Augustæ Vindel. et Ulmæ, 1732. For the ancient belief regarding giants, see Leopardi, *Saggio*. For accounts of the views of Mazurier and Scheuchzer, see Cuvier; also Büchner, *Man in Past, Present, and Future*, English translation, pp. 235, 236. For Increase Mather's views, see *Philosophical Transactions*, vol. xxiv, p. 85. As to similar fossils sent from New York to the Royal Society as remains of giants, see Weld, *History of the Royal Society*, vol. i, p. 421. For Father Torrubia and his *Gigantologia Española*, see D'Archiac, *Introduction à l'Étude de la Paléontologie Stratigraphique*, Paris, 1864, p. 201. For admirable summaries, see Lyell, *Principles of Geology*, London, 1867; D'Archiac, *Géologie et Paléontologie*, Paris, 1866; Pictet, *Traité de Paléontologie*, Paris, 1853; Vezian, *Prodrome de la Géologie*, Paris, 1863; Haeckel, *History of Creation*, English translation, New York, 1876, chap. iii; and for recent progress, Prof. O. S. Marsh's *Address on the History and Methods of Paleontology*.

In the midst all this came an episode very comical but very instructive; for it shows that the attempt to shape the deductions of science to meet the exigencies of dogma may mislead heterodoxy as absurdly as orthodoxy.

About the year 1760 news of the discovery of marine fossils in various elevated districts of Europe reached Voltaire. He, too, had a theologic system to support, though his system was opposed to that of the sacred books of the Hebrews; and, fearing that these new discoveries might be used to support the Mosaic accounts of the Deluge, all his wisdom and wit were compacted into arguments to prove that the fossil fishes were remains of fishes intended for food, but spoiled and thrown away by travellers; that the fossil shells were accidentally dropped by crusaders and pilgrims returning from the Holy Land; and that the fossil bones found between Paris and Étampes were parts of a skeleton belonging to the cabinet of some ancient philosopher. Through chapter after chapter, Voltaire, obeying the supposed necessities of his theology, fought desperately the growing results of the geologic investigations of his time.*

But far more prejudicial to Christianity was the continued effort on the other side to show that the fossils were caused by the Deluge of Noah.

No supposition was too violent to support this theory, which was considered vital to the Bible. By taking the mere husks and rinds of biblical truth for truth itself, by taking sacred poetry as prose, and by giving a literal interpretation of it, the followers of Burnet, Whiston, and Woodward built up systems which bear to real geology much the same relation that the *Christian Topography* of Cosmas bears to real geography. In vain were exhibited the absolute geological, zoölogical, astronomical proofs that no universal deluge, or deluge covering any large part of the earth, had taken place within the last six thousand or sixty thousand years; in vain did so enlightened a churchman as Bishop Clayton declare that the Deluge could not have extended

---

* See Voltaire, *Dissertation sur les Changements arrivés dans notre Globe*; also Voltaire, *Les Singularités de la Nature*, chap. xii; also Jevons, *Principles of Science*, vol. ii, p. 328.

beyond that district where Noah lived before the Flood; in vain did others, like Bishop Croft and Bishop Stillingfleet, and the nonconformist Matthew Poole, show that the Deluge might not have been and probably was not universal; in vain was it shown that, even if there had been a universal deluge, the fossils were not produced by it: the only answers were the citation of the text, "And all the high mountains which were under the whole heaven were covered," and, to clinch the matter, Worthington and men like him insisted that any argument to show that fossils were not remains of animals drowned at the Deluge of Noah was "infidelity." In England, France, and Germany, belief that the fossils were produced by the Deluge of Noah was widely insisted upon as part of that faith essential to salvation.*

But the steady work of science went on: not all the force of the Church—not even the splendid engravings in Scheuchzer's Bible—could stop it, and the foundations of this theological theory began to crumble away. The process was, indeed, slow; it required a hundred and twenty years for the searchers of God's truth, as revealed in Nature—such men as Hooke, Linnæus, Whitehurst, Daubenton, Cuvier, and William Smith—to push their works under this fabric of error, and, by statements which could not be resisted, to undermine it. As we arrive at the beginning of the nineteenth century, science is becoming irresistible in this field. Blumenbach, Von Buch, and Schlotheim led the way, but most important on the Continent was the work of Cuvier. In the early years of the present century his researches among fossils began to throw new light into the whole subject of geology. He was, indeed, very conservative, and even more wary and diplomatic; seeming, like Voltaire, to feel that "among wolves one must howl a little." It was a time of reaction. Napoleon had made peace with the Church, and to disturb

---

* For a candid summary of the proofs from geology, astronomy, and zoölogy, that the Noachian Deluge was not universally or widely extended, see McClintock and Strong, *Cyclopædia of Biblical Theology and Ecclesiastical Literature*, article *Deluge*. For general history, see Lyell, D'Archiac, and Vezian. For special cases showing the bitterness of the conflict, see the Rev. Mr. Davis's *Life of Rev. Dr. Pye Smith*, *passim*. For a late account, see Prof. Huxley on *The Lights of the Church and the Light of Science*, in the *Nineteenth Century* for July, 1890.

that peace was akin to treason. By large but vague concessions Cuvier kept the theologians satisfied, while he undermined their strongest fortress. The danger was instinctively felt by some of the champions of the Church, and typical among these was Chateaubriand, who in his best-known work, once so great, now so little—the *Genius of Christianity* —grappled with the questions of creation by insisting upon a sort of general deception "in the beginning," under which everything was created by a sudden fiat, but with appearances of pre-existence. His words are as follows: "It was part of the perfection and harmony of the nature which was displayed before men's eyes that the deserted nests of last year's birds should be seen on the trees, and that the seashore should be covered with shells which had been the abode of fish, and yet the world was quite new, and nests and shells had never been inhabited." * But the real victory was with Brongniart, who, about 1820, gave forth his work on fossil plants, and thus built a barrier against which the enemies of science raged in vain.†

Still the struggle was not ended, and, a few years later, a forlorn hope was led in England by Granville Penn.

His fundamental thesis was that "our globe has undergone only two revolutions, the Creation and the Deluge, and both by the immediate fiat of the Almighty"; he insisted that the Creation took place in exactly six days of ordinary time, each made up of "the evening and the morning"; and he ended with a piece of that peculiar presumption so familiar to the world, by calling on Cuvier and all other geologists to "ask for the old paths and walk therein until they shall simplify their system and reduce their numerous revolutions to the two events or epochs only—the six days of Creation and the Deluge." ‡ The geologists showed no disposition to yield to this peremptory summons; on the contrary, the President of the British Geological Society, and even so eminent a churchman and geologist as Dean Buckland, soon acknowledged that facts obliged them to give up

---

* *Génie du Christianisme*, chap. v, pp. 1–14, cited by Reusch, vol. i, p. 250.

† For admirable sketches of Brongniart and other paleobotanists, see Ward, as above.

‡ See the works of Granville Penn, vol. ii, p. 273.

the theory that the fossils of the coal measures were deposited at the Deluge of Noah, and to deny that the Deluge was universal.

The defection of Buckland was especially felt by the orthodox party. His ability, honesty, and loyalty to his profession, as well as his position as Canon of Christ Church and Professor of Geology at Oxford, gave him great authority, which he exerted to the utmost in soothing his brother ecclesiastics. In his inaugural lecture he had laboured to show that geology confirmed the accounts of Creation and the Flood as given in Genesis, and in 1823, after his cave explorations had revealed overwhelming evidences of the vast antiquity of the earth, he had still clung to the Flood theory in his *Reliquiæ Diluvianæ*.

This had not, indeed, fully satisfied the anti-scientific party, but as a rule their attacks upon him took the form not so much of abuse as of humorous disparagement. An epigram by Shuttleworth, afterward Bishop of Chichester, in imitation of Pope's famous lines upon Newton, ran as follows:

> "Some doubts were once expressed about the Flood:
> Buckland arose, and all was clear as mud."

On his leaving Oxford for a journey to southern Europe, Dean Gaisford was heard to exclaim: "Well, Buckland is gone to Italy; so, thank God, we shall have no more of this geology!"

Still there was some comfort as long as Buckland held to the Deluge theory; but, on his surrender, the combat deepened: instead of epigrams and caricatures came bitter attacks, and from the pulpit and press came showers of missiles. The worst of these were hurled at Lyell. As we have seen, he had published in 1830 his *Principles of Geology*. Nothing could have been more cautious. It simply gave an account of the main discoveries up to that time, drawing the necessary inferences with plain yet convincing logic, and it remains to this day one of those works in which the Anglo-Saxon race may most justly take pride,—one of the landmarks in the advance of human thought.

But its tendency was inevitably at variance with the Chaldean and other ancient myths and legends regarding the Creation and Deluge which the Hebrews had received from the older civilizations among their neighbours, and had incorporated into the sacred books which they transmitted to the modern world; it was therefore extensively "refuted."

Theologians and men of science influenced by them insisted that his minimizing of geological changes, and his laying stress on the gradual action of natural causes still in force, endangered the sacred record of Creation and left no place for miraculous intervention; and when it was found that he had entirely cast aside their cherished idea that the great geological changes of the earth's surface and the multitude of fossil remains were due to the Deluge of Noah, and had shown that a far longer time was demanded for Creation than any which could possibly be deduced from the Old Testament genealogies and chronicles, orthodox indignation burst forth violently; eminent dignitaries of the Church attacked him without mercy and for a time he was under social ostracism.

As this availed little, an effort was made on the scientific side to crush him beneath the weighty authority of Cuvier; but the futility of this effort was evident when it was found that thinking men would no longer listen to Cuvier and persisted in listening to Lyell. The great orthodox text-book, Cuvier's *Theory of the Earth*, became at once so discredited in the estimation of men of science that no new edition of it was called for, while Lyell's work speedily ran through twelve editions and remained a firm basis of modern thought.*

As typical of his more moderate opponents we may take Fairholme, who in 1837 published his *Mosaic Deluge*, and argued that no early convulsions of the earth, such as those supposed by geologists, could have taken place, because there could have been no deluge "before moral guilt could possibly have been incurred"—that is to say, before the creation of mankind. In touching terms he bewailed the defection of the President of the Geological Society and Dean Buckland—protesting against geologists who "persist

* For Buckland and the various forms of attack upon him, see Gordon, *Life of Buckland*, especially pp. 10, 26, 136. For the attack on Lyell and his book, see Huxley, *The Lights of the Church and the Light of Science.*

in closing their eyes upon the solemn declarations of the Almighty."

Still the geologists continued to seek truth: the germs planted especially by William Smith, "the Father of English Geology," were developed by a noble succession of investigators, and the victory was sure. Meanwhile those theologians who felt that denunciation of science as "godless" could accomplish little, laboured upon schemes for reconciling geology with Genesis. Some of these show amazing ingenuity, but an eminent religious authority, going over them with great thoroughness, has well characterized them as "daring and fanciful." Such attempts have been variously classified, but the fact regarding them all is that each mixes up more or less of science with more or less of Scripture, and produces a result more or less absurd. Though a few men here and there have continued these exercises, the capitulation of the party which set the literal account of the Deluge of Noah against the facts revealed by geology was at last clearly made.*

One of the first evidences of the completeness of this surrender has been so well related by the eminent physiologist, Dr. W. B. Carpenter, that it may best be given in his own words: "You are familiar with a book of considerable value, Dr. W. Smith's *Dictionary of the Bible*. I happened to know the influences under which that dictionary was framed. The idea of the publisher and of the editor was to give as much scholarship and such results of modern criticism as should be compatible with a very judicious conservatism. There was to be no objection to geology, but the universality of the Deluge was to be strictly maintained. The editor committed the article *Deluge* to a man of very considerable ability, but when the article came to him he found that it was so excessively heretical that he could not venture to put it in. There was not time for a second article under that head, and if you look in that dictionary you will find under the word *Deluge* a reference to *Flood*. Before *Flood* came, a sec-

---

* For Fairholme, see his *Mosaic Deluge*, London, 1837, p. 358. For a very just characterization of various schemes of "reconciliation," see Shields, *The Final Philosophy*, p. 340.

ond article had been commissioned from a source that was believed safely conservative; but when the article came in it was found to be worse than the first. A third article was then commissioned, and care was taken to secure its 'safety.' If you look for the word *Flood* in the dictionary, you will find a reference to *Noah*. Under that name you will find an article written by a distinguished professor of Cambridge, of which I remember that Bishop Colenso said to me at the time, 'In a very guarded way the writer concedes the whole thing.' You will see by this under what trammels scientific thought has laboured in this department of inquiry." *

A similar surrender was seen when from a new edition of Horne's *Introduction to the Scriptures*, the standard text-book of orthodoxy, its accustomed use of fossils to prove the universality of the Deluge was quietly dropped. †

A like capitulation in the United States was foreshadowed in 1841, when an eminent Professor of Biblical Literature and Interpretation in the most important theological seminary of the Protestant Episcopal Church, Dr. Samuel Turner, showed his Christian faith and courage by virtually accepting the new view; and the old contention was utterly cast away by the thinking men of another great religious body when, at a later period, two divines among the most eminent for piety and learning in the Methodist Episcopal Church inserted in the *Biblical Cyclopædia*, published under their supervision, a candid summary of the proofs from geology, astronomy, and zoölogy that the Deluge of Noah was not universal, or even widely extended, and this without protest from any man of note in any branch of the American Church. ‡

The time when the struggle was relinquished by enlight-

---

* See *Official Report of the National Conference of Unitarian and other Christian Churches held at Saratoga, 1882*, p. 97.

† This was about 1856; see Tylor, *Early History of Mankind*, p. 329.

‡ For Dr. Turner, see his *Companion to the Book of Genesis*, London and New York, 1841, pp. 216–219. For McClintock and Strong, see their *Cyclopædia of Biblical Knowledge*, etc., article *Deluge*. For similar surrenders of the Deluge in various other religious encyclopædias and commentaries, see Huxley, *Essays on Controverted Questions*, chap. xiii.

ened theologians of the Roman Catholic Church may be fixed at about 1862, when Reusch, Professor of Theology at Bonn, in his work on *The Bible and Nature*, cast off the old diluvial theory and all its supporters, accepting the conclusions of science.*

But, though the sacred theory with the Deluge of Noah as a universal solvent for geological difficulties was evidently dying, there still remained in various quarters a touching fidelity to it. In Roman Catholic countries the old theory was widely though quietly cherished, and taught from the religious press, the pulpit, and the theological professor's chair. Pope Pius IX was doubtless in sympathy with this feeling when, about 1850, he forbade the scientific congress of Italy to meet at Bologna. †

In 1856 Father Debreyne congratulated the theologians of France on their admirable attitude: "Instinctively," he says, "they still insist upon deriving the fossils from Noah's Flood." ‡ In 1875 the Abbé Choyer published at Paris and Angers a text-book widely approved by Church authorities, in which he took similar ground; and in 1877 the Jesuit father Bosizio published at Mayence a treatise on *Geology and the Deluge*, endeavouring to hold the world to the old solution of the problem, allowing, indeed, that the "days" of Creation were long periods, but making atonement for this concession by sneers at Darwin.#

In the Russo-Greek Church, in 1869, Archbishop Macarius, of Lithuania, urged the necessity of believing that Creation in six days of ordinary time and the Deluge of Noah are the only causes of all that geology seeks to explain; and, as late as 1876, another eminent theologian of the same Church went even farther, and refused to allow the faithful to believe that any change had taken place since "the beginning" mentioned in Genesis, when the strata of the earth were laid, tilted, and twisted, and the fossils scat-

* See Reusch, *Bibel und Natur*, chap. xxi.

† See Whiteside, *Italy in the Nineteenth Century*, vol. iii, chap. xiv.

‡ See Zoeckler, vol. ii, p. 472.

See Zoeckler, vol. ii, p. 478, and Bosizio, *Geologie und die Sündfluth*, Mayence, 1877, preface, p. xiv.

tered among them by the hand of the Almighty during six ordinary days.*

In the Lutheran branch of the Protestant Church we also find echoes of the old belief. Keil, eminent in scriptural interpretation at the University of Dorpat, gave forth in 1860 a treatise insisting that geology is rendered futile and its explanations vain by two great facts: the Curse which drove Adam and Eve out of Eden, and the Flood that destroyed all living things save Noah, his family, and the animals in the ark. In 1867, Phillippi, and in 1869, Dieterich, both theologians of eminence, took virtually the same ground in Germany, the latter attempting to beat back the scientific hosts with a phrase apparently pithy, but really hollow—the declaration that "modern geology observes what is, but has no right to judge concerning the beginning of things." As late as 1876, Zugler took a similar view, and a multitude of lesser lights, through pulpit and press, brought these anti-scientific doctrines to bear upon the people at large—the only effect being to arouse grave doubts regarding Christianity among thoughtful men, and especially among young men, who naturally distrusted a cause using such weapons.

For just at this time the traditional view of the Deluge received its death-blow, and in a manner entirely unexpected. By the investigations of George Smith among the Assyrian tablets of the British Museum, in 1872, and by his discoveries just afterward in Assyria, it was put beyond a reasonable doubt that a great mass of accounts in Genesis are simply adaptations of earlier and especially of Chaldean myths and legends. While this proved to be the fact as regards the accounts of Creation and the fall of man, it was seen to be most strikingly so as regards the Deluge. The eleventh of the twelve tablets, on which the most important of these inscriptions was found, was almost wholly preserved, and it revealed in this legend, dating from a time far earlier than that of Moses, such features peculiar to the childhood of the world as the building of the great ship or ark to escape the flood, the careful caulking of its seams, the saving of a

* See Zoeckler, vol. ii, pp. 472, 571, and elsewhere; also citations in Reusch and Shields.

man beloved of Heaven, his selecting and taking with him into the vessel animals of all sorts in couples, the impressive final closing of the door, the sending forth different birds as the flood abated, the offering of sacrifices when the flood had subsided, the joy of the Divine Being who had caused the flood as the odour of the sacrifice reached his nostrils; while throughout all was shown that partiality for the Chaldean sacred number seven which appears so constantly in the Genesis legends and throughout the Hebrew sacred books.

Other devoted scholars followed in the paths thus opened —Sayce in England, Lenormant in France, Schrader in Germany—with the result that the Hebrew account of the Deluge, to which for ages theologians had obliged all geological research to conform, was quietly relegated, even by most eminent Christian scholars, to the realm of myth and legend.*

Sundry feeble attempts to break the force of this discovery, and an evidently widespread fear to have it known, have certainly impaired not a little the legitimate influence of the Christian clergy.

And yet this adoption of Chaldean myths into the Hebrew Scriptures furnishes one of the strongest arguments for the value of our Bible as a record of the upward growth of man; for, while the Chaldean legend primarily ascribes the Deluge to the mere arbitrary caprice of one among many gods (Bel), the Hebrew development of the legend ascribes it to the justice, the righteousness, of the Supreme God; thus showing the evolution of a higher and nobler sentiment which demanded a moral cause adequate to justify such a catastrophe.

Unfortunately, thus far, save in a few of the broader and nobler minds among the clergy, the policy of ignoring such new revelations has prevailed, and the results of this policy, both in Roman Catholic and in Protestant countries, are not far to seek. What the condition of thought is among the middle classes of France and Italy needs not to be stated

---

* For George Smith, see his *Chaldean Account of Genesis*, New York, 1876, especially pp. 36, 263, 286; also his special work on the subject. See also Lenormant, *Les Origines de l'Histoire*, Paris, 1880, chap. viii. For Schrader, see his *The Cuneiform Inscriptions and the Old Testament*, Whitehouse's translation, London, 1885, vol. i, pp. 47–49 and 58–60, and elsewhere.

here. In Germany, as a typical fact, it may be mentioned that there was in the year 1881 church accommodation in the city of Berlin for but two per cent of the population, and that even this accommodation was more than was needed. This fact is not due to the want of a deep religious spirit among the North Germans: no one who has lived among them can doubt the existence of such a spirit; but it is due mainly to the fact that, while the simple results of scientific investigation have filtered down among the people at large, the dominant party in the Lutheran Church has steadily refused to recognise this fact, and has persisted in imposing on Scripture the fetters of literal and dogmatic interpretation which Germany has largely outgrown. A similar danger threatens every other country in which the clergy pursue a similar policy. No thinking man, whatever may be his religious views, can fail to regret this. A thoughtful, reverent, enlightened clergy is a great blessing to any country; and anything which undermines their legitimate work of leading men out of the worship of material things to the consideration of that which is highest is a vast misfortune.*

IV. FINAL EFFORTS AT COMPROMISE.—THE VICTORY OF SCIENCE COMPLETE.

Before concluding, it may be instructive to note a few especially desperate attempts at truces or compromises, such as always appear when the victory of any science has become absolutely sure. Typical among the earliest of these may be mentioned the effort of Carl von Raumer in 1819. With much pretension to scientific knowledge, but with aspirations bounded by the limits of Prussian orthodoxy, he made a laboured attempt to produce a statement which, by its vagueness, haziness, and "depth," should obscure the real questions at issue. This statement appeared in the

* For the foregoing statements regarding Germany the writer relies on his personal observation as a student at the University of Berlin in 1856, as a traveller at various periods afterward, and as Minister of the United States in 1879, 1880, and 1881.

shape of an argument, used by Bertrand and others in the previous century, to prove that fossil remains of plants in the coal measures had never existed as living plants, but had been simply a "result of the development of imperfect plant embryos"; and the same misty theory was suggested to explain the existence of fossil animals without supposing the epochs and changes required by geological science.

In 1837 Wagner sought to uphold this explanation; but it was so clearly a mere hollow phrase, unable to bear the weight of the facts to be accounted for, that it was soon given up.

Similar attempts were made throughout Europe, the most noteworthy appearing in England. In 1853 was issued an anonymous work having as its title *A Brief and Complete Refutation of the Anti-Scriptural Theory of Geologists*: the author having revived an old idea, and put a spark of life into it—this idea being that "all the organisms found in the depths of the earth were made on the first of the six creative days, as models for the plants and animals to be created on the third, fifth, and sixth days." *

But while these attempts to preserve the old theory as to fossil remains of lower animals were thus pressed, there appeared upon the geological field a new scientific column far more terrible to the old doctrines than any which had been seen previously.

For, just at the close of the first quarter of the nineteenth century, geologists began to examine the caves and beds of drift in various parts of the world; and within a few years from that time a series of discoveries began in France, in Belgium, in England, in Brazil, in Sicily, in India, in Egypt, and in America, which established the fact that a period of time much greater than any which had before been thought of had elapsed since the first human occupation of the earth. The chronologies of Archbishop Usher, Petavius, Bossuet, and the other great authorities on which theology had securely leaned, were found worthless. It was clearly seen that, no matter how well based upon the Old Testament genealogies and lives of the patriarchs, all these systems

---

* See Zoeckler, vol. ii, p. 475.

must go for nothing. The most conservative geologists were gradually obliged to admit that man had been upon the earth not merely six thousand, or sixty thousand, or one hundred and sixty thousand years. And when, in 1863, Sir Charles Lyell, in his book on *The Antiquity of Man*, retracted solemnly his earlier view—yielding with a reluctance almost pathetic, but with a thoroughness absolutely convincing—the last stronghold of orthodoxy in this field fell.*

The supporters of a theory based upon the letter of Scripture, who had so long taken the offensive, were now obliged to fight upon the defensive and at fearful odds. Various lines of defence were taken; but perhaps the most pathetic effort was that made in the year 1857, in England, by Gosse. As a naturalist he had rendered great services to zoölogical science, but he now concentrated his energies upon one last effort to save the literal interpretation of Genesis and the theological structure built upon it. In his work entitled *Omphalos* he developed the theory previously urged by Granville Penn, and asserted a new principle called "prochronism." In accordance with this, all things were created by the Almighty hand literally within the six days, each made up of "the evening and the morning," and each great branch of creation was brought into existence in an instant. Accepting a declaration of Dr. Ure, that "neither reason nor revelation will justify us in extending the origin of the material system beyond six thousand years from our own days," Gosse held that all the evidences of convulsive changes and long epochs in strata, rocks, minerals, and fossils are simply "*appearances*"—only that and nothing more. Among these mere "appearances," all created simultaneously, were the glacial furrows and scratches on rocks, the marks of retreat on rocky masses, as at Niagara, the tilted and twisted strata, the piles of lava from extinct volcanoes, the fossils of every sort in every part of the earth, the foot-tracks of birds and reptiles, the half-digested remains of weaker animals found in the fossilized bodies of the

* See Prof. Marsh's address as President of the Society for the Advancement of Science, in 1879; and for a development of the matter, see the chapters on *The Antiquity of Man and Egyptology* and *The Fall of Man and Anthropology*, in this work.

stronger, the marks of hyenas' teeth on fossilized bones found in various caves, and even the skeleton of the Siberian mammoth at St. Petersburg with lumps of flesh bearing the marks of wolves' teeth—all these, with all gaps and imperfections, he urged mankind to believe came into being in an instant. The preface of the work is especially touching, and it ends with the prayer that science and Scripture may be reconciled by his theory, and "that the God of truth will deign so to use it, and if he do, to him be all the glory." * At the close of the whole book Gosse declared: "The field is left clear and undisputed for the one witness on the opposite side, whose testimony is as follows: 'In six days Jehovah made heaven and earth, the sea, and all that in them is.'" This quotation he placed in capital letters, as the final refutation of all that the science of geology had built.

In other parts of Europe desperate attempts were made even later to save the letter of our sacred books by the revival of a theory in some respects more striking. To shape this theory to recent needs, vague reminiscences of a text in Job regarding fire beneath the earth, and vague conceptions of speculations made by Humboldt and Laplace, were mingled with Jewish tradition. Out of the mixture thus obtained Schubert developed the idea that the Satanic "principalities and powers" formerly inhabiting our universe plunged it into the chaos from which it was newly created by a process accurately described in Genesis. Rougemont made the earth one of the "morning stars" of Job, reduced to chaos by Lucifer and his followers, and thence developed in accordance with the nebular hypothesis. Kurtz evolved from this theory an opinion that the geological disturbances were caused by the opposition of the devil to the rescue of our universe from chaos by the Almighty. Delitzsch put a similar idea into a more scholastic jargon; but most desperate of all were the statements of Dr. Anton Westermeyer, of Munich, in *The Old Testament vindicated from Modern Infidel Objections*. The following passage will serve to show his

* See Gosse, *Omphalos*, London, 1857, p. 5, and *passim*; and for a passage giving the keynote of the whole, with a most farcical note on coprolites, see pp. 353, 354.

ideas: "By the fructifying brooding of the Divine Spirit on the waters of the deep, creative forces began to stir; the devils who inhabited the primeval darkness and considered it their own abode saw that they were to be driven from their possessions, or at least that their place of habitation was to be contracted, and they therefore tried to frustrate God's plan of creation and exert all that remained to them of might and power to hinder or at least to mar the new creation." So came into being "the horrible and destructive monsters, these caricatures and distortions of creation," of which we have fossil remains. Dr. Westermeyer goes on to insist that "whole generations called into existence by God succumbed to the corruption of the devil, and for that reason had to be destroyed"; and that "in the work of the six days God caused the devil to feel his power in all earnest, and made Satan's enterprise appear miserable and vain." *

Such was the last important assault upon the strongholds of geological science in Germany; and, in view of this and others of the same kind, it is little to be wondered at that when, in 1870, Johann Silberschlag made an attempt to again base geology upon the Deluge of Noah, he found such difficulties that, in a touching passage, he expressed a desire to get back to the theory that fossils were "sports of Nature." †

But the most noted among efforts to keep geology well within the letter of Scripture is of still more recent date. In the year 1885 Mr. Gladstone found time, amid all his labours and cares as the greatest parliamentary leader in England, to take the field in the struggle for the letter of Genesis against geology.

On the face of it his effort seemed Quixotic, for he confessed at the outset that in science he was "utterly destitute of that kind of knowledge which carries authority," and his argument soon showed that this confession was entirely true.

---

* See Shields's *Final Philosophy*, pp. 340 *et seq.*, and Reusch's *Nature and the Bible* (English translation, 1886), vol. i, pp. 318–320.

† See Reusch, vol. i, p. 264.

But he had some other qualities of which much might be expected: great skill in phrase-making, great shrewdness in adapting the meanings of single words to conflicting necessities in discussion, wonderful power in erecting showy structures of argument upon the smallest basis of fact, and a facility almost preternatural in "explaining away" troublesome realities. So striking was his power in this last respect, that a humorous London chronicler once advised a bigamist, as his only hope, to induce Mr. Gladstone to explain away one of his wives.

At the basis of this theologico-geological structure Mr. Gladstone placed what he found in the text of Genesis: "A grand fourfold division" of animated Nature "set forth in an orderly succession of times." And he arranged this order and succession of creation as follows: "First, the water population; secondly, the air population; thirdly, the land population of animals; fourthly, the land population consummated in man."

His next step was to slide in upon this basis the apparently harmless proposition that this division and sequence "is understood to have been so affirmed in our time by natural science that it may be taken as a demonstrated conclusion and established fact."

Finally, upon these foundations he proceeded to build an argument out of the coincidences thus secured between the record in the Hebrew sacred books and the truths revealed by science as regards this order and sequence, and he easily arrived at the desired conclusion with which he crowned the whole structure, namely, as regards the writer of Genesis, that "his knowledge was divine." *

Such was the skeleton of the structure; it was abundantly decorated with the rhetoric in which Mr. Gladstone is so skilful an artificer, and it towered above "the average man" as a structure beautiful and invincible—like some Chinese fortress in the nineteenth century, faced with porcelain and defended with crossbows.

Its strength was soon seen to be unreal. In an essay ad-

* See Mr. Gladstone's *Dawn of Creation and Worship*, a reply to Dr. Réville, in the *Nineteenth Century* for November, 1885.

mirable in its temper, overwhelming in its facts, and absolutely convincing in its argument, Prof. Huxley, late President of the Royal Society, and doubtless the most eminent contemporary authority on the scientific questions concerned, took up the matter.

Mr. Gladstone's first proposition, that the sacred writings give us a great "fourfold division" created "in an orderly succession of times," Prof. Huxley did not presume to gainsay.

As to Mr. Gladstone's second proposition, that "this great fourfold division . . . created in an orderly succession of times . . . has been so affirmed in our own time by natural science that it may be taken as a demonstrated conclusion and established fact," Prof. Huxley showed that, as a matter of fact, no such "fourfold division" and "orderly succession" exist; that, so far from establishing Mr. Gladstone's assumption that the population of water, air, and land followed each other in the order given, "all the evidence we possess goes to prove that they did not"; that the distribution of fossils through the various strata proves that some land animals originated before sea animals; that there has been a mixing of sea, land, and air "population" utterly destructive to the "great fourfold division" and to the creation "in an orderly succession of times"; that, so far is the view presented in the sacred text, as stated by Mr. Gladstone, from having been "so affirmed in our own time by natural science, that it may be taken as a demonstrated conclusion and established fact" that Mr. Gladstone's assertion is "directly contradictory to facts known to every one who is acquainted with the elements of natural science"; that Mr. Gladstone's only geological authority, Cuvier, had died more than fifty years before, when geological science was in its infancy [and he might have added, when it was necessary to make every possible concession to the Church]; and, finally, he challenged Mr. Gladstone to produce any contemporary authority in geological science who would support his so-called scriptural view. And when, in a rejoinder, Mr. Gladstone attempted to support his view on the authority of Prof. Dana, Prof. Huxley had no difficulty in showing from Prof. Dana's works that Mr. Gladstone's inference was utterly unfounded.

But, while the fabric reared by Mr. Gladstone had been thus undermined by Huxley on the scientific side, another opponent began an attack from the biblical side. The Rev. Canon Driver, professor at Mr. Gladstone's own University of Oxford, took up the question in the light of scriptural interpretation. In regard to the comparative table drawn up by Sir J. W. Dawson, showing the supposed correspondence between the scriptural and the geological order of creation, Canon Driver said: "The two series are evidently at variance. The geological record contains no evidence of clearly defined periods corresponding to the 'days' of Genesis. In Genesis, vegetation is complete two days before animal life appears. Geology shows that they appear simultaneously—even if animal life does not appear first. In Genesis, birds appear together with aquatic creatures, and precede all land animals; according to the evidence of geology, birds are unknown till a period much later than that at which aquatic creatures (including fishes and amphibia) abound, and they are preceded by numerous species of land animals—in particular, by insects and other 'creeping things.'" Of the Mosaic account of the existence of vegetation before the creation of the sun, Canon Driver said, "No reconciliation of this representation with the data of science has yet been found"; and again: "From all that has been said, however reluctant we may be to make the admission, only one conclusion seems possible. Read without prejudice or bias, the narrative of Genesis i. creates an impression at variance with the facts revealed by science." The eminent professor ends by saying that the efforts at reconciliation are "different modes of obliterating the characteristic features of Genesis, and of reading into it a view which it does not express."

Thus fell Mr. Gladstone's fabric of coincidences between the "great fourfold division" in Genesis and the facts ascertained by geology. Prof. Huxley had shattered the scientific parts of the structure, Prof. Driver had removed its biblical foundations, and the last great fortress of the opponents of unfettered scientific investigation was in ruins.

In opposition to all such attempts we may put a noble

utterance by a clergyman who has probably done more to save what is essential in Christianity among English-speaking people than any other ecclesiastic of his time. The late Dean of Westminster, Dr. Arthur Stanley, was widely known and beloved on both continents. In his memorial sermon after the funeral of Sir Charles Lyell he said: "It is now clear to diligent students of the Bible that the first and second chapters of Genesis contain two narratives of the creation side by side, differing from each other in almost every particular of time and place and order. It is well known that, when the science of geology first arose, it was involved in endless schemes of attempted reconciliation with the letter of Scripture. There were, there are perhaps still, two modes of reconciliation of Scripture and science, which have been each in their day attempted, *and each has totally and deservedly failed.* One is the endeavour to wrest the words of the Bible from their natural meaning and *force it to speak the language of science.*" And again, speaking of the earliest known example, which was the interpolation of the word "not" in Leviticus xi, 6, he continues: "This is the earliest instance of *the falsification of Scripture to meet the demands of science;* and it has been followed in later times by the various efforts which have been made to twist the earlier chapters of the book of Genesis into *apparent* agreement with the last results of geology—representing days not to be days, morning and evening not to be morning and evening, the Deluge not to be the Deluge, and the ark not to be the ark."

After a statement like this we may fitly ask, Which is the more likely to strengthen Christianity for its work in the twentieth century which we are now about to enter—a large, manly, honest, fearless utterance like this of Arthur Stanley, or hair-splitting sophistries, bearing in their every line the germs of failure, like those attempted by Mr. Gladstone?

The world is finding that the scientific revelation of creation is ever more and more in accordance with worthy conceptions of that great Power working in and through the universe. More and more it is seen that inspiration has never ceased, and that its prophets and priests are not those who work to fit the letter of its older literature to the needs of dogmas and sects, but those, above all others, who patiently, fearlessly, and reverently devote themselves to the search for truth as truth, in the faith that there is a Power in the universe wise enough to make truth-seeking safe and good enough to make truth-telling useful.*

---

* For the Huxley-Gladstone controversy, see *The Nineteenth Century* for 1885–'86. For Canon Driver, see his article, *The Cosmogony of Genesis*, in *The Expositor* for January, 1886.

CHAPTER VI.

THE ANTIQUITY OF MAN, EGYPTOLOGY, AND ASSYRIOLOGY.

I. THE SACRED CHRONOLOGY.

In the great ranges of investigation which bear most directly upon the origin of man, there are two in which Science within the last few years has gained final victories. The significance of these in changing, and ultimately in reversing, one of the greatest currents of theological thought, can hardly be overestimated; not even the tide set in motion by Cusa, Copernicus, and Galileo was more powerful to bring in a new epoch of belief.

The first of these conquests relates to the antiquity of man on the earth.

The fathers of the early Christian Church, receiving all parts of our sacred books as equally inspired, laid little, if any, less stress on the myths, legends, genealogies, and tribal, family, and personal traditions contained in the Old and the New Testaments, than upon the most powerful appeals, the most instructive apologues, and the most lofty poems of prophets, psalmists, and apostles. As to the age of our planet and the life of man upon it, they found in the Bible a carefully recorded series of periods, extending from Adam to the building of the Temple at Jerusalem, the length of each period being explicitly given.

Thus they had a biblical chronology—full, consecutive, and definite—extending from the first man created to an event of known date well within ascertained profane history; as a result, the early Christian commentators arrived at conclusions varying somewhat, but in the main agreeing. Some, like Origen, Eusebius, Lactantius, Clement of Alexandria, and the great fathers generally of the first three centuries, dwelling especially upon the Septuagint version of the Scriptures, thought that man's creation took place about six thousand years before the Christian era. Strong confirmation of this view was found in a simple piece of purely theological reasoning: for, just as the seven candlesticks of the Apocalypse were long held to prove the existence of seven heavenly bodies revolving about the earth, so it was felt that the six days of creation prefigured six thousand years during which the earth in its first form was to endure; and that, as the first Adam came on the sixth day, Christ, the second Adam, had come at the sixth millennial period. Theophilus, Bishop of Antioch, in the second century clinched this argument with the text, "One day is with the Lord as a thousand years."

On the other hand, Eusebius and St. Jerome, dwelling more especially upon the Hebrew text, which we are brought up to revere, thought that man's origin took place at a somewhat shorter period before the Christian era; and St. Jerome's overwhelming authority made this the dominant view throughout western Europe during fifteen centuries.

The simplicity of these great fathers as regards chronology is especially reflected from the tables of Eusebius. In these, Moses, Joshua, and Bacchus,—Deborah, Orpheus, and the Amazons,—Abimelech, the Sphinx, and Œdipus, appear together as personages equally real, and their positions in chronology equally ascertained.

At times great bitterness was aroused between those holding the longer and those holding the shorter chronology, but after all the difference between them, as we now see, was trivial; and it may be broadly stated that in the early Church, "always, everywhere, and by all," it was held as certain, upon the absolute warrant of Scripture, that man was created from four to six thousand years before the Christian era.

To doubt this, and even much less than this, was to risk damnation. St. Augustine insisted that belief in the antipodes and in the longer duration of the earth than six thousand years were deadly heresies, equally hostile to Scripture.

Philastrius, the friend of St. Ambrose and St. Augustine, whose fearful catalogue of heresies served as a guide to intolerance throughout the Middle Ages, condemned with the same holy horror those who expressed doubt as to the orthodox number of years since the beginning of the world, and those who doubted an earthquake to be the literal voice of an angry God, or who questioned the plurality of the heavens, or who gainsaid the statement that God brings out the stars from his treasures and hangs them up in the solid firmament above the earth every night.

About the beginning of the seventh century Isidore of Seville, the great theologian of his time, took up the subject. He accepted the dominant view not only of Hebrew but of all other chronologies, without anything like real criticism. The childlike faith of his system may be imagined from his summaries which follow. He tells us:

"Joseph lived one hundred and five years. Greece began to cultivate grain."

"The Jews were in slavery in Egypt one hundred and forty-four years. Atlas discovered astrology."

"Joshua ruled for twenty-seven years. Ericthonius yoked horses together."

"Othniel, forty years. Cadmus introduced letters into Greece."

"Deborah, forty years. Apollo discovered the art of medicine and invented the cithara."

"Gideon, forty years. Mercury invented the lyre and gave it to Orpheus."

Reasoning in this general way, Isidore kept well under the longer date; and, the great theological authority of southern Europe having thus spoken, the question was virtually at rest throughout Christendom for nearly a hundred years.

Early in the eighth century the Venerable Bede took up the problem. Dwelling especially upon the received Hebrew text of the Old Testament, he soon entangled himself in very serious difficulties; but, in spite of the great fathers of the first three centuries, he reduced the antiquity of man on the earth by nearly a thousand years, and, in spite of mutterings against him as coming dangerously near a limit which made the theological argument from the six days of creation to the six ages of the world look doubtful, his authority had great weight, and did much to fix western Europe in its allegiance to the general system laid down by Eusebius and Jerome.

In the twelfth century this belief was re-enforced by a tide of thought from a very different quarter. Rabbi Moses Maimonides and other Jewish scholars, by careful study of the Hebrew text, arrived at conclusions diminishing the antiquity of man still further, and thus gave strength throughout the Middle Ages to the shorter chronology: it was incorporated into the sacred science of Christianity; and Vincent of Beauvais, in his great *Speculum Historiale*, forming part of that still more enormous work intended to sum up all the knowledge possessed by the ages of faith, placed the creation of man at about four thousand years before our era.*

At the Reformation this view was not disturbed. The same manner of accepting the sacred text which led Luther, Melanchthon, and the great Protestant leaders generally, to oppose the Copernican theory, fixed them firmly in this biblical chronology; the keynote was sounded for them by Luther when he said, "We know, on the authority of Moses, that longer ago than six thousand years the world did not exist." Melanchthon, more exact, fixed the creation of man at 3963 B. C.

---

* For a table summing up the periods, from Adam to the building of the Temple, explicitly given in the Scriptures, see the admirable paper on *The Pope and the Bible*, in *The Contemporary Review* for April, 1893. For the date of man's creation as given by leading chronologists in various branches of the Church, see *L'Art de Vérifier les Dates*, Paris, 1819, vol. i, pp. 27 *et seq*. In this edition there are sundry typographical errors; compare with Wallace, *True Age of the World*, London, 1844. As to preference for the longer computation by the fathers of the Church, see Clinton, *Fasti Hellenici*, vol. ii, p. 291. For the sacred significance of the six days of creation in ascertaining the antiquity of man, see especially Eicken, *Geschichte der mittelalterlichen Weltanschauung*; also Wallace, *True Age of the World*, pp. 2, 3. For the views of St. Augustine, see Topinard, *Anthropologie*, citing the *De Civ. Dei.*, lib. xvi, c. viii, lib. xii, c. x. For the views of Philastrius, see the *De Hæresibus*, c. 102, 112, *et passim*, in Migne, tome xii. For Eusebius's simple credulity, see the tables in Palmer's *Egyptian Chronicles*, vol. ii, pp. 828, 829. For Bede, see Usher's *Chronologia Sacra*, cited in Wallace, *True Age of the World*, p. 35. For Isidore of Seville, see the *Etymologia*, lib. v, c. 39; also lib. iii, in Migne, tome lxxxii.

But the great Christian scholars continued the old endeavour to make the time of man's origin more precise: there seems to have been a sort of fascination in the subject which developed a long array of chronologists, all weighing the minutest indications in our sacred books, until the Protestant divine De Vignolles, who had given forty years to the study of biblical chronology, declared in 1738 that he had gathered no less than two hundred computations based upon Scripture, and no two alike.

As to the Roman Church, about 1580 there was published, by authority of Pope Gregory XIII, the Roman Martyrology, and this, both as originally published and as revised in 1640 under Pope Urban VIII, declared that the creation of man took place 5199 years before Christ.

But of all who gave themselves up to these chronological studies, the man who exerted the most powerful influence upon the dominant nations of Christendom was Archbishop Usher. In 1650 he published his *Annals of the Ancient and New Testaments*, and it at once became the greatest authority for all English-speaking peoples. Usher was a man of deep and wide theological learning, powerful in controversy; and his careful conclusion, after years of the most profound study of the Hebrew Scriptures, was that man was created 4004 years before the Christian era. His verdict was widely received as final; his dates were inserted in the margins of the authorized version of the English Bible, and were soon practically regarded as equally inspired with the sacred text itself: to question them seriously was to risk preferment in the Church and reputation in the world at large.

The same adhesion to the Hebrew Scriptures which had influenced Usher brought leading men of the older Church to the same view: men who would have burned each other at the stake for their differences on other points, agreed on this: Melanchthon and Tostatus, Lightfoot and Jansen, Salmeron and Scaliger, Petavius and Kepler, inquisitors and reformers, Jesuits and Jansenists, priests and rabbis, stood together in the belief that the creation of man was proved by Scripture to have taken place between 3900 and 4004 years before Christ.

In spite of the severe pressure of this line of authorities, extending from St. Jerome and Eusebius to Usher and Petavius, in favour of this scriptural chronology, even devoted Christian scholars had sometimes felt obliged to revolt. The first great source of difficulty was increased knowledge regarding the Egyptian monuments. As far back as the last years of the sixteenth century Joseph Scaliger had done what he could to lay the foundations of a more scientific treatment of chronology, insisting especially that the historical indications in Persia, in Babylon, and above all in Egypt, should be brought to bear on the question. More than that, he had the boldness to urge that the chronological indications of the Hebrew Scriptures should be fully and critically discussed in the light of Egyptian and other records, without any undue bias from theological considerations. His idea may well be called inspired; yet it had little effect as regards a true view of the antiquity of man, even upon himself, for the theological bias prevailed above all his reasonings, even in his own mind. Well does a brilliant modern writer declare that, "among the multitude of strong men in modern times abdicating their reason at the command of their prejudices, Joseph Scaliger is perhaps the most striking example."

Early in the following century Sir Walter Raleigh, in his *History of the World* (1603–1616), pointed out the danger of adhering to the old system. He, too, foresaw one of the results of modern investigation, stating it in these words, which have the ring of prophetic inspiration: "For in Abraham's time all the then known parts of the world were developed. . . . Egypt had many magnificent cities, . . . and these not built with sticks, but of hewn stone, . . . which magnificence needed a parent of more antiquity than these other men have supposed.". In view of these considerations Raleigh followed the chronology of the Septuagint version, which enabled him to give to the human race a few more years than were usually allowed.

About the middle of the seventeenth century Isaac Vossius, one of the most eminent scholars of Christendom, attempted to bring the prevailing belief into closer accordance with ascertained facts, but, save by a chosen few, his efforts were rejected. In some parts of Europe a man holding new views on chronology was by no means safe from bodily harm. As an example of the extreme pressure exerted by the old theological system at times upon honest scholars, we may take the case of La Peyrère, who about the middle of the seventeenth century put forth his book on the Pre-Adamites—an attempt to reconcile sundry well-known difficulties in Scripture by claiming that man existed on earth before the time of Adam. He was taken in hand at once; great theologians rushed forward to attack him from all parts of Europe; within fifty years thirty-six different refutations of his arguments had appeared; the Parliament of Paris burned the book, and the Grand Vicar of the archdiocese of Mechlin threw him into prison and kept him there until he was forced, not only to retract his statements, but to abjure his Protestantism.

In England, opposition to the growing truth was hardly less earnest. Especially strong was Pearson, afterward Master of Trinity and Bishop of Chester. In his treatise on the Creed, published in 1659, which has remained a theologic classic, he condemned those who held the earth to be more than fifty-six hundred years old, insisted that the first man was created just six days later, declared that the Egyptian records were forged, and called all Christians to turn from them to "the infallible annals of the Spirit of God."

But, in spite of warnings like these, we see the new idea cropping out in various parts of Europe. In 1672, Sir John Marsham published a work in which he showed himself bold and honest. After describing the heathen sources of Oriental history, he turns to the Christian writers, and, having used the history of Egypt to show that the great Church authorities were not exact, he ends one important argument with the following words: "Thus the most interesting antiquities of Egypt have been involved in the deepest obscurity by the very interpreters of her chronology, who have jumbled everything up (*qui omnia susque deque permiscuerunt*), so as to make them match with their own reckonings of Hebrew chronology. Truly a very bad example, and quite unworthy of religious writers."

This sturdy protest of Sir John against the dominant system and against the "jumbling" by which Eusebius had endeavoured to cut down ancient chronology within safe and sound orthodox limits, had little effect. Though eminent chronologists of the eighteenth century, like Jackson, Hales, and Drummond, gave forth multitudes of ponderous volumes pleading for a period somewhat longer than that generally allowed, and insisting that the received Hebrew text was grossly vitiated as regards chronology, even this poor favour was refused them; the mass of believers found it more comfortable to hold fast the faith committed to them by Usher, and it remained settled that man was created about four thousand years before our era.

To those who wished even greater precision, Dr. John Lightfoot, Vice-Chancellor of the University of Cambridge, the great rabbinical scholar of his time, gave his famous demonstration from our sacred books that "heaven and earth, centre and circumference, were created together, in the same instant, and clouds full of water," and that "this work took place and man was created by the Trinity on the twenty-third of October, 4004 B. C., at nine o'clock in the morning."

This tide of theological reasoning rolled on through the eighteenth century, swollen by the biblical researches of leading commentators, Catholic and Protestant, until it came in much majesty and force into our own nineteenth century. At the very beginning of the century it gained new strength from various great men in the Church, among whom may be especially named Dr. Adam Clarke, who declared that, "to preclude the possibility of a mistake, the unerring Spirit of God directed Moses in the selection of his facts and the ascertaining of his dates."

All opposition to the received view seemed broken down, and as late as 1835—indeed, as late as 1850—came an announcement in the work of one of the most eminent Egyptologists, Sir J. G. Wilkinson, to the effect that he had modified the results he had obtained from Egyptian monuments, in order that his chronology might not interfere with the received date of the Deluge of Noah.*

* For Lightfoot, see his *Prolegomena* relating to the age of the world at the birth of Christ; see also in the edition of his works, London, 1822, vol. iv, pp. 64, 112. For Scaliger, see the *De Emendatione Temporum*, 1583; also Mark Pattison, *Es-*

II. THE NEW CHRONOLOGY.

But all investigators were not so docile as Wilkinson, and there soon came a new train of scientific thought which rapidly undermined all this theological chronology. Not to speak of other noted men, we have early in the present century Young, Champollion, and Rosellini, beginning a new epoch in the study of the Egyptian monuments. Nothing could be more cautious than their procedure, but the evidence was soon overwhelming in favour of a vastly longer existence of man in the Nile Valley than could be made to agree with even the longest duration then allowed by theologians.

For, in spite of all the suppleness of men like Wilkinson, it became evident that, whatever system of scriptural chronology was adopted, Egypt was the seat of a flourishing civilization at a period before the "Flood of Noah," and that no such flood had ever interrupted it. This was bad, but worse remained behind: it was soon clear that the civilization of Egypt began earlier than the time assigned for the creation of man, even according to the most liberal of the sacred chronologists.

As time went on, this became more and more evident. The long duration assigned to human civilization in the fragments of Manetho, the Egyptian scribe at Thebes in the third century B. C., was discovered to be more accordant with truth than the chronologies of the great theologians; and, as the

---

says, Oxford, 1889, vol. i, pp. 162 *et seq.* For Raleigh's misgivings, see his *History of the World*, London, 1614, p. 227, book ii of part i, section 7 of chapter i; also Clinton's *Fasti Hellenici*, vol. ii, p. 293. For Usher, see his *Annales Vet. et Nov. Test.*, London, 1650. For Pearson, see his *Exposition of the Creed*, sixth edition, London, 1692, pp. 59 *et seq.* For Marsham, see his *Chronicus Canon Ægypticus, Ebraicus, Græcus, et Disquisitiones*, London, 1672. For La Peyrère, see especially Quatrefages, in *Revue des Deux Mondes* for 1861; also other chapters in this work. For Jackson, Hales, and others, see Wallace's *True Age of the World*. For Wilkinson, see various editions of his work on Egypt. For Vignolles, see Leblois, vol. iii, p. 617. As to the declarations in favour of the recent origin of man, sanctioned by Popes Gregory XIII and Urban VIII, see Strauchius, cited in Wallace, p. 97. For the general agreement of Church authorities, as stated, see *L'Art de Vérifier les Dates*, as above. As to difficulties of scriptural chronology, see Ewald, *History of Israel*, English translation, London, 1883, pp. 204 *et seq.*

present century has gone on, scientific results have been reached absolutely fatal to the chronological view based by the universal Church upon Scripture for nearly two thousand years.

As is well known, the first of the Egyptian kings of whom mention is made upon the monuments of the Nile Valley is Mena, or Menes. Manetho had given a statement, according to which Mena must have lived nearly six thousand years before the Christian era. This was looked upon for a long time as utterly inadmissible, as it was so clearly at variance with the chronology of our own sacred books; but, as time went on, large fragments of the original work of Manetho were more carefully studied and distinguished from corrupt transcriptions, the lists of kings at Karnak, Sacquarah, and the two temples at Abydos were brought to light, and the lists of court architects were discovered. Among all these monuments the scholar who visits Egypt is most impressed by the sculptured tablets giving the lists of kings. Each shows the monarch of the period doing homage to the long line of his ancestors. Each of these sculptured monarchs has near him a tablet bearing his name. That great care was always taken to keep these imposing records correct is certain; the loyalty of subjects, the devotion of priests, and the family pride of kings were all combined in this; and how effective this care was, is seen in the fact that kings now known to be usurpers are carefully omitted. The lists of court architects, extending over the period from Seti to Darius, throw a flood of light over the other records.

Comparing, then, all these sources, and applying an average from the lengths of the long series of well-known reigns to the reigns preceding, the most careful and cautious scholars have satisfied themselves that the original fragments of Manetho represent the work of a man honest and well informed, and, after making all allowances for discrepancies and the overlapping of reigns, it has become clear that the period known as the reign of Mena must be fixed at more than three thousand years B. C. In this the great Egyptologists of our time concur. Mariette, the eminent French authority, puts the date at 5004 B. C.; Brugsch, the leading German authority, puts it at about 4500 B. C.; and Meyer, the latest and most cautious of the historians of antiquity, declares 3180 B. C. the latest possible date that can be assigned it. With these dates the foremost English authorities, Sayce and Flinders Petrie, substantially agree. This view is also confirmed on astronomical grounds by Mr. Lockyer, the Astronomer Royal. We have it, then, as the result of a century of work by the most acute and trained Egyptologists, and with the inscriptions upon the temples and papyri before them, both of which are now read with as much facility as many mediæval manuscripts, that the reign of Mena must be placed more than five thousand years ago.

But the significance of this conclusion can not be fully understood until we bring into connection with it some other facts revealed by the Egyptian monuments.

The first of these is that which struck Sir Walter Raleigh, that, even in the time of the first dynasties in the Nile Valley, a high civilization had already been developed. Take, first, man himself: we find sculptured upon the early monuments types of the various races—Egyptians, Israelites, negroes, and Libyans—as clearly distinguishable in these paintings and sculptures of from four to six thousand years ago as the same types are at the present day. No one can look at these sculptures upon the Egyptian monuments, or even the drawings of them, as given by Lepsius or Prisse d'Avennes, without being convinced that they indicate, even at that remote period, a difference of races so marked that long previous ages must have been required to produce it.

The social condition of Egypt revealed in these early monuments of art forces us to the same conclusion. Those earliest monuments show that a very complex society had even then been developed. We not only have a separation between the priestly and military orders, but agriculturists, manufacturers, and traders, with a whole series of subdivisions in each of these classes. The early tombs show us sculptured and painted representations of a daily life which even then had been developed into a vast wealth and variety of grades, forms, and usages.

Take, next, the political and military condition. One fact out of many reveals a policy which must have been the result of long experience. Just as now, at the end of the nineteenth century, the British Government, having found that they can not rely upon the native Egyptians for the protection of the country, are drilling the negroes from the interior of Africa as soldiers, so the celebrated inscription of Prince Una, as far back as the sixth dynasty, speaks of the Maksi or negroes levied and drilled by tens of thousands for the Egyptian army.

Take, next, engineering. Here we find very early operations in the way of canals, dikes, and great public edifices, so bold in conception and thorough in execution as to fill our greatest engineers of these days with astonishment. The quarrying, conveyance, cutting, jointing, and polishing of the enormous blocks in the interior of the Great Pyramid alone are the marvel of the foremost stone-workers of our century.

As regards architecture, we find not only the pyramids, which date from the very earliest period of Egyptian history, and which are to this hour the wonder of the world for size, for boldness, for exactness, and for skilful contrivance, but also the temples, with long ranges of colossal columns wrought in polished granite, with wonderful beauty of ornamentation, with architraves and roofs vast in size and exquisite in adjustment, which by their proportions tax the imagination, and lead the beholder to ask whether all this can be real.

As to sculpture, we have not only the great Sphinx of Gizeh, so marvellous in its boldness and dignity, dating from the very first period of Egyptian history, but we have ranges of sphinxes, heroic statues, and bas-reliefs, showing that even in the early ages this branch of art had reached an amazing development.

As regards the perfection of these, Lübke, the most eminent German authority on plastic art, referring to the early works in the tombs about Memphis, declares that, "as monuments of the period of the fourth dynasty, they are an evidence of the high perfection to which the sculpture of the Egyptians had attained." Brugsch declares that "every artistic production of those early days, whether picture, writing, or sculpture, bears the stamp of the highest perfection in art." Maspero, the most eminent French authority in this field, while expressing his belief that the Sphinx was sculptured even before the time of Mena, declares that "the art which conceived and carved this prodigious statue was a finished art—an art which had attained self-mastery and was sure of its effects"; while, among the more eminent English authorities, Sayce tells us that "art is at its best in the age of the pyramid-builders," and Sir James Fergusson declares, "We are startled to find Egyptian art nearly as perfect in the oldest periods as in any of the later."

The evidence as to the high development of Egyptian sculpture in the earlier dynasties becomes every day more overwhelming. What exquisite genius the early Egyptian sculptors showed in their lesser statues is known to all who have seen those most precious specimens in the museum at Cairo, which were wrought before the conventional type was adopted in obedience to religious considerations.

In decorative and especially in ceramic art, as early as the fourth and fifth dynasties, we have vases, cups, and other vessels showing exquisite beauty of outline and a general sense of form almost if not quite equal to Etruscan and Grecian work of the best periods.

Take, next, astronomy. Going back to the very earliest period of Egyptian civilization, we find that the four sides of the Great Pyramid are adjusted to the cardinal points with the utmost precision. "The day of the equinox can be taken by observing the sun set across the face of the pyramid, and the neighbouring Arabs adjust their astronomical dates by its shadow." Yet this is but one out of many facts which prove that the Egyptians, at the earliest period of which their monuments exist, had arrived at knowledge and skill only acquired by long ages of observation and thought. Mr. Lockyer, Astronomer Royal of Great Britain, has recently convinced himself, after careful examination of various ruined temples at Thebes and elsewhere, that they were placed with reference to observations of stars. To state his conclusion in his own words: "There seems a very high probability that three thousand, and possibly four thousand, years before Christ the Egyptians had among them men with some knowledge of astronomy, and that six thousand years ago the course of the sun through the year was practically very well known, and methods had been invented by means of which in time it might be better known; and that, not very long after that, they not only considered questions relating to the sun, but began to take up other questions relating to the position and movement of the stars."

The same view of the antiquity of man in the Nile valley is confirmed by philologists. To use the words of Max Duncker: "The oldest monuments of Egypt—and they are the oldest monuments in the world—exhibit the Egyptian in possession of the art of writing." It is found also, by the inscriptions of the early dynasties, that the Egyptian language had even at that early time been developed in all essential particulars to the highest point it ever attained. What long periods it must have required for such a development every scholar in philology can imagine.

As regards medical science, we have the Berlin papyrus, which, although of a later period, refers with careful specification to a medical literature of the first dynasty.

As regards archæology, the earliest known inscriptions point to still earlier events and buildings, indicating a long sequence in previous history.

As to all that pertains to the history of civilization, no man of fair and open mind can go into the museums of Cairo or the Louvre or the British Museum and look at the monuments of those earlier dynasties without seeing in them the results of a development in art, science, laws, customs, and language, which must have required a vast period before the time of Mena. And this conclusion is forced upon us all the more invincibly when we consider the slow growth of ideas in the earlier stages of civilization as compared with the later—a slowness of growth which has kept the natives of many parts of the world in that earliest civilization to this hour. To this we must add the fact that Egyptian civilization was especially immobile: its development into castes is but one among many evidences that it was the very opposite of a civilization developed rapidly.

As to the length of the period before the time of Mena, there is, of course, nothing exact. Manetho gives lists of great personages before that first dynasty, and these extend over twenty-four thousand years. Bunsen, one of the most learned of Christian scholars, declares that not less than ten thousand years were necessary for the development of civilization up to the point where we find it in Mena's time. No one can claim precision for either of these statements, but they are valuable as showing the impression of vast antiquity made upon the most competent judges by the careful study of those remains: no unbiased judge can doubt that an immensely long period of years must have been required for the development of civilization up to the state in which we there find it.

The investigations in the bed of the Nile confirm these views. That some unwarranted conclusions have at times been announced is true; but the fact remains that again and again rude pottery and other evidences of early stages of civilization have been found in borings at places so distant from each other, and at depths so great, that for such a range of concurring facts, considered in connection with the rate of earthy deposit by the Nile, there is no adequate explanation save the existence of man in that valley thousands on thousands of years before the longest time admitted by our sacred chronologists.

Nor have these investigations been of a careless character. Between the years 1851 and 1854, Mr. Horner, an extremely cautious English geologist, sank ninety-six shafts in four rows at intervals of eight English miles, at right angles to the Nile, in the neighbourhood of Memphis. In these pottery was brought up from various depths, and beneath the statue of Rameses II at Memphis from a depth of thirty-nine feet. At the rate of the Nile deposit a careful estimate has declared this to indicate a period of over eleven thousand years. So eminent a German authority in geography as Peschel characterizes objections to such deductions as groundless. However this may be, the general results of these investigations, taken in connection with the other results of research, are convincing.

And, finally, as if to make assurance doubly sure, a series of archæologists of the highest standing, French, German, English, and American, have within the past twenty years discovered relics of a savage period, of vastly earlier date than the time of Mena, prevailing throughout Egypt. These relics have been discovered in various parts of the country, from Cairo to Luxor, in great numbers. They are the same sort of prehistoric implements which prove to us the early existence of man in so many other parts of the world at a geological period so remote that the figures given by our sacred chronologists are but trivial. The last and most convincing of these discoveries, that of flint implements in the drift, far down below the tombs of early kings at Thebes, and upon high terraces far above the present bed of the Nile, will be referred to later.

But it is not in Egypt alone that proofs are found of the utter inadequacy of the entire chronological system derived from our sacred books. These results of research in Egypt are strikingly confirmed by research in Assyria and Babylonia. Prof. Sayce exhibits various proofs of this. To use his own words regarding one of these proofs: "On the shelves of the British Museum you may see huge sun-dried bricks, on which are stamped the names and titles of kings who erected or repaired the temples where they have been found. . . . They must . . . have reigned before the time when, according to the margins of our Bibles, the Flood of Noah was covering the earth and reducing such bricks as these to their primeval slime."

This conclusion was soon placed beyond a doubt. The lists of kings and collateral inscriptions recovered from the temples of the great valley between the Tigris and Euphrates, and the records of astronomical observations in that region, showed that there, too, a powerful civilization had grown up at a period far earlier than could be made consistent with our sacred chronology. The science of Assyriology was thus combined with Egyptology to furnish one more convincing proof that, precious as are the moral and religious truths in our sacred books and the historical indications which they give us, these truths and indications are necessarily inclosed in a setting of myth and legend.*

---

* As to Manetho, see, for a very full account of his relations to other chronologists, Palmer, *Egyptian Chronicles*, vol. i, chap. ii. For a more recent and readable account, see Brugsch, *Egypt under the Pharaohs*, English edition, London, 1879, chap. iv. For lists of kings at Abydos and elsewhere, also the lists of architects, see Brugsch, Palmer, Mariette, and others; also illustrations in Lepsius. For

proofs that the dynasties given were consecutive and not contemporaneous, as was once so fondly argued by those who tried to save Archbishop Usher's chronology, see Mariette; also Sayce's *Herodotus*, appendix, p. 316. For the various race types given on early monuments, see the coloured engravings in Lepsius, *Denkmäler*; also Prisse d'Avennes, and the frontispiece in the English edition of Brugsch; see also statement regarding the same subject in Tylor, *Anthropology*, chap. i. For the fulness of development in Egyptian civilization in the earliest dynasties, see Rawlinson's *Egypt*, London, 1881, chap. xiii; also Brugsch and other works cited. For the perfection of Egyptian engineering, I rely not merely upon my own observation, but on what is far more important, the testimony of my friend the Hon. J. G. Batterson, probably the largest and most experienced worker in granite in the United States, who acknowledges, from personal observation, that the early Egyptian work is, in boldness and perfection, far beyond anything known since, and a source of perpetual wonder to him. As to the perfection of Egyptian architecture, see very striking statements in Fergusson, *History of Architecture*, book i, chap. i. As to the pyramids, showing a very high grade of culture already reached under the earliest dynasties, see Lübke, *Gesch. der Arch.*, book i. For Sayce's views, see his *Herodotus*, appendix, p. 348. As to sculpture, see for representations photographs published by the Boulak Museum, and such works as the *Description de l'Égypte*, Lepsius's *Denkmäler*, and Prisse d'Avennes; see also a most valuable small work, easy of access, Maspero, *Archæology*, translated by Miss A. B. Edwards, New York and London, 1887, chaps. i and ii. See especially in Prisse, vol. ii, the statue of Chafré the Scribe, and the group of "Tea" and his wife. As to the artistic value of the Sphinx, see Maspero, as above, pp. 202, 203. See also similar ideas in Lübke's *History of Sculpture*, vol. i, p. 24. As to astronomical knowledge evidenced by the Great Pyramid, see Tylor, as above, p. 21; also Lockyer, *On Some Points in the Early History of Astronomy*, in *Nature* for 1891, and especially in the issues of June 4th and July 2d; also his *Dawn of Astronomy, passim*. For a recent and conservative statement as to the date of Mena, see Flinders Petrie, *History of Egypt*, London, 1894, chap. ii. For delineations of vases, etc., showing Grecian proportion and beauty of form under the fourth and fifth dynasties, see Prisse, vol. ii, *Art Industriel*. As to the philological question, and the development of language in Egypt, with the hieroglyphic system of writing, see Rawlinson's *Egypt*, London, 1881, chap. xiii; also Lenormant; also Max Düncker, *Geschichte des Alterthums*, Abbott's translation, 1877. As to the medical papyrus of Berlin, see Brugsch, vol. i, p. 58, but especially the *Papyrus Ebers*. As to the corruption of later copies of Manetho and fidelity of originals as attested by the monuments, see Brugsch, chap. iv. On the accuracy of the present Egyptian chronology as regards long periods, see ibid., vol. i, p. 32. As to the pottery found deep in the Nile and the value of Horner's discovery, see Peschel, *Races of Man*, New York, 1876, pp. 42–44. For succinct statement, see also Laing, *Problems of the Future*, p. 94. For confirmatory proofs from Assyriology, see Sayce, *Lectures on the Religion of the Babylonians* (Hibbert Lectures for 1887), London, 1887, introductory chapter, and especially pp. 21–25. See also Laing, *Human Origins*, chap. ii, for an excellent summary. For an account of flint implements recently found in gravel terraces fifteen hundred feet above the present level of the Nile, and showing evidences of an age vastly greater even than those dug out of the gravel at Thebes, see article by Flinders Petrie in *London Times* of April 18th, 1895.

CHAPTER VII.

THE ANTIQUITY OF MAN AND PREHISTORIC ARCHÆOLOGY.

I. THE THUNDER-STONES.

WHILE the view of chronology based upon the literal acceptance of Scripture texts was thus shaken by researches in Egypt, another line of observation and thought was slowly developed, even more fatal to the theological view.

From a very early period there had been dug from the earth, in various parts of the world, strangely shaped masses of stone, some rudely chipped, some polished: in ancient times the larger of these were very often considered as thunderbolts, the smaller as arrows, and all of them as weapons which had been hurled by the gods and other supernatural personages. Hence a sort of sacredness attached to them. In Chaldea, they were built into the wall of temples; in Egypt, they were strung about the necks of the dead; in India, fine specimens are to this day seen upon altars, receiving prayers and sacrifices.

Naturally these beliefs were brought into the Christian mythology and adapted to it. During the Middle Ages many of these well-wrought stones were venerated as weapons, which during the "war in heaven" had been used in driving forth Satan and his hosts; hence in the eleventh century an Emperor of the East sent to the Emperor of the West a "heaven axe"; and in the twelfth century a Bishop of Rennes asserted the value of thunder-stones as a divinely-appointed means of securing success in battle, safety on the sea, security against thunder, and immunity from unpleasant dreams. Even as late as the seventeenth century a French ambassador brought a stone hatchet, which still exists in the museum at Nancy, as a present to the Prince-Bishop of Verdun, and claimed for it health-giving virtues.

In the last years of the sixteenth century Michael Mercati tried to prove that the "thunder-stones" were weapons or implements of early races of men; but from some cause his book was not published until the following century, when other thinkers had begun to take up the same idea, and then it had to contend with a theory far more accordant with theologic modes of reasoning in science. This was the theory of the learned Tollius, who in 1649 told the world that these chipped or smoothed stones were "generated in the sky by a fulgurous exhalation conglobed in a cloud by the circumposed humour."

But about the beginning of the eighteenth century a fact of great importance was quietly established. In the year 1715 a large pointed weapon of black flint was found in contact with the bones of an elephant, in a gravel bed near Gray's Inn Lane, in London. The world in general paid no heed to this: if the attention of theologians was called to it, they dismissed it summarily with a reference to the Deluge of Noah; but the specimen was labelled, the circumstances regarding it were recorded, and both specimen and record carefully preserved.

In 1723 Jussieu addressed the French Academy on *The Origin and Uses of Thunder-stones.* He showed that recent travellers from various parts of the world had brought a number of weapons and other implements of stone to France, and that they were essentially similar to what in Europe had been known as "thunder-stones." A year later this fact was clinched into the scientific mind of France by the Jesuit Lafitau, who published a work showing the similarity between the customs of aborigines then existing in other lands and those of the early inhabitants of Europe. So began, in these works of Jussieu and Lafitau, the science of Comparative Ethnography.

But it was at their own risk and peril that thinkers drew from these discoveries any conclusions as to the antiquity of man. Montesquieu, having ventured to hint, in an early edition of his *Persian Letters*, that the world might be much older than had been generally supposed, was soon made to feel danger both to his book and to himself, so that in succeeding editions he suppressed the passage.

In 1730 Mahudel presented a paper to the French Academy of Inscriptions on the so-called "thunder-stones," and also presented a series of plates which showed that these were stone implements, which must have been used at an early period in human history.

In 1778 Buffon, in his *Époques de la Nature*, intimated his belief that "thunder-stones" were made by early races of men; but he did not press this view, and the reason for his reserve was obvious enough: he had already one quarrel with the theologians on his hands, which had cost him dear—public retraction and humiliation. His declaration, therefore, attracted little notice.

In the year 1800 another fact came into the minds of thinking men in England. In that year John Frere presented to the London Society of Antiquaries sundry flint implements found in the clay beds near Hoxne: that they were of human make was certain, and, in view of the undisturbed depths in which they were found, the theory was suggested that the men who made them must have lived at a very ancient geological epoch; yet even this discovery and theory passed like a troublesome dream, and soon seemed to be forgotten.

About twenty years later Dr. Buckland published a discussion of the subject, in the light of various discoveries in the drift and in caves. It received wide attention, but theology was soothed by his temporary concession that these striking relics of human handiwork, associated with the remains of various extinct animals, were proofs of the Deluge of Noah.

In 1823 Boué, of the Vienna Academy of Sciences, showed to Cuvier sundry human bones found deep in the alluvial deposits of the upper Rhine, and suggested that they were of an early geological period; this Cuvier virtually, if not explicitly, denied. Great as he was in his own field, he was not a great geologist; he, in fact, led geology astray for many years. Moreover, he lived in a time of reaction; it was the period of the restored Bourbons, of the Voltairean

King Louis XVIII, governing to please orthodoxy. Boué's discovery was, therefore, at first opposed, then enveloped in studied silence.

Cuvier evidently thought, as Voltaire had felt under similar circumstances, that "among wolves one must howl a little"; and his leading disciple, Élie de Beaumont, who succeeded him in the sway over geological science in France, was even more opposed to the new view than his great master had been. Boué's discoveries were, therefore, apparently laid to rest forever.*

In 1825 Kent's Cavern, near Torquay, was explored by the Rev. Mr. McEnery, a Roman Catholic clergyman, who seems to have been completely overawed by orthodox opinion in England and elsewhere; for, though he found human bones and implements mingled with remains of extinct animals, he kept his notes in manuscript, and they were only brought to light more than thirty years later by Mr. Vivian.

The coming of Charles X, the last of the French Bourbons, to the throne, made the orthodox pressure even greater. It was the culmination of the reactionary period—the time in France when a clerical committee, sitting at the Tuileries, took such measures as were necessary to hold in check all science that was not perfectly "safe"; the time in Austria when Kaiser Franz made his famous declaration to sundry professors, that what he wanted of them was simply to train obedient subjects, and that those who did not make this their purpose would be dismissed; the time in Germany when Nicholas of Russia and the princelings and ministers under his control, from the King of Prussia downward, put forth all their might in behalf of "scriptural science"; the time in Italy when a scientific investigator, arriving at any conclu-

* For the general history of early views regarding stone implements, see the first chapters in Cartailhac, *La France Préhistorique*; also Joly, *L'Homme avant les Métaux*; also Lyell, Lubbock, and Evans. For lightning-stones in China and elsewhere, see citation from a Chinese encyclopædia of 1662, in Tylor, *Early History of Mankind*, p. 209. On the universality of this belief, on the surviving use of stone implements even into civilized times, and on their manufacture to-day, see ibid., chapter viii. For the treatment of Boué's discovery, see especially Mortillet, *Le Préhistorique*, Paris, 1885, p. 11. For the suppression of the passage in Montesquieu's *Persian Letters*, see Letter 113, cited in Schlosser's *History of the Eighteenth Century* (English translation), vol. i, p. 135.

sion distrusted by the Church, was sure of losing his place and in danger of losing his liberty; the time in England when what little science was taught was held in due submission to Archdeacon Paley; the time in the United States when the first thing essential in science was, that it be adjusted to the ideas of revival exhorters.

Yet men devoted to scientific truth laboured on; and in 1828 Tournal, of Narbonne, discovered in the cavern of Bize specimens of human industry, with a fragment of a human skeleton, among bones of extinct animals. In the following year Christol published accounts of his excavations in the caverns of Gard; he had found in position, and under conditions which forbade the idea of after-disturbance, human remains mixed with bones of the extinct hyena of the early Quaternary period. Little general notice was taken of this, for the reactionary orthodox atmosphere involved such discoveries in darkness.

But in the French Revolution of 1830 the old politico-theological system collapsed: Charles X and his advisers fled for their lives; the other continental monarchs got glimpses of new light; the priesthood in charge of education were put on their good behaviour for a time, and a better era began.

Under the constitutional monarchy of the house of Orleans in France and Belgium less attention was therefore paid by Government to the saving of souls; and we have in rapid succession new discoveries of remains of human industry, and even of human skeletons so mingled with bones of extinct animals as to give additional proofs that the origin of man was at a period vastly earlier than any which theologians had dreamed of.

A few years later the reactionary clerical influence against science in this field rallied again. Schmerling in 1833 had explored a multitude of caverns in Belgium, especially at Engis and Engihoul, and had found human skulls and bones closely associated with bones of extinct animals, such as the cave bear, hyena, elephant, and rhinoceros, while mingled with these were evidences of human workmanship in the shape of chipped flint implements; discoveries of a similar sort had been made by De Serres in France and by

Lund in Brazil; but, at least as far as continental Europe was concerned, these discoveries were received with much coolness both by Catholic leaders of opinion in France and Belgium and by Protestant leaders in England and Holland. Schmerling himself appears to have been overawed, and gave forth a sort of apologetic theory, half scientific, half theologic, vainly hoping to satisfy the clerical side.

Nor was it much better in England. Sir Charles Lyell, so devoted a servant of prehistoric research thirty years later, was still holding out against it on the scientific side; and, as to the theological side, it was the period when that great churchman, Dean Cockburn, was insulting geologists from the pulpit of York Minster, and the Rev. Mellor Brown denouncing geology as "a black art," "a forbidden province"; and when, in America, Prof. Moses Stuart and others like him were belittling the work of Benjamin Silliman and Edward Hitchcock.

In 1840 Godwin Austin presented to the Royal Geological Society an account of his discoveries in Kent's Cavern, near Torquay, and especially of human bones and implements mingled with bones of the elephant, rhinoceros, cave bear, hyena, and other extinct animals; yet this memoir, like that of McEnery fifteen years before, found an atmosphere so unfavourable that it was not published.

II. THE FLINT WEAPONS AND IMPLEMENTS.

At the middle of the nineteenth century came the beginning of a new epoch in science—an epoch when all these earlier discoveries were to be interpreted by means of investigations in a different field: for, in 1847, a man previously unknown to the world at large, Boucher de Perthes, published at Paris the first volume of his work on *Celtic and Antediluvian Antiquities*, and in this he showed engravings of typical flint implements and weapons, of which he had discovered thousands upon thousands in the high drift beds near Abbeville, in northern France.

The significance of this discovery was great indeed—far greater than Boucher himself at first supposed. The very title of his book showed that he at first regarded these implements and weapons as having belonged to men overwhelmed at the Deluge of Noah; but it was soon seen that they were something very different from proofs of the literal exactness of Genesis: for they were found in terraces at great heights above the river Somme, and, under any possible theory having regard to fact, must have been deposited there at a time when the river system of northern France was vastly different from anything known within the historic period. The whole discovery indicated a series of great geological changes since the time when these implements were made, requiring cycles of time compared to which the space allowed by the orthodox chronologists was as nothing.

His work was the result of over ten years of research and thought. Year after year a force of men under his direction had dug into these high-terraced gravel deposits of the river Somme, and in his book he now gave, in the first full form, the results of his labour. So far as France was concerned, he was met at first by what he calls "a conspiracy of silence," and then by a contemptuous opposition among orthodox scientists, at the head of whom stood Élie de Beaumont.

This heavy, sluggish opposition seemed immovable: nothing that Boucher could do or say appeared to lighten the pressure of the orthodox theological opinion behind it; not even his belief that these fossils were remains of men drowned at the Deluge of Noah, and that they were proofs of the literal exactness of Genesis seemed to help the matter. His opponents felt instinctively that such discoveries boded danger to the accepted view, and they were right: Boucher himself soon saw the folly of trying to account for them by the orthodox theory.

And it must be confessed that not a little force was added to the opposition by certain characteristics of Boucher de Perthes himself. Gifted, far-sighted, and vigorous as he was, he was his own worst enemy. Carried away by his own discoveries, he jumped to the most astounding conclusions. The engravings in the later volume of his great work, showing what he thought to be human features and inscriptions upon some of the flint implements, are worthy of a comic almanac; and at the National Museum of Archæology at St. Germain, beneath the shelves bearing the remains which he discovered, which mark the beginning of a new epoch in science, are drawers containing specimens hardly worthy of a penny museum, but from which he drew the most unwarranted inferences as to the language, religion, and usages of prehistoric man.

Boucher triumphed none the less. Among his bitter opponents at first was Dr. Rigollot, who in 1855, searching earnestly for materials to refute the innovator, dug into the deposits of St. Acheul—and was converted: for he found implements similar to those of Abbeville, making still more certain the existence of man during the Drift period. So, too, Gaudry a year later made similar discoveries.

But most important was the evidence of the truth which now came from other parts of France and from other countries. The French leaders in geological science had been held back not only by awe of Cuvier but by recollections of Scheuchzer. Ridicule has always been a serious weapon in France, and the ridicule which finally overtook the supporters of the attempt of Scheuchzer, Mazurier, and others, to square geology with Genesis, was still remembered. From the great body of French geologists, therefore, Boucher secured at first no aid. His support came from the other side of the Channel. The most eminent English geologists, such as Falconer, Prestwich, and Lyell, visited the beds at Abbeville and St. Acheul, convinced themselves that the discoveries of Boucher, Rigollot, and their colleagues were real, and then quietly but firmly told England the truth.

And now there appeared a most effective ally in France. The arguments used against Boucher de Perthes and some of the other early investigators of bone caves had been that the implements found might have been washed about and turned over by great floods, and therefore that they might be of a recent period; but in 1861 Edward Lartet published an account of his own excavations at the Grotto of Aurignac, and the proof that man had existed in the time of the Quaternary animals was complete. This grotto had been carefully sealed in prehistoric times by a stone at its entrance; no interference from disturbing currents of water had been possible; and Lartet found, in place, bones of eight out of nine of the main species of animals which characterize the Quaternary period in Europe; and upon them marks of cutting implements, and in the midst of them coals and ashes.

Close upon these came the excavations at Eyzies by Lartet and his English colleague, Christy. In both these men there was a carefulness in making researches and a sobriety in stating results which converted many of those who had been repelled by the enthusiasm of Boucher de Perthes. The two colleagues found in the stony deposits made by the water dropping from the roof of the cave at Eyzies the bones of numerous animals extinct or departed to arctic regions—one of these a vertebra of a reindeer with a flint lance-head still fast in it, and with these were found evidences of fire.

Discoveries like these were thoroughly convincing; yet there still remained here and there gainsayers in the supposed interest of Scripture, and these, in spite of the convincing array of facts, insisted that in some way, by some combination of circumstances, these bones of extinct animals of vastly remote periods might have been brought into connection with all these human bones and implements of human make in all these different places, refusing to admit that these ancient relics of men and animals were of the same period. Such gainsayers virtually adopted the reasoning of quaint old Persons, who, having maintained that God created the world "about five thousand sixe hundred and odde yeares agoe," added, "And if they aske what God was doing before this short number of yeares, we answere with St. Augustine replying to such curious questioners, that He was framing Hell for them." But a new class of discoveries came to silence this opposition. At La Madeleine in France, at the Kessler cave in Switzerland, and at various other places, were found rude but striking carvings and engravings on bone and stone representing sundry specimens of those long-vanished species; and these specimens, or casts of them, were soon to be seen in all the principal museums. They showed the hairy mammoth, the cave bear, and various other animals of the Quaternary period, carved rudely but vigorously by contemporary men; and, to complete the significance of these discoveries, travellers returning from the icy regions of North America brought similar carvings of animals now existing in those regions, made by the Eskimos during their long arctic winters to-day.*

As a result of these discoveries and others like them, showing that man was not only contemporary with long-extinct animals of past geological epochs, but that he had already developed into a stage of culture above pure savagery, the tide of thought began to turn. Especially was this seen in 1863, when Lyell published the first edition of his *Geological Evidence of the Antiquity of Man*; and the fact that he had so long opposed the new ideas gave force to the clear and conclusive argument which led him to renounce his early scientific beliefs.

Research among the evidences of man's existence in the early Quaternary, and possibly in the Tertiary period, was now pressed forward along the whole line. In 1864 Gabriel Mortillet founded his review devoted to this subject; and in 1865 the first of a series of scientific congresses devoted to such researches was held in Italy. These investigations went on vigorously in all parts of France and spread rapidly

---

* For the explorations in Belgium, see Dupont, *Le Temps Préhistorique en Belgique*. For the discoveries by McEnery and Godwin Austin, see Lubbock, *Prehistoric Times*, London, 1869, chap. x; also Cartailhac, Joly, and others above cited. For Boucher de Perthes, see his *Antiquités Celtiques et Antédiluviennes*, Paris, 1847–'64, vol. iii, pp. 526 *et seq*. For sundry extravagances of Boucher de Perthes, see Reinach, *Description raisonnée du Musée de St.-Germain-en-Laye*, Paris, 1889, vol. i, pp. 16 *et seq*. For the mixture of sound and absurd results in Boucher's work, see Cartailhac as above, p. 19. Boucher had published in 1838 a work entitled *De la Création*, but it seems to have dropped dead from the press, For the attempts of Scheuchzer to reconcile geology and Genesis by means of the *Homo diluvii testis*, and similar "diluvian fossils," see the chapter on *Geology* in this series. The original specimens of those prehistoric engravings upon bone and stone may be best seen at the Archæological Museum of St.-Germain and the British Museum. For engravings of some of the most recent, see especially Dawkins's *Early Man in Britain*, chap. vii, and the *Description du Musée de St.-Germain*. As to the Kessler etchings and their antiquity, see D. G. Brinton, in *Science*, August 12, 1892. For comparison of this prehistoric work with that produced to-day by the Eskimos and others, see Lubbock, *Prehistoric Times*, chapters x and xiv. For very striking exhibitions of this same artistic gift in a higher field to-day by descendants of the barbarian tribes of northern America, see the very remarkable illustrations in Rink, *Danish Greenland*, London, 1877, especially those in chap. xiv.

to other countries. The explorations which Dupont began in 1864, in the caves of Belgium, gave to the museum at Brussels eighty thousand flint implements, forty thousand bones of animals of the Quaternary period, and a number of human skulls and bones found mingled with these remains. From Germany, Italy, Spain, America, India, and Egypt similar results were reported.

Especially noteworthy were the further explorations of the caves and drift throughout the British Islands. The discovery by Colonel Wood, in 1861, of flint tools in the same strata with bones of the earlier forms of the rhinoceros, was but typical of many. A thorough examination of the caverns of Brixham and Torquay, by Pengelly and others, made it still more evident that man had existed in the early Quaternary period. The existence of a period before the Glacial epoch or between different glacial epochs in England, when the Englishman was a savage, using rude stone tools, was then fully ascertained, and, what was more significant, there were clearly shown a gradation and evolution even in the history of that period. It was found that this ancient Stone epoch showed progress and development. In the upper layers of the caves, with remains of the reindeer, who, although he has migrated from these regions, still exists in more northern climates, were found stone implements revealing some little advance in civilization; next below these, sealed up in the stalagmite, came, as a rule, another layer, in which the remains of reindeer were rare and those of the mammoth more frequent, the implements found in this stratum being less skilfully made than those in the upper and more recent layers; and, finally, in the lowest levels, near the floors of these ancient caverns, with remains of the cave bear and others of the most ancient extinct animals, were found stone implements evidently of a yet ruder and earlier stage of human progress. No fairly unprejudiced man can visit the cave and museum at Torquay without being convinced that there were a gradation and an evolution in these beginnings of human civilization. The evidence is complete; the masses of breccia taken from the cave, with the various soils, implements, and bones carefully kept in place, put this progress beyond a doubt.

All this indicated a great antiquity for the human race, but in it lay the germs of still another great truth, even more important and more serious in its consequences to the older theologic view, which will be discussed in the following chapter.

But new evidences came in, showing a yet greater antiquity of man. Remains of animals were found in connection with human remains, which showed not only that man was living in times more remote than the earlier of the new investigators had dared dream, but that some of these early periods of his existence must have been of immense length, embracing climatic changes betokening different geological periods; for with remains of fire and human implements and human bones were found not only bones of the hairy mammoth and cave bear, woolly rhinoceros, and reindeer, which could only have been deposited there in a time of arctic cold, but bones of the hyena, hippopotamus, sabre-toothed tiger, and the like, which could only have been deposited when there was in these regions a torrid climate. The conjunction of these remains clearly showed that man had lived in England early enough and long enough to pass through times when there was arctic cold and times when there was torrid heat; times when great glaciers stretched far down into England and indeed into the continent, and times when England had a land connection with the European continent, and the European continent with Africa, allowing tropical animals to migrate freely from Africa to the middle regions of England.

The question of the origin of man at a period vastly earlier than the sacred chronologists permitted was thus absolutely settled, but among the questions regarding the existence of man at a period yet more remote, the Drift period, there was one which for a time seemed to give the champions of science some difficulty. The orthodox leaders in the time of Boucher de Perthes, and for a considerable time afterward, had a weapon of which they made vigorous use: the statement that no human bones had yet been discovered in the drift. The supporters of science naturally answered that few if any other bones as small as those of man had been found, and that this fact was an additional proof of the great length of the period since man had lived with the extinct animals; for, since specimens of human workmanship proved man's existence as fully as remains of his bones could do, the absence or even rarity of human and other small bones simply indicated the long periods of time required for dissolving them away.

Yet Boucher, inspired by the genius he had already shown, and filled with the spirit of prophecy, declared that human bones would yet be found in the midst of the flint implements, and in 1863 he claimed that this prophecy had been fulfilled by the discovery at Moulin Quignon of a portion of a human jaw deep in the early Quaternary deposits. But his triumph was short-lived: the opposition ridiculed his discovery; they showed that he had offered a premium to his workmen for the discovery of human remains, and they naturally drew the inference that some tricky labourer had deceived him. The result of this was that the men of science felt obliged to acknowledge that the Moulin Quignon discovery was not proven.

But ere long human bones were found in the deposits of the early Quaternary period, or indeed of an earlier period, in various other parts of the world, and the question regarding the Moulin Quignon relic was of little importance.

We have seen that researches regarding the existence of prehistoric man in England and on the Continent were at first mainly made in the caverns; but the existence of man in the earliest Quaternary period was confirmed on both sides of the English Channel, in a way even more striking, by the close examination of the drift and early gravel deposits. The results arrived at by Boucher de Perthes were amply confirmed in England. Rude stone implements were found in terraces a hundred feet and more above the levels at which various rivers of Great Britain now flow, and under circumstances which show that, at the time when they were deposited, the rivers of Great Britain in many cases were entirely different from those of the present period, and formed parts of the river system of the European continent. Researches in the high terraces above the Thames and the Ouse, as well as at other points in Great Britain, placed beyond a doubt the fact that man existed on the British

Islands at a time when they were connected by solid land with the Continent, and made it clear that, within the period of the existence of man in northern Europe, a large portion of the British Islands had been sunk to depths between fifteen hundred and twenty-five hundred feet beneath the Northern Ocean,—had risen again from the water,—had formed part of the continent of Europe, and had been in unbroken connection with Africa, so that elephants, bears, tigers, lions, the rhinoceros and hippopotamus, of species now mainly extinct, had left their bones in the same deposits with human implements as far north as Yorkshire. Moreover, connected with this fact came in the new conviction, forced upon geologists by the more careful examination of the earth and its changes, that such elevations and depressions of Great Britain and other parts of the world were not necessarily the results of sudden cataclysms, but generally of slow processes extending through vast cycles of years—processes such as are now known to be going on in various parts of the world. Thus it was that the six or seven thousand years allowed by the most liberal theologians of former times were seen more and more clearly to be but a mere nothing in the long succession of ages since the appearance of man.

Confirmation of these results was received from various other parts of the world. In Africa came the discovery of flint implements deep in the hard gravel of the Nile Valley at Luxor and on the high hills behind Esneh. In America the discoveries at Trenton, N. J., and at various places in Delaware, Ohio, Minnesota, and elsewhere, along the southern edge of the drift of the Glacial epochs, clinched the new scientific truth yet more firmly; and the statement made by an eminent American authority is, that "man was on this continent when the climate and ice of Greenland extended to the mouth of New York harbour." The discoveries of prehistoric remains on the Pacific coast, and especially in British Columbia, finished completely the last chance at a reasonable contention by the adherents of the older view. As to these investigations on the Pacific slope of the United States, the discoveries of Whitney and others in California had been so made and announced that the judgment of scientific men regarding them was suspended until the visit of perhaps the greatest living authority in his department, Alfred Russel Wallace, in 1887. He confirmed the view of Prof. Whitney and others with the statement that "both the actual remains and works of man found deep under the lava-flows of Pliocene age show that he existed in the New World at least as early as in the Old." To this may be added the discoveries in British Columbia, which prove that, since man existed in these regions, "valleys have been filled up by drift from the waste of mountains to a depth in some cases of fifteen hundred feet; this covered by a succession of tuffs, ashes, and lava-streams from volcanoes long since extinct, and finally cut down by the present rivers through beds of solid basalt, and through this accumulation of lavas and gravels." The immense antiquity of the human remains in the gravels of the Pacific coast is summed up by a most eminent English authority and declared to be proved, "first, by the present river systems being of subsequent date, sometimes cutting through them and their superincumbent lava-cap to a depth of two thousand feet; secondly, by the great denudation that has taken place since they were deposited, for they sometimes lie on the summits of mountains six thousand feet high; thirdly, by the fact that the Sierra Nevada has been partly elevated since their formation." *

* For the general subject of investigations in British prehistoric remains, see especially Boyd Dawkins, *Early Man in Britain and his Place in the Tertiary Period*, London, 1880. For Boucher de Perthes's account of his discovery of the human jaw at Moulin Quignon, see his *Antiquités Celtiques et Antédiluviennes*, vol. iii, p. 542 *et seq.*, Appendix. For an excellent account of special investigations in the high terraces above the Thames, see J. Allen Brown, F. G. S., *Palæolithic Man in Northwest Middlesex*, London, 1887. For discoveries in America, and the citation regarding them, see Wright, *The Ice Age in North America*, New York, 1889, chap. xxi. Very remarkable examples of these specimens from the drift at Trenton may be seen in Prof. Abbott's collections at the University of Pennsylvania. For an admirable statement, see Prof. Henry W. Haynes, in Wright, as above. For proofs of the vast antiquity of man upon the Pacific coast, cited in the text, see Skertchley, F. G. S., in the *Journal of the Anthropological Institute* for 1887, p. 336; see also Wallace, *Darwinism*, London, 1890, chap. xv; and for a summary, as cited, Laing, *Problems of the Future*, London, 1889. For a striking summary of the evidence that man lived before the last submergence of Britain, see Brown, *Palæolithic Man in Northwest Middlesex*, as above cited. For proofs that man existed in a period when the streams were flowing hundreds of feet above their present level, see ibid., p. 33. As to the evidence of the action of the sea and of glacial ac-

As an important supplement to these discoveries of ancient implements came sundry comparisons made by eminent physiologists between human skulls and bones found in different places and under circumstances showing vast antiquity.

Human bones had been found under such circumstances as early as 1835 at Cannstadt near Stuttgart, and in 1856 in the Neanderthal near Düsseldorf; but in more recent searches they had been discovered in a multitude of places, especially in Germany, France, Belgium, England, the Caucasus, Africa, and North and South America. Comparison of these bones showed that even in that remote Quaternary period there were great differences of race, and here again came in an argument for the yet earlier existence of man on the earth; for long previous periods must have been required

tion in the Welsh bone caves after the remains of extinct animals and weapons of human workmanship had been deposited, see ibid., p. 198. For a good statement of the slowness of the submergence and emergence of Great Britain, with an illustration from the rising of the shore of Finland, see ibid., pp. 47, 48. As to the flint implements of Palæolithic man in the high terraced gravels throughout the Thames Valley, associated with bones of the mammoth, woolly rhinoceros, etc., see Brown, p. 31 For still more conclusive proofs that man inhabited North Wales before the last submergence of the greater part of the British Islands to a depth of twelve hundred to fourteen hundred feet, see ibid., pp. 199, 200. For maps showing the connection of the British river system with that of the Continent, see Boyd Dawkins, *Early Man in Britain*, London, 1880, pp. 18, 41, 73; also Lyell, *Antiquity of Man*, chap. xiv. As to the long continuance of the early Stone period, see James Geikie, *The Great Ice Age*, New York, 1888, p. 402. As to the impossibility of the animals of arctic and torrid regions living together or visiting the same place at different times in the same year, see Geikie, as above, pp. 421 *et seq.*; and for a conclusive argument that the animals of the period assigned lived in England not since, but before, the Glacial period, or in the interglacial period, see ibid., p. 459. For a very candid statement by perhaps the foremost leader of the theological rear-guard, admitting the insuperable difficulties presented by the Old Testament chronology as regards the Creation and the Deluge, see the Duke of Argyll's *Primeval Man*, pp. 90–100, and especially pp. 93, 124. For a succinct statement on the general subject, see Laing, *Problems of the Future*, London, 1889, chapters v and vi. For discoveries of prehistoric implements in India, see notes by Bruce Foote, F. G. S., in the *British Journal of the Anthropological Institute* for 1886 and 1887. For similar discoveries in South Africa, see Gooch, in *Journal of the Anthropological Institute of Great Britain and Ireland*, vol. xi, pp. 124 *et seq.* For proofs of the existence of Palæolithic man in Egypt, see Mook, Haynes, Pitt-Rivers, Flinders-Petrie, and others, cited at length in the next chapter. For the corroborative and concurrent testimony of ethnology, philology, and history to the vast antiquity of man, see Tylor, *Anthropology*, chap. i.

to develop such racial differences. Considerations of this kind gave a new impulse to the belief that man's existence might even date back into the Tertiary period. The evidence for this earlier origin of man was ably summed up, not only by its brilliant advocate, Mortillet, but by a former opponent, one of the most conservative of modern anthropologists, Quatrefages; and the conclusion arrived at by both was, that man did really exist in the Tertiary period. The acceptance of this conclusion was also seen in the more recent work of Alfred Russel Wallace, who, though very cautious and conservative, placed the origin of man not only in the Tertiary period, but in an earlier stage of it than most had dared assign—even in the Miocene.

The first thing raising a strong presumption, if not giving proof, that man existed in the Tertiary, was the fact that from all explored parts of the world came in more and more evidence that in the earlier Quaternary man existed in different, strongly marked races and in great numbers. From all regions which geologists had explored, even from those the most distant and different from each other, came this same evidence—from northern Europe to southern Africa; from France to China; from New Jersey to British Columbia; from British Columbia to Peru. The development of man in such numbers and in so many different regions, with such differences of race and at so early a period, must have required a long previous time.

This argument was strengthened by discoveries of bones bearing marks apparently made by cutting instruments, in the Tertiary formations of France and Italy, and by the discoveries of what were claimed to be flint implements by the Abbé Bourgeois in France, and of implements and human bones by Prof. Capellini in Italy.

On the other hand, some of the more cautious men of science are still content to say that the existence of man in the Tertiary period is not yet proven. As to his existence throughout the Quaternary epoch, no new proofs are needed; even so determined a supporter of the theological side as the Duke of Argyll has been forced to yield to the evidence.

Of attempts to make an exact chronological statement throwing light on the length of the various prehistoric periods, the most notable have been those by M. Morlot, on the accumulated strata of the Lake of Geneva; by Gilliéron, on the silt of Lake Neufchâtel; by Horner, in the delta deposits of Egypt; and by Riddle, in the delta of the Mississippi. But while these have failed to give anything like an exact result, all these investigations together point to the central truth, so amply established, of the vast antiquity of man, and the utter inadequacy of the chronology given in our sacred books. The period of man's past life upon our planet, which has been fixed by the universal Church, "always, everywhere, and by all," is thus perfectly proved to be insignificant compared with those vast geological epochs during which man is now known to have existed.*

---

* As to the evidence of man in the Tertiary period, see works already cited, especially Quatrefages, Cartailhac, and Mortillet. For an admirable summary, see Laing, *Human Origins*, chap. viii. See also, for a summing up of the evidence in favour of man in the Tertiary period, Quatrefages, *Histoire Générale des Races Humaines*, in the *Bibliothèque Ethnologique*, Paris, 1887, chap. iv. As to the earlier view, see Vogt, *Lectures on Man*, London, 1864, lecture xi. For a thorough and convincing refutation of Sir J. W. Dawson's attempt to make the old and new Stone periods coincide, see H. W. Haynes, in chap. vi of the *History of America*, edited by Justin Winsor. For development of various important points in the relation of anthropology to the human occupancy of our planet, see Topinard, *Anthropology*, London, 1890, chap. ix.

CHAPTER VIII.

THE "FALL OF MAN" AND ANTHROPOLOGY.

IN the previous chapters we have seen how science, especially within the eighteenth and nineteenth centuries, has thoroughly changed the intelligent thought of the world in regard to the antiquity of man upon our planet; and how the fabric built upon the chronological indications in our sacred books—first, by the early fathers of the Church, afterward by the mediæval doctors, and finally by the reformers and modern orthodox chronologists—has virtually disappeared before an entirely different view forced upon us, especially by Egyptian and Assyrian studies, as well as by geology and archæology.

In this chapter I purpose to present some outlines of the work of Anthropology, especially as assisted by Ethnology, in showing what the evolution of human civilization has been.

Here, too, the change from the old theological view based upon the letter of our sacred books to the modern scientific view based upon evidence absolutely irrefragable is complete. Here, too, we are at the beginning of a vast change in the basis and modes of thought upon man—a change even more striking than that accomplished by Copernicus and Galileo, when they substituted for a universe in which sun and planets revolved about the earth a universe in which the earth is but the merest grain or atom revolving with other worlds, larger and smaller, about the sun; and all these forming but one among innumerable systems.

Ever since the beginning of man's effective thinking upon the great problems around him, two antagonistic views have existed regarding the life of the human race upon earth.

The first of these is the belief that man was created "in the beginning" a perfect being, endowed with the highest moral and intellectual powers, but that there came a "fall," and, as its result, the entrance into the world of evil, toil, sorrow, and death.

Nothing could be more natural than such an explanation of the existence of evil, in times when men saw everywhere miracle and nowhere law. It is, under such circumstances, by far the most easy of explanations, for it is in accordance with the appearances of things: men adopted it just as naturally as they adopted the theory that the Almighty hangs up the stars as lights in the solid firmament above the earth, or hides the sun behind a mountain at night, or wheels the planets around the earth, or flings comets as "signs and wonders" to scare a wicked world, or allows evil spirits to control thunder, lightning, and storm, and to cause diseases of body and mind, or opens the "windows of heaven" to let down "the waters that be above the heavens," and thus to give rain upon the earth.

A belief, then, in a primeval period of innocence and perfection—moral, intellectual, and physical—from which men for some fault fell, is perfectly in accordance with what we should expect.

Among the earliest known records of our race we find this view taking shape in the Chaldean legends of war between the gods, and of a fall of man; both of which seemed necessary to explain the existence of evil.

In Greek mythology perhaps the best-known statement was made by Hesiod: to him it was revealed, regarding the men of the most ancient times, that they were at first "a golden race," that "as gods they were wont to live, with a life void of care, without labour and trouble; nor was wretched old age at all impending; but ever did they delight themselves out of the reach of all ills, and they died as if overcome by sleep; all blessings were theirs: of its own will the fruitful field would bear them fruit, much and ample, and they gladly used to reap the labours of their hands in quietness along with many good things, being rich in flocks and true to the blessed gods." But there came a "fall," caused by human curiosity. Pandora, the first woman created, received a vase which, by divine command, was to remain closed; but she was tempted to open it, and troubles, sorrow, and disease escaped into the world, hope alone remaining.

So, too, in Roman mythological poetry the well-known picture by Ovid is but one among the many exhibitions of this same belief in a primeval golden age—a Saturnian cycle; one of the constantly recurring attempts, so universal and so natural in the early history of man, to account for the existence of evil, care, and toil on earth by explanatory myths and legends.

This view, growing out of the myths, legends, and theologies of earlier peoples, we also find embodied in the sacred tradition of the Jews, and especially in one of the documents which form the impressive poem beginning the books attributed to Moses. As to the Christian Church, no word of its Blessed Founder indicates that it was committed by him to this theory, or that he even thought it worthy of his attention. How, like so many other dogmas never dreamed of by Jesus of Nazareth and those who knew him best, it was developed, it does not lie within the province of this chapter to point out; nor is it worth our while to dwell upon its evolution in the early Church, in the Middle Ages, at the Reformation, and in various branches of the Protestant Church: suffice it that, though among English-speaking nations by far the most important influence in its favour has come from Milton's inspiration rather than from that of older sacred books, no doctrine has been more universally accepted, "always, everywhere, and by all," from the earliest fathers of the Church down to the present hour.

On the other hand appeared at an early period the opposite view—that mankind, instead of having fallen from a high intellectual, moral, and religious condition, has slowly risen from low and brutal beginnings. In Greece, among the philosophers contemporary with Socrates, we find Critias depicting a rise of man, from a time when he was beastlike and lawless, through a period when laws were developed, to a time when morality received enforcement from religion; but among all the statements of this theory the most noteworthy is that given by Lucretius in his great poem on *The Nature of Things.* Despite its errors, it remains among the most remarkable examples of prophetic insight in the history of our race. The inspiration of Lucretius gave him almost miraculous glimpses of truth; his view of the development of civilization from the rudest beginnings to the height of its achievements is a wonderful growth, rooted in observation and thought, branching forth into a multitude of striking facts and fancies; and among these is the statement regarding the sequence of inventions:

> "Man's earliest arms were fingers, teeth, and nails,
> And stones and fragments from the branching woods;
> Then copper next; and last, as latest traced,
> The tyrant, iron."

Thus did the poet prophesy one of the most fruitful achievements of modern science: the discovery of that series of epochs which has been so carefully studied in our century.

Very striking, also, is the statement of Horace, though his idea is evidently derived from Lucretius. He dwells upon man's first condition on earth as low and bestial, and pictures him lurking in caves, progressing from the use of his fists and nails, first to clubs, then to arms which he had learned to forge, and, finally, to the invention of the names of things, to literature, and to laws.*

During the mediæval ages of faith this view was almost

---

* For the passage in Hesiod, as given, see the *Works and Days*, lines 109–120, in Banks's translation. As to Horace, see the *Satires*, i, 3, 99. As to the relation of the poetic account of the Fall in Genesis to Chaldean myths, see Smith, *Chaldean Account of Genesis*, pp. 13, 17. For a very instructive separation of the Jehovistic and Elohistic parts of Genesis, with the account of the "Fall" as given in the former, see Lenormant, *La Génèse*, Paris, 1883, pp. 166–168; also Bacon, *Genesis of Genesis*. Of the lines of Lucretius—

> "Arma antiqua, manus, ungues, dentesque fuerunt,
> Et lapides, et item sylvarum fragmina rami,
> Posterius ferri vis est, aerisque reperta,
> Sed prior aeris erat, quam ferri cognitus usus"—

the translation given is that of Good. For a more exact prose translation, see Munro's Lucretius, fourth edition, which is much more careful, at least in the proof-reading, than the first edition. As regards Lucretius's prophetic insight into some of the greatest conclusions of modern science, see Munro's translation and notes, fourth edition, book v, notes ii, p. 335. On the relation of several passages in Horace to the ideas of Lucretius, see Munro as above. For the passage from Luther, see the *Table Talk*, Hazlitt's translation, p. 242.

entirely obscured, and at the Reformation it seemed likely to remain so. Typical of the simplicity of belief in "the Fall" cherished among the Reformers is Luther's declaration regarding Adam and Eve. He tells us, "they entered into the garden about noon, and having a desire to eat, she took the apple; then came the fall—according to our account at about two o'clock." But in the revival of learning the old eclipsed truth reappeared, and in the first part of the seventeenth century we find that, among the crimes for which Vanini was sentenced at Toulouse to have his tongue torn out and to be burned alive, was his belief that there is a gradation extending upward from the lowest to the highest form of created beings.

Yet, in the same century, the writings of Bodin, Bacon, Descartes, and Pascal were evidently undermining the old idea of "the Fall." Bodin especially, brilliant as were his services to orthodoxy, argued lucidly against the doctrine of general human deterioration.

Early in the eighteenth century Vico presented the philosophy of history as an upward movement of man out of animalism and barbarism. This idea took firm hold upon human thought, and in the following centuries such men as Lessing and Turgot gave new force to it.

The investigations of the last forty years have shown that Lucretius and Horace were inspired prophets: what they saw by the exercise of reason illumined by poetic genius, has been now thoroughly based upon facts carefully ascertained and arranged—until Thomsen and Nilsson, the northern archæologists, have brought these prophecies to evident fulfilment, by presenting a scientific classification dividing the age of prehistoric man in various parts of the world between an old stone period, a new stone period, a period of beaten copper, a period of bronze, and a period of iron, and arraying vast masses of facts from all parts of the world, fitting thoroughly into each other, strengthening each other, and showing beyond a doubt that, instead of a *fall*, there has been a *rise* of man, from the earliest indications in the Quaternary, or even, possibly, in the Tertiary period.*

---

* For Vanini, see Topinard, *Éléments d'Anthropologie*, p. 52. For a brief and careful summary of the agency of Eccard in Germany, Goguet in France, Hoare in England, and others in various parts of Europe, as regards this development of the

The first blow at the fully developed doctrine of "the Fall" came, as we have seen, from geology. According to that doctrine, as held quite generally from its beginnings among the fathers and doctors of the primitive Church down to its culmination in the minds of great Protestants like John Wesley, the statement in our sacred books that "death entered the world by sin" was taken as a historic fact, necessitating the conclusion that, before the serpent persuaded Eve to eat of the forbidden fruit, death on our planet was unknown. Naturally, when geology revealed, in the strata of a period long before the coming of man on earth, a vast multitude of carnivorous tribes fitted to destroy their fellow-creatures on land and sea, and within the fossilized skeletons of many of these the partially digested remains of animals, this doctrine was too heavy to be carried, and it was quietly dropped.

But about the middle of the nineteenth century the doctrine of the rise of man as opposed to the doctrine of his "fall" received a great accession of strength from a source most unexpected. As we saw in the last chapter, the facts proving the great antiquity of man foreshadowed a new and even more remarkable idea regarding him. We saw, it is true, that the opponents of Boucher de Perthes, while they could not deny his discovery of human implements in the drift, were successful in securing a verdict of "Not proven" as regarded his discovery of human bones; but their triumph was short-lived. Many previous discoveries, little thought of up to that time, began to be studied, and others were added which resulted not merely in confirming the truth regarding the antiquity of man, but in establishing another doctrine which the opponents of science regarded with vastly greater dislike—the doctrine that man has not fallen from an

---

scientific view during the eighteenth century, see Mortillet, *Le Préhistorique*, Paris, 1885, chap. i. For the agency of Bodin, Bacon, Descartes, and Pascal, see Flint, *Philosophy of History*, introduction, pp. 28 *et seq.* For a shorter summary, see Lubbock, *Prehistoric Man.* For the statements by the northern archæologists, see Nilsson, Worsaae, and the other main works cited in this article. For a generous statement regarding the great services of the Danish archæologists in this field, see Quatrefages, introduction to Cartailhac, *Les Ages Préhistoriques de l'Espagne et du Portugal.*

original high estate in which he was created about six thousand years ago, but that, from a period vastly earlier than any warranted by the sacred chronologists, he has been, in spite of lapses and deteriorations, rising.

A brief review of this new growth of truth may be useful. As early as 1835 Prof. Jaeger had brought out from a quantity of Quaternary remains dug up long before at Cannstadt, near Stuttgart, a portion of a human skull, apparently of very low type. A battle raged about it for a time, but this finally subsided, owing to uncertainties arising from the circumstances of the discovery.

In 1856, in the Neanderthal, near Düsseldorf, among Quaternary remains gathered on the floor of a grotto, another skull was found bearing the same evidence of a low human type. As in the case of the Cannstadt skull, this again was fiercely debated, and finally the questions regarding it were allowed to remain in suspense. But new discoveries were made: at Eguisheim, at Brux, at Spy, and elsewhere, human skulls were found of a similarly low type; and, while each of the earlier discoveries was open to debate, and either, had no other been discovered, might have been considered an abnormal specimen, the combination of all these showed conclusively that not only had a race of men existed at that remote period, but that it was of a type as low as the lowest, perhaps below the lowest, now known.

Research was now redoubled, and, as a result, human skulls and complete skeletons of various types began to be discovered in the ancient deposits of many other parts of the world, and especially in France, Belgium, Germany, the Caucasus, Africa, and North and South America.

But soon began to emerge from all these discoveries a fact of enormous importance. The skulls and bones found at Cro Magnon, Solutré, Furfooz, Grenelle, and elsewhere, were compared, and it was thus made certain that various races had already appeared and lived in various grades of civilization, even in those exceedingly remote epochs; that even then there were various strata of humanity ranging from races of a very low to those of a very high type; and that upon any theory—certainly upon the theory of the origin of mankind from a single pair—two things were evident: first, that long, slow processes during vast periods of time must have been required for the differentiation of these races, and for the evolution of man up to the point where the better specimens show him, certainly in the early Quaternary and perhaps in the Tertiary period; and, secondly, that there had been from the first appearance of man, of which we have any traces, an *upward* tendency.*

This second conclusion, the upward tendency of man from low beginnings, was made more and more clear by bringing into relations with these remains of human bodies and of extinct animals the remains of human handiwork. As stated in the last chapter, the river drift and bone caves in Great Britain, France, and other parts of the world, revealed a progression, even in the various divisions of the earliest Stone period; for, beginning at the very lowest strata of these remains, on the floors of the caverns, associated mainly with the bones of extinct animals, such as the cave bear, the hairy elephant, and the like, were the rudest implements; then, in strata above these, sealed in the stalagmite of the cavern floors, lying with the bones of animals extinct but more recent, stone implements were found, still rude, but, as a rule, of an improved type; and, finally, in a still higher stratum, associated with bones of animals like the reindeer and bison, which, though not extinct, have departed to other climates, were rude stone implements, on the whole of a still better workmanship. Such was the foreshadowing, even at that early rude Stone period, of the proofs that the tendency

* For Wesley's statement of the amazing consequences of the entrance of death into the world by sin, see citations from his sermon on *The Fall of Man* in the chapter on Geology. For Boucher de Perthes, see his *Life* by Ledieu, especially chapters v and xix; also letters in the appendix; also *Les Antiquités Celtiques et Antédiluviennes*, as cited in previous chapters of this work. For an account of the Neanderthal man and other remains mentioned, see Quatrefages, *Human Species*, chap. xxvi; also Mortillet, *Le Préhistorique*, Paris, 1885, pp. 232 *et seq.*; also other writers cited in this chapter. For the other discoveries mentioned, see the same sources. For an engraving of the skull and the restored human face of the Neanderthal man, see Reinach, *Antiquités Nationales*, etc., vol. i, p. 138. For the vast regions over which that early race spread, see Quatrefages as above, p. 307. See also the same author, *Histoire Générale des Races Humaines*, in the *Bibliothèque Ethnologique*, Paris, 1887, p. 4. In the vast mass of literature bearing on this sub ject, see Quatrefages, Dupont, Reinach, Joly, Mortillet, Tylor, and Lubbock, in works cited through these chapters.

of man has been from his earliest epoch and in all parts of the world, as a rule, upward.

But this rule was to be much further exemplified. About 1850, while the French and English geologists were working more especially among the relics of the drift and cave periods, noted archæologists of the North—Forchammer, Steenstrup, and Worsaae—were devoting themselves to the investigation of certain remains upon the Danish Peninsula. These remains were of two kinds: first, there were vast shell-heaps or accumulations of shells and other refuse cast aside by rude tribes which at some unknown age in the past lived on the shores of the Baltic, principally on shellfish. That these shell-heaps were very ancient was evident: the shells of oysters and the like found in them were far larger than any now found on those coasts; their size, so far from being like that of the corresponding varieties which now exist in the brackish waters of the Baltic, was in every case like that of those varieties which only thrive in the waters of the open salt sea. Here was a clear indication that at the time when man formed these shell-heaps those coasts were in far more direct communication with the salt sea than at present, and that sufficient time must have elapsed since that period to have wrought enormous changes in sea and land throughout those regions.

Scattered through these heaps were found indications of a grade of civilization when man still used implements of stone, but implements and weapons which, though still rude, showed a progress from those of the drift and early cave period, some of them being of polished stone.

With these were other evidences that civilization had progressed. With implements rude enough to have survived from early periods, other implements never known in the drift and bone caves began to appear, and, though there were few if any bones of other domestic animals, the remains of dogs were found; everything showed that there had been a progress in civilization between the former Stone epoch and this.

The second series of discoveries in Scandinavia was made in the peat-beds: these were generally formed in hollows or bowls varying in depth from ten to thirty feet, and a section of them, like a section of the deposits in the bone caverns, showed a gradual evolution of human culture. The lower strata in these great bowls were found to be made up chiefly of mosses and various plants matted together with the trunks of fallen trees, sometimes of very large diameter; and the botanical examination of the lowest layer of these trees and plants in the various bowls revealed a most important fact: for this layer, the first in point of time, was always of the Scotch fir—which now grows nowhere in the Danish islands, and can not be made to grow anywhere in them—and of plants which are now extinct in these regions, but have retreated within the arctic circle. Coming up from the bottom of these great bowls there was found above the first layer a second, in which were matted together masses of oak trees of different varieties; these, too, were relics of a bygone epoch, since the oak has almost entirely disappeared from Denmark. Above these came a third stratum made up of fallen beech trees; and the beech is now, and has been since the beginning of recorded history, the most common tree of the Danish Peninsula.

Now came a second fact of the utmost importance as connected with the first. Scattered, as a rule, through the lower of these deposits, that of the extinct fir trees and plants, were found implements and weapons of smooth stone; in the layer of oak trees were found implements of bronze; and among the layer of beeches were found implements and weapons of iron.

The general result of these investigations in these two sources, the shell mounds and the peat deposits, was the same: the first civilization evidenced in them was marked by the use of stone implements more or less smooth, showing a progress from the earlier rude Stone period made known by the bone caves; then came a later progress to a higher civilization, marked by the use of bronze implements; and, finally, a still higher development when iron began to be used.

The labours of the Danish archæologists have resulted in the formation of a great museum at Copenhagen, and on the specimens they have found, coupled with those of the drift and bone caves, is based the classification between the main periods or divisions in the evolution of the human race above referred to.

It was not merely in Scandinavian lands that these results were reached; substantially the same discoveries were made in Ireland and France, in Sardinia and Portugal, in Japan and in Brazil, in Cuba and in the United States; in fact, as a rule, in nearly every part of the world which was thoroughly examined.*

But from another quarter came a yet more striking indication of this same evolution. As far back as the year 1829 there were discovered, in the Lake of Zurich, piles and other antiquities indicating a former existence of human dwellings, standing in the water at some distance from the shore; but the usual mixture of thoughtlessness and dread of new ideas seems to have prevailed, and nothing was done until about 1853, when new discoveries of the same kind were followed up vigorously, and Rütimeyer, Keller, Troyon, and others showed not only in the Lake of Zurich, but in many other lakes in Switzerland, remains of former habitations, and, in the midst of these, great numbers of relics, exhibiting the grade of civilization which those lake-dwellers had attained.

Here, too, were accumulated proofs of the upward tendency of the human race. Implements of polished stone, bone, leather, pottery of various grades, woven cloth, bones

* For the general subject, see Mortillet, *Le Préhistorique*, p. 498, *et passim*. For examples of the rude stone implements, improving as we go from earlier to later layers in the bone caves, see Boyd Dawkins, *Early Man in Britain*, chap. vii, p. 186; also Quatrefages, *Human Species*, New York, 1879, pp. 305 *et seq*. An interesting gleam of light is thrown on the subject in De Baye, *Grottes Préhistoriques de la Marne*, pp. 31 *et seq.*; also Evans, as cited in the previous chapter. For the more recent investigations in the Danish shell-heaps, see Boyd Dawkins, *Early Man in Britain*, pp. 303, 304. For these evidences of advanced civilization in the shell-heaps, see Mortillet, p. 498. He, like Nilsson, says that only the bones of the dog were found; but compare Dawkins, p. 305. For the very full list of these discoveries, with their bearing on each other, see Mortillet, p. 499. As to those in Scandinavian countries, see Nilsson, *The Primitive Inhabitants of Scandinavia*, third edition, with Introduction by Lubbock, London, 1868; also the *Pre-History of the North*, by Worsaae, English translation, London, 1886. For shell-mounds and their contents in the Spanish Peninsula, see Cartailhac's greater work already cited. For summary of such discoveries throughout the world, see Mortillet, *Le Préhistorique*, pp. 497 *et seq*.

of several kinds of domestic animals, various sorts of grain, bread which had been preserved by charring, and a multitude of evidences of progress never found among the earlier, ruder relics of civilization, showed yet more strongly that man had arrived here at a still higher stage than his predecessor of the drift, cave, and shell-heap periods, and had gone on from better to better.

Very striking evidences of this upward tendency were found in each class of implements. As by comparing the chipped flint implements of the lower and earlier strata in the cave period with those of the later and upper strata we saw progress, so, in each of the periods of polished stone, bronze, and iron, we see, by similar comparisons, a steady progress from rude to perfected implements; and especially is this true in the remains of the various lake-dwellings, for among these can be traced out constant increase in the variety of animals domesticated, and gradual improvements in means of subsistence and in ways of living.

Incidentally, too, a fact, at first sight of small account, but on reflection exceedingly important, was revealed. The earlier bronze implements were frequently found to imitate in various minor respects implements of stone; in other words, forms were at first given to bronze implements natural in working stone, but not natural in working bronze. This showed the *direction* of the development—that it was upward from stone to bronze, not downward from bronze to stone; that it was progress rather than decline.

These investigations were supplemented by similar researches elsewhere. In many other parts of the world it was found that lake-dwellers had existed in different grades of civilization, but all within a certain range, intermediate between the cave-dwellers and the historic period. To explain this epoch of the lake-dwellers History came in with the account given by Herodotus of the lake-dwellings on Lake Prasias, which gave protection from the armies of Persia. Still more important, Comparative Ethnography showed that to-day, in various parts of the world, especially in New Guinea and West Africa, races of men are living in lake-dwellings built upon piles, and with a range of implements and weapons strikingly like many of those discovered in these ancient lake deposits of Switzerland.

In Great Britain, France, Germany, Italy, Ireland, Scotland, and other countries, remains of a different sort were also found, throwing light on this progress. The cromlechs, cranogs, mounds, and the like, though some of them indicate the work of weaker tribes pressed upon by stronger, show, as a rule, the same upward tendency.

At a very early period in the history of these discoveries, various attempts were made—nominally in the interest of religion, but really in the interest of sundry creeds and catechisms framed when men knew little or nothing of natural laws—to break the force of such evidences of the progress and development of the human race from lower to higher. Out of all the earlier efforts two may be taken as fairly typical, for they exhibit the opposition to science as developed under two different schools of theology, each working in its own way. The first of these shows great ingenuity and learning, and is presented by Mr. Southall in his book, published in 1875, entitled *The Recent Origin of the World.* In this he grapples first of all with the difficulties presented by the early date of Egyptian civilization, and the keynote of his argument is the statement made by an eminent Egyptologist, at a period before modern archæological discoveries were well understood, that "Egypt laughs the idea of a rude Stone age, a polished Stone age, a Bronze age, an Iron age, to scorn."

Mr. Southall's method was substantially that of the late excellent Mr. Gosse in geology. Mr. Gosse, as the readers of this work may remember, felt obliged, in the supposed interest of Genesis, to urge that safety to men's souls might be found in believing that, six thousand years ago, the Almighty, for some inscrutable purpose, suddenly set Niagara pouring very near the spot where it is pouring now; laid the various strata, and sprinkled the fossils through them like plums through a pudding; scratched the glacial grooves upon the rocks, and did a vast multitude of things, subtle and cunning, little and great, in all parts of the world, required to delude geologists of modern times into the conviction that all these things were the result of a steady progress through long epochs. On a similar plan, Mr. Southall proposed, at the very beginning of his book, as a final solution of the problem, the declaration that Egypt, with its high civilization in the time of Mena, with its races, classes, institutions, arrangements, language, monuments—all indicating an evolution through a vast previous history—was a sudden creation which came fully made from the hands of the Creator. To use his own words, "The Egyptians had no Stone age, and were born civilized."

There is an old story that once on a time a certain jovial King of France, making a progress through his kingdom, was received at the gates of a provincial town by the mayor's deputy, who began his speech on this wise: "May it please your Majesty, there are just thirteen reasons why His Honour the Mayor can not be present to welcome you this morning. The first of these reasons is that he is dead." On this the king graciously declared that this first reason was sufficient, and that he would not trouble the mayor's deputy for the twelve others.

So with Mr. Southall's argument: one simple result of scientific research out of many is all that it is needful to state, and this is, that in these later years we have a new and convincing evidence of the existence of prehistoric man in Egypt in his earliest, rudest beginnings; the very same evidence which we find in all other parts of the world which have been carefully examined. This evidence consists of stone implements and weapons which have been found in Egypt in such forms, at such points, and in such positions that when studied in connection with those found in all other parts of the world, from New Jersey to California, from France to India, and from England to the Andaman Islands, they force upon us the conviction that civilization in Egypt, as in all other parts of the world, was developed by the same slow process of evolution from the rudest beginnings.

It is true that men learned in Egyptology had discouraged the idea of an earlier Stone age in Egypt, and that among these were Lepsius and Brugsch; but these men were not trained in prehistoric archæology; their devotion to the study of the monuments of Egyptian civilization had evidently drawn them away from sympathy, and indeed from acquaintance, with the work of men like Boucher de Perthes, Lartet, Nilsson, Troyon, and Dawkins. But a new era was beginning. In 1867 Worsaae called attention to the prehistoric implements found on the borders of Egypt; two years later Arcelin discussed such stone implements found beneath the soil of Sakkara and Gizeh, the very focus of the earliest Egyptian civilization; in the same year Hamy and Lenormant found such implements washed out from the depths higher up the Nile at Thebes, near the tombs of the kings; and in the following year they exhibited more flint implements found at various other places. Coupled with these discoveries was the fact that Horner and Linant found a copper knife at twenty-four feet, and pottery at sixty feet, below the surface. In 1872 Dr. Reil, director of the baths at Helouan, near Cairo, discovered implements of chipped flint; and in 1877 Dr. Jukes Brown made similar discoveries in that region. In 1878 Oscar Fraas, summing up the question, showed that the stone implements were mainly such as are found in the prehistoric deposits of other countries, and that, Zittel having found them in the Libyan Desert, far from the oases, there was reason to suppose that these implements were used before the region became a desert and before Egypt was civilized. Two years later Dr. Mook, of Würzburg, published a work giving the results of his investigations, with careful drawings of the rude stone implements discovered by him in the upper Nile Valley, and it was evident that, while some of these implements differed slightly from those before known, the great mass of them were of the character so common in the prehistoric deposits of other parts of the world.

A yet more important contribution to this mass of facts was made by Prof. Henry Haynes, of Boston, who in the winter of 1877 and 1878 began a very thorough investigation of the subject, and discovered, a few miles east of Cairo, many flint implements. The significance of Haynes's discoveries was twofold: First, there were, among these, stone axes like those found in the French drift beds of St. Acheul, showing that the men who made or taught men how to make these in Egypt were passing through the same phase of savagery as that of Quaternary France; secondly, he found a workshop for making these implements, proving that these flint implements were not brought into Egypt by invaders, but were made to meet the necessities of the country. From this first field Prof. Haynes went to Helouan, north of Cairo, and there found, as Dr. Reil had done, various worked flints, some of them like those discovered by M. Rivière in the caves of southern France; thence he went up the Nile to Luxor, the site of ancient Thebes, began a thorough search in the Tertiary limestone hills, and found multitudes of chipped stone implements, some of them, indeed, of original forms, but most of forms common in other parts of the world under similar circumstances, some of the chipped stone axes corresponding closely to those found in the drift beds of northern France.

All this seemed to show conclusively that, long ages before the earliest period of Egyptian civilization of which the monuments of the first dynasties give us any trace, mankind in the Nile Valley was going through the same slow progress from the period when, standing just above the brutes, he defended himself with implements of rudely chipped stone.

But in 1881 came discoveries which settled the question entirely. In that year General Pitt-Rivers, a Fellow of the Royal Society and President of the Anthropological Institute, and J. F. Campbell, Fellow of the Royal Geographical Society of England, found implements not only in alluvial deposits, associated with the bones of the zebra, hyena, and other animals which have since retreated farther south, but, at Djebel Assas, near Thebes, they found implements of chipped flint in the hard, stratified gravel, from six and a half to ten feet below the surface; relics evidently, as Mr. Campbell says, "beyond calculation older than the oldest Egyptian temples and tombs." They certainly proved that Egyptian civilization had not issued in its completeness, and all at once, from the hand of the Creator in the time of Mena. Nor was this all. Investigators of the highest character and ability—men like Hull and Flinders Petrie—revealed geological changes in Egypt requiring enormous periods of time, and traces of man's handiwork dating from a period when the waters in the Nile Valley extended hundreds of feet above the present level. Thus was ended the contention of Mr. Southall.

Still another attack upon the new scientific conclusions came from France, when in 1883 the Abbé Hamard, Priest of the Oratory, published his *Age of Stone and Primitive Man.* He had been especially vexed at the arrangement of prehistoric implements by periods at the Paris Exposition of 1878; he bitterly complains of this as having an anti-Christian tendency, and rails at science as "the idol of the day." He attacks Mortillet, one of the leaders in French archæology, with a great display of contempt; speaks of the "venom" in books on prehistoric man generally; complains that the Church is too mild and gentle with such monstrous doctrines; bewails the concessions made to science by some eminent preachers; and foretells his own martyrdom at the hands of men of science.

Efforts like this accomplished little, and a more legitimate attempt was made to resist the conclusions of archæology by showing that knives of stone were used in obedience to a sacred ritual in Egypt for embalming, and in Judea for circumcision, and that these flint knives might have had this later origin. But the argument against the conclusions drawn from this view was triple: First, as we have seen, not only stone knives, but axes and other implements of stone similar to those of a prehistoric period in western Europe were discovered; secondly, these implements were discovered in the hard gravel drift of a period evidently far earlier than that of Mena; and, thirdly, the use of stone implements in Egyptian and Jewish sacred functions within the historic period, so far from weakening the force of the arguments for the long and slow development of Egyptian civilization from the men who used rude flint implements to the men who built and adorned the great temples of the early dynasties, is really an argument in favour of that long evolution. A study of comparative ethnology has made it clear that the sacred stone knives and implements of the Egyptian and Jewish priestly ritual were natural survivals of that previous period. For sacrificial or ritual purposes, the knife of stone was considered more sacred than the knife of bronze or iron, simply because it was ancient; just as to-day, in India, Brahman priests kindle the sacred fire not with matches or flint and steel, but by a process found in the earliest, lowest stages of human culture—by violently boring a pointed stick into another piece of wood until a spark comes; and just as to-day, in Europe and America, the architecture of the Middle Ages survives as a special religious form in the erection of our most recent churches, and to such an extent that thousands on thousands of us feel that we can not worship fitly unless in the midst of windows, decorations, vessels, implements, vestments, and ornaments, no longer used for other purposes, but which have survived in sundry branches of the Christian Church, and derived a special sanctity from the fact that they are of ancient origin.

Taking, then, the whole mass of testimony together, even though a plausible or very strong argument against single evidences may be made here and there, the force of its combined mass remains, and leaves both the vast antiquity of man and the evolution of civilization from its lowest to its highest forms, as proved by the prehistoric remains of Egypt and so many other countries in all parts of the world, beyond a reasonable doubt. Most important of all, the recent discoveries in Assyria have thrown a new light upon the evolution of the dogma of "the fall of man." Reverent scholars like George Smith, Sayce, Delitzsch, Jensen, Schrader, and their compeers have found in the Ninevite records the undoubted source of that form of the fall legend which was adopted by the Hebrews and by them transmitted to Christianity.*

---

* For Mr. Southall's views, see his *Recent Origin of Man*, p. 20, and elsewhere. For Mr. Gosse's views, see his *Omphalos* as cited in the chapter on Geology in this work. For a summary of the work of Arcelin, Hamy, Lenormant, Richard, Lubbock, Mook, and Haynes, see Mortillet, *Le Préhistorique*, *passim*. As to Zittel's discovery, see Oscar Fraas's *Aus dem Orient*, Stuttgart, 1878. As to the striking similarities of the stone implements found in Egypt with those found in the drift and bone caves, see Mook's monograph, Würzburg, 1880, cited in the next chapter, especially Plates IX, XI, XII. For even more striking reproductions of photographs showing this remarkable similarity between Egyptian and European chipped stone remains, see H. W. Haynes, *Palæolithic Implements in Upper Egypt*, Boston, 1881. See also Evans, *Ancient Stone Implements*, chap. i, pp. 8, 9, 44, 102, 316, 329. As to stone implements used by priests of Jehovah, priests of Baal, priests

of Moloch, priests of Odin, and Egyptian priests, as religious survivals, see Cartailhac, as above, 6 and 7; also Lartet, in De Luynes, *Expedition to the Dead Sea*; also Nilsson, *Primitive Inhabitants of Scandinavia*, pp. 96, 97; also Sayce, *Herodotus*, p. 171, note. For the discoveries by Pitt-Rivers, see the *Journal of the Anthropological Institute of Great Britain and Ireland* for 1882, vol. xi, pp. 382 *et seq.*; and for Campbell's decision regarding them, see ibid., pp. 396, 397. For facts summed up in the words, "It is most probable that Egypt at a remote period passed like many other countries through its stone period," see Hilton Price, F. S. A., F. G. S., paper in the *Journal of the Archæological Institute of Great Britain and Ireland* for 1884, p. 56. Specimens of palæolithic implements from Egypt—knives, arrowheads, spearheads, flakes, and the like, both of peculiar and ordinary forms—may be seen in various museums, but especially in that of Prof. Haynes, of Boston. Some interesting light is also thrown into the subject by the specimens obtained by General Wilson and deposited in the Smithsonian Institution at Washington. For the Abbé Hamard's attack, see his *L'Age de la Pierre et l'Homme Primitif*, Paris, 1883—especially his preface. For the stone weapon found in the high drift behind Esneh, see Flinders Petrie, *History of Egypt*, chap. i. Of these discoveries by Pitt-Rivers and others Maspero appears to know nothing.

CHAPTER IX.

THE "FALL OF MAN" AND ETHNOLOGY.

We have seen that, closely connected with the main lines of investigation in archæology and anthropology, there were other researches throwing much light on the entire subject. In a previous chapter we saw especially that Lafitau and Jussieu were among the first to collect and compare facts bearing on the natural history of man, gathered by travellers in various parts of the earth, thus laying foundations for the science of comparative ethnology. It was soon seen that ethnology had most important bearings upon the question of the material, intellectual, moral, and religious evolution of the human race; in every civilized nation, therefore, appeared scholars who began to study the characteristics of various groups of men as ascertained from travellers, and to compare the results thus gained with each other and with those obtained by archæology.

Thus, more and more clear became the evidences that the tendency of the race has been upward from low beginnings. It was found that groups of men still existed possessing characteristics of those in the early periods of development to whom the drift and caves and shell-heaps and pile-dwellings bear witness; groups of men using many of the same implements and weapons, building their houses in the same way, seeking their food by the same means, enjoying the same amusements, and going through the same general stages of culture; some being in a condition corresponding to the earlier, some to the later, of those early periods.

From all sides thus came evidence that we have still upon the earth examples of all the main stages in the development of human civilization; that from the period when man appears little above the brutes, and with little if any religion in any accepted sense of the word, these examples can be arranged in an ascending series leading to the highest planes which humanity has reached; that philosophic observers may among these examples study existing beliefs, usages, and institutions back through earlier and earlier forms, until, as a rule, the whole evolution can be easily divined if not fully seen. Moreover, the basis of the whole structure became more and more clear: the fact that "the lines of intelligence have always been what they are, and have always operated as they do now; that man has progressed from the simple to the complex, from the particular to the general."

As this evidence from ethnology became more and more strong, its significance to theology aroused attention, and naturally most determined efforts were made to break its force. On the Continent the two great champions of the Church in this field were De Maistre and De Bonald; but the two attempts which may be especially recalled as the most influential among English-speaking peoples were those of Whately, Archbishop of Dublin, and the Duke of Argyll.

First in the combat against these new deductions of science was Whately. He was a strong man, whose breadth of thought and liberality in practice deserve all honour; but these very qualities drew upon him the distrust of his orthodox brethren; and, while his writings were powerful in the first half of the present century to break down many bulwarks of unreason, he seems to have been constantly in fear of losing touch with the Church, and therefore to have promptly attacked some scientific reasonings, which, had he been a layman, not holding a brief for the Church, he would probably have studied with more care and less prejudice. He was not slow to see the deeper significance of archæology and ethnology in their relations to the theological conception of "the Fall," and he set the battle in array against them.

His contention was, to use his own words, that "no community ever did or ever can emerge unassisted by external helps from a state of utter barbarism into anything that can be called civilization"; and that, in short, all imperfectly civilized, barbarous, and savage races are but fallen descendants of races more fully civilized. This view was urged with his usual ingenuity and vigour, but the facts proved too strong for him: they made it clear, first, that many races were without simple possessions, instruments, and arts which never, probably, could have been lost if once acquired—as, for example, pottery, the bow for shooting, various domesticated animals, spinning, the simplest principles of agriculture, household economy, and the like; and, secondly, it was shown as a simple matter of fact that various savage and barbarous tribes *had* raised themselves by a development of means which no one from outside could have taught them; as in the cultivation and improvement of various indigenous plants, such as the potato and Indian corn among the Indians of North America; in the domestication of various animals peculiar to their own regions, such as the llama among the Indians of South America; in the making of sundry fabrics out of materials and by processes not found among other nations, such as the bark cloth of the Polynesians; and in the development of weapons peculiar to sundry localities, but known in no others, such as the boomerang in Australia.

Most effective in bringing out the truth were such works as those of Sir John Lubbock and Tylor; and so conclusive were they that the arguments of Whately were given up as untenable by the other of the two great champions above referred to, and an attempt was made by him to form the diminishing number of thinking men supporting the old theological view on a new line of defence.

This second champion, the Duke of Argyll, was a man of wide knowledge and strong powers in debate, whose high moral sense was amply shown in his adhesion to the side of the American Union in the struggle against disunion and slavery, despite the overwhelming majority against him in the high aristocracy to which he belonged. As an honest man and close thinker, the duke was obliged to give up completely the theological view of the antiquity of man. The whole biblical chronology as held by the universal Church, "always, everywhere, and by all," he sacrificed, and gave all his powers in this field to support the theory of "the Fall." *Noblesse oblige*: the duke and his ancestors had been for centuries the chief pillars of the Church of Scotland, and it was too much to expect that he could break away from a tenet which forms really its "chief cornerstone."

Acknowledging the insufficiency of Archbishop Whately's argument, the duke took the ground that the lower, barbarous, savage, brutal races were the remains of civilized races which, in the struggle for existence, had been pushed and driven off to remote and inclement parts of the earth, where the conditions necessary to a continuance in their early civilization were absent; that, therefore, the descendants of primeval, civilized men degenerated and sank in the scale of culture. To use his own words, the weaker races were "driven by the stronger to the woods and rocks," so that they became "mere outcasts of the human race."

In answer to this, while it was conceded, first, that there have been examples of weaker tribes sinking in the scale of culture after escaping from the stronger into regions unfavourable to civilization, and, secondly, that many powerful nations have declined and decayed, it was shown that the men in the most remote and unfavourable regions have not always been the lowest in the scale; that men have been frequently found "among the woods and rocks" in a higher state of civilization than on the fertile plains, such examples being cited as Mexico, Peru, and even Scotland; and that, while there were many examples of special and local decline, overwhelming masses of facts point to progress as a rule.

The improbability, not to say impossibility, of many of the conclusions arrived at by the duke appeared more and more strongly as more became known of the lower tribes of mankind. It was necessary on his theory to suppose many things which our knowledge of the human race absolutely forbids us to believe: for example, it was necessary to suppose that the Australians or New Zealanders, having once possessed so simple and convenient an art as that of the potter, had lost every trace of it; and that the same tribes, having once had so simple a means of saving labour as the spindle or small stick weighted at one end for spinning, had given it up and gone back to twisting threads with the hand. In fact, it was necessary to suppose that one of the main occupations of man from "the beginning" had been the forgetting of simple methods, processes, and implements which all experience in the actual world teaches us are never entirely forgotten by peoples who have once acquired them.

Some leading arguments of the duke were overthrown by simple statements of fact. Thus, his instance of the Eskimo as pushed to the verge of habitable America, and therefore living in the lowest depths of savagery, which, even if it were true, by no means proved a general rule, was deprived of its force by the simple fact that the Eskimos are by no means the lowest race on the American continent, and that various tribes far more centrally and advantageously placed, as, for instance, those in Brazil, are really inferior to them in the scale of culture. Again, his statement that "in Africa there appear to be no traces of any time when the natives were not acquainted with the use of iron," is met by the fact that from the Nile Valley to the Cape of Good Hope we find, wherever examination has been made, the same early stone implements which in all other parts of the world precede the use of iron, some of which would not have been made had their makers possessed iron. The duke also tried to show that there were no distinctive epochs of stone, bronze, and iron, by adducing the fact that some stone implements are found even in some high civilizations. This is indeed a fact. We find some few European peasants to-day using stone mallet-heads; but this proves simply that the old stone mallet-heads have survived as implements cheap and effective.

The argument from Comparative Ethnology in support of the view that the tendency of mankind is upward has received strength from many sources. Comparative Philology shows that in the less civilized, barbarous, and savage races childish forms of speech prevail—frequent reduplications and the like, of which we have survivals in the later and even in the most highly developed languages. In various languages, too, we find relics of ancient modes of thought in the simplest words and expressions used for arithmetical calculations. Words and phrases for this purpose are frequently found to be derived from the words for hands, feet, fingers, and toes, just as clearly as in our own language some of our simplest measures of length are shown by their names to have been measures of parts of the human body, as the cubit, the foot, and the like, and therefore to date from a time when exactness was not required. To add another out of many examples, it is found to-day that various rude nations go through the simplest arithmetical processes by means of pebbles. Into our own language, through the Latin, has come a word showing that our distant progenitors reckoned in this way: the word *calculate* gives us an absolute proof of this. According to the theory of the Duke of Argyll, men ages ago used pebbles (*calculi*) in performing the simplest arithmetical calculations because we to-day "*calculate*." No reduction to absurdity could be more thorough. The simple fact must be that we "calculate" because our remote ancestors used pebbles in their arithmetic.

Comparative Literature and Folklore also show among peoples of a low culture to-day childish modes of viewing nature, and childish ways of expressing the relations of man to nature, such as clearly survive from a remote ancestry; noteworthy among these are the beliefs in witches and fairies, and multitudes of popular and poetic expressions in the most civilized nations.

So, too, Comparative Ethnography, the basis of Ethnology, shows in contemporary barbarians and savages a childish love of playthings and games, of which we have many survivals.

All these facts, which were at first unobserved or observed as matters of no significance, have been brought into connection with a fact in biology acknowledged alike by all important schools; by Agassiz on one hand and by Darwin on the other—namely, as stated by Agassiz, that "the young states of each species and group resemble older forms of the same group," or, as stated by Darwin, that "in two or more groups of animals, however much they may at first differ from each other in structure and habits, if they pass through closely similar embryonic stages, we may feel almost assured that they have descended from the same parent form, and are therefore closely related." *

* For the stone forms given to early bronze axes, etc., see Nilsson, *Primitive Inhabitants of Scandinavia*, London, 1868, Lubbock's *Introduction*, p. 31; and

for plates, see Lubbock's *Prehistoric Man*, chap. ii ; also Cartailhac, *Les Ages Préhistoriques de l'Espagne et du Portugal*, p. 227 ; also Keller, *Lake Dwellings* ; also Troyon, *Habitations Lacustres* ; also Boyd Dawkins, *Early Man in Great Britain*, p. 292 ; also Lubbock, p. 6 ; also Lyell, *Antiquity of Man*, chap. ii. For the cranogs, etc., in the north of Europe, see Munro, *Ancient Scottish Lake Dwellings*, Edinburgh, 1882. For mounds and greater stone constructions in the extreme south of Europe, see Cartailhac's work on Spain and Portugal above cited, part iii, chap. iii. For the source of Mr. Southall's contention, see Brugsch, *Egypt of the Pharaohs*. For the two sides of the question whether in the lowest grades of savagery there is really any recognition of a superior power, or anything which can be called, in any accepted sense, religion, compare Quatrefages with Lubbock, in works already cited. For a striking but rather *ad captandum* effort to show that there is a moral and religious sense in the very lowest Australian tribes, see one of the discourses of Archbishop Vaughan on *Science and Religion*, Baltimore, 1879. For one out of multitudes of striking and instructive resemblances in ancient stone implements and those now in use among sundry savage tribes, see comparison between old Scandinavian arrowheads and those recently brought from Tierra del Fuego, in Nilsson as above, especially in Plate V. For a brief and admirable statement of the arguments on both sides, see Sir J. Lubbock's Dundee paper, given in the appendix to the American edition of his *Origin of Civilization*, etc. For the general argument referred to between Whately and the Duke of Argyll on one side and Lubbock on the other, see Lubbock's Dundee paper as above cited ; Tylor, *Early History of Mankind*, especially p. 193 ; and the Duke of Argyll, *Primeval Man*, part iv. For difficulties of savages in arithmetic, see Lubbock, as above, pp. 459 *et seq*. For a very temperate and judicial view of the whole question, see Tylor as above, chaps. vii and xiii. For a brief summary of the scientific position regarding the stagnation and deterioration of races, resulting in the statement that such deterioration "in no way contradicts the theory that civilization itself is developed from low to high stages," see Tylor, *Anthropology*, chap. i. For striking examples of the testimony of language to upward progress, see Tylor, chap. xii.

CHAPTER X.

THE "FALL OF MAN" AND HISTORY.

THE history of art, especially as shown by architecture, in the noblest monuments of the most enlightened nations of antiquity, gives abundant proofs of the upward tendency of man from the rudest and simplest beginnings. Many columns of early Egyptian temples or tombs are but bundles of Nile reeds slightly conventionalized in stone; the temples of Greece, including not only the earliest forms, but the Parthenon itself, while in parts showing an evolution out of Egyptian and Assyrian architecture, exhibit frequent reminiscences and even imitations of earlier constructions in wood; the mediæval cathedrals, while evolved out of Roman and Byzantine structures, constantly show unmistakable survivals of prehistoric construction.*

So, too, general history has come in, illustrating the unknown from the known: the development of man in the prehistoric period from his development within historic times. Nothing is more evident from history than the fact that weaker bodies of men driven out by stronger do not necessarily relapse into barbarism, but frequently rise, even under the most unfavourable circumstances, to a civilization

* As to evolution in architecture, and especially of Greek forms and ornaments out of Egyptian and Assyrian, with survivals in stone architecture of forms obtained in Egypt when reeds were used, and in Greece when wood construction prevailed, see Fergusson's *Handbook of Architecture*, vol. i, pp. 100, 228, 233, and elsewhere; also Otfried Müller, *Ancient Art and its Remains*, English translation, London, 1852, pp. 219, *passim*. For a very brief but thorough statement, see A. Mangnard's paper in the *Proceedings of the American Oriental Society*, October, 1889, entitled *Reminiscences of Egypt in Doric Architecture*. On the general subject, see Hommel, *Babylonien*, ch. i, and Meyer, *Alterthum*, i, § 199.

equal or superior to that from which they have been banished. Out of very many examples showing this law of upward development, a few may be taken as typical. The Slavs, who sank so low under the pressure of stronger races that they gave the modern world a new word to express the most hopeless servitude, have developed powerful civilizations peculiar to themselves; the barbarian tribes who ages ago took refuge amid the sand-banks and morasses of Holland, have developed one of the world's leading centres of civilization; the wretched peasants who about the fifth century took refuge from invading hordes among the lagoons and mud banks of Venetia, developed a power in art, arms, and politics which is among the wonders of human history; the Puritans, driven from the civilization of Great Britain to the unfavourable climate, soil, and circumstances of early New England,—the Huguenots, driven from France, a country admirably fitted for the highest growth of civilization, to various countries far less fitted for such growth,—the Irish peasantry, driven in vast numbers from their own island to other parts of the world on the whole less fitted to them—all are proofs that, as a rule, bodies of men once enlightened, when driven to unfavourable climates and brought under the most depressing circumstances, not only retain what enlightenment they have, but go on increasing it. Besides these, we have such cases as those of criminals banished to various penal colonies, from whose descendants has been developed a better morality; and of pirates, like those of the Bounty, whose descendants, in a remote Pacific island, became sober, steady citizens. Thousands of examples show the prevalence of this same rule—that men in masses do not forget the main gains of their civilization, and that, in spite of deteriorations, their tendency is upward.

Another class of historic facts also testifies in the most striking manner to this same upward tendency: the decline and destruction of various civilizations brilliant but hopelessly vitiated. These catastrophes are seen more and more to be but steps in this development. The crumbling away of the great ancient civilizations based upon despotism, whether the despotism of monarch, priest, or mob—the decline and fall of Roman civilization, for example, which, in his most remarkable generalization, Guizot has shown to have been necessary to the development of the richer civilization of modern Europe; the terrible struggle and loss of the Crusades, which once appeared to be a mere catastrophe, but are now seen to have brought in, with the downfall of feudalism, the beginnings of the centralizing, civilizing monarchical period; the French Revolution, once thought a mere outburst of diabolic passion, but now seen to be an unduly delayed transition from the monarchical to the constitutional epoch: all show that even widespread deterioration and decline—often, indeed, the greatest political and moral catastrophes—so far from leading to a fall of mankind, tend in the long run to raise humanity to higher planes.

Thus, then, Anthropology and its handmaids, Ethnology, Philology, and History, have wrought out, beyond a doubt, proofs of the upward evolution of humanity since the appearance of man upon our planet.

Nor have these researches been confined to progress in man's material condition. Far more important evidences have been found of upward evolution in his family, social, moral, intellectual, and religious relations. The light thrown on this subject by such men as Lubbock, Tylor, Herbert Spencer, Buckle, Draper, Max Müller, and a multitude of others, despite mistakes, haltings, stumblings, and occasional following of delusive paths, is among the greatest glories of the century now ending. From all these investigators in their various fields, holding no brief for any system sacred or secular, but seeking truth as truth, comes the same general testimony of the evolution of higher out of lower. The process has been indeed slow and painful, but this does not prove that it may not become more rapid and less fruitful in sorrow as humanity goes on.*

While, then, it is not denied that many instances of retrogression can be found, the consenting voice of unbiased investigators in all lands has declared more and more that the beginnings of our race must have been low and brutal, and that the tendency has been upward. To combat this

* As to the good effects of migration, see Waitz, *Introduction to Anthropology*, London, 1863, p. 345.

conclusion by examples of decline and deterioration here and there has become impossible: as well try to prove that, because in the Mississippi there are eddies in which the currents flow northward, there is no main stream flowing southward; or that, because trees decay and fall, there is no law of upward growth from germ to trunk, branches, foliage, and fruit.

A very striking evidence that the theological theory had become untenable was seen when its main supporter in the scientific field, Von Martius, in the full ripeness of his powers, publicly declared his conversion to the scientific view.

Yet, while the tendency of enlightened human thought in recent times is unmistakable, the struggle against the older view is not yet ended. The bitterness of the Abbé Hamard in France has been carried to similar and even greater extremes among sundry Protestant bodies in Europe and America. The simple truth of history makes it a necessity, unpleasant though it be, to chronicle two typical examples in the United States.

In the year 1875 a leader in American industrial enterprise endowed at the capital of a Southern State a university which bore his name. It was given into the hands of one of the religious sects most powerful in that region, and a bishop of that sect became its president. To its chair of Geology was called Alexander Winchell, a scholar who had already won eminence as a teacher and writer in that field, a professor greatly beloved and respected in the two universities with which he had been connected, and a member of the sect which the institution of learning above referred to represented.

But his relations to this Southern institution were destined to be brief. That his lectures at the Vanderbilt University were learned, attractive, and stimulating, even his enemies were forced to admit; but he was soon found to believe that there had been men earlier than the period assigned to Adam, and even that all the human race are not descended from Adam. His desire was to reconcile science and Scripture, and he was now treated by a Methodist Episcopal Bishop in Tennessee just as, two centuries before, La Peyrère had been treated, for a similar effort, by a Roman

Catholic vicar-general in Belgium. The publication of a series of articles on the subject, contributed by the professor to a Northern religious newspaper at its own request, brought matters to a climax; for, the articles having fallen under the notice of a leading Southwestern organ of the denomination controlling the Vanderbilt University, the result was a most bitter denunciation of Prof. Winchell and of his views. Shortly afterward the professor was told by Bishop McTyeire that "our people are of the opinion that such views are contrary to the plan of redemption," and was requested by the bishop to quietly resign his chair. To this the professor made the fitting reply: "If the board of trustees have the manliness to dismiss me for cause, and declare the cause, I prefer that they should do it. No power on earth could persuade me to resign."

"We do not propose," said the bishop, with quite gratuitous suggestiveness, "to treat you as the Inquisition treated Galileo."

"But what you propose is the same thing," rejoined Dr. Winchell. "It is ecclesiastical proscription for an opinion which must be settled by scientific evidence."

Twenty-four hours later Dr. Winchell was informed that his chair had been abolished, and its duties, with its salary, added to those of a colleague; the public were given to understand that the reasons were purely economic; the banished scholar was heaped with official compliments, evidently in hope that he would keep silence.

Such was not Dr. Winchell's view. In a frank letter to the leading journal of the university town he stated the whole matter. The intolerance-hating press of the country, religious and secular, did not hold its peace. In vain the authorities of the university waited for the storm to blow over. It was evident, at last, that a defence must be made, and a local organ of the sect, which under the editorship of a fellow-professor had always treated Dr. Winchell's views with the luminous inaccuracy which usually characterizes a professor's ideas of a rival's teachings, assumed the task. In the articles which followed, the usual scientific hypotheses as to the creation were declared to be "absurd," "vague and unintelligible," "preposterous and gratuitous." This new champion stated that "the objections drawn from the fossiliferous strata and the like are met by reference to the analogy of Adam and Eve, who presented the phenomena of adults when they were but a day old, and by the Flood of Noah and other cataclysms, which, with the constant change of Nature, are sufficient to account for the phenomena in question"!

Under inspiration of this sort the Tennessee Conference of the religious body in control of the university had already, in October, 1878, given utterance to its opinion of unsanctified science as follows: "This is an age in which scientific atheism, having divested itself of the habiliments that most adorn and dignify humanity, walks abroad in shameless denudation. The arrogant and impertinent claims of this 'science, falsely so called,' have been so boisterous and persistent, that the unthinking mass have been sadly deluded; but our university alone has had the courage to lay its young but vigorous hand upon the mane of untamed Speculation and say, 'We will have no more of this.'"

It is a consolation to know how the result, thus devoutly sought, has been achieved; for in the "ode" sung at the laying of the corner-stone of a new theological building of the same university, in May, 1880, we read:

> "Science and Revelation here
> In perfect harmony appear,
> Guiding young feet along the road
> Through grace and Nature up to God."

It is also pleasing to know that, while an institution calling itself a university thus violated the fundamental principles on which any institution worthy of the name must be based, another institution which has the glory of being the first in the entire North to begin something like a university organization—the State University of Michigan—recalled Dr. Winchell at once to his former professorship, and honoured itself by maintaining him in that position, where, unhampered, he was thereafter able to utter his views in the midst of the largest body of students on the American continent.

Disgraceful as this history was to the men who drove

out Dr. Winchell, they but succeeded, as various similar bodies of men making similar efforts have done, in advancing their supposed victim to higher position and more commanding influence.*

A few years after this suppression of earnest Christian thought at an institution of learning in the western part of our Southern States, there appeared a similar attempt in sundry seaboard States of the South.

As far back as the year 1857 the Presbyterian Synod of Mississippi passed the following resolution:

"*Whereas,* We live in an age in which the most insidious attacks are made on revealed religion through the natural sciences, and as it behooves the Church at all times to have men capable of defending the faith once delivered to the saints;

"*Resolved,* That this presbytery recommend the endowment of a professorship of Natural Science as connected with revealed religion in one or more of our theological seminaries."

Pursuant to this resolution such a chair was established in the theological seminary at Columbia, S. C., and James Woodrow was appointed professor. Dr. Woodrow seems to have been admirably fitted for the position—a devoted Christian man, accepting the Presbyterian standards of faith in which he had been brought up, and at the same time giving every effort to acquaint himself with the methods and conclusions of science. To great natural endowments he added constant labours to arrive at the truth in this field. Visiting Europe, he made the acquaintance of many of the

* For Dr. Winchell's original statements, see *Adamites and Pre-Adamites*, Syracuse, N. Y., 1878. For the first important denunciation of his views, see the *St. Louis Christian Advocate*, May 22, 1878. For the conversation with Bishop McTyeire, see Dr. Winchell's own account in the *Nashville American*, June 16, 1878. For the curious reply from Dr. Winchell's colleague, see the *Nashville Christian Advocate*, July 12, 1878; and for the further development of the matter, see the *Nashville American* of July 19, 1878. For the further course of the attack in the denominational organ of Dr. Winchell's oppressors, see the *Nashville Christian Advocate*, April 26, 1879. For the oratorical declaration of the Tennessee Conference upon the matter, see the *Nashville American*, October 15, 1878; and for the "ode" regarding the "harmony of science and revelation" as supported at the university, see the same journal for May 2, 1880.

foremost scientific investigators, became a student in university lecture rooms and laboratories, an interested hearer in scientific conventions, and a correspondent of leading men of science at home and abroad. As a result, he came to the conclusion that the hypothesis of evolution is the only one which explains various leading facts in natural science. This he taught, and he also taught that such a view is not incompatible with a true view of the sacred Scriptures.

In 1882 and 1883 the board of directors of the theological seminary, in fear that "scepticism in the world is using alleged discoveries in science to impugn the Word of God," requested Prof. Woodrow to state his views in regard to evolution. The professor complied with this request in a very powerful address, which was published and widely circulated, to such effect that the board of directors shortly afterward passed resolutions declaring the theory of evolution as defined by Prof. Woodrow not inconsistent with perfect soundness in the faith.

In the year 1884 alarm regarding Dr. Woodrow's teachings began to show itself in larger proportions, and a minority report was introduced into the Synod of South Carolina declaring that "the synod is called upon to decide not upon the question whether the said views of Dr. Woodrow contradict the Bible in its highest and absolute sense, but upon the question whether they contradict the interpretation of the Bible by the Presbyterian Church in the United States."

Perhaps a more self-condemnatory statement was never presented, for it clearly recognized, as a basis for intolerance, at least a possible difference between "the interpretation of the Bible by the Presbyterian Church" and the teachings of "the Bible in its highest and absolute sense."

This hostile movement became so strong that, in spite of the favourable action of the directors of the seminary, and against the efforts of a broad-minded minority in the representative bodies having ultimate charge of the institution, the delegates from the various synods raised a storm of orthodoxy and drove Dr. Woodrow from his post. Happily, he was at the same time professor in the University of South Carolina in the same city of Columbia, and from his chair in that institution he continued to teach natural science with the approval of the great majority of thinking men in that region; hence, the only effect of the attempt to crush him was, that his position was made higher, respect for him deeper, and his reputation wider.

In spite of attempts by the more orthodox to prevent students of the theological seminary from attending his lectures at the university, they persisted in hearing him; indeed, the reputation of heresy seemed to enhance his influence.

It should be borne in mind that the professor thus treated had been one of the most respected and beloved university instructors in the South during more than a quarter of a century, and that he was turned out of his position with no opportunity for careful defence, and, indeed, without even the formality of a trial. Well did an eminent but thoughtful divine of the Southern Presbyterian Church declare that "the method of procedure to destroy evolution by the majority in the Church is vicious and suicidal," and that "logical dynamite has been used to put out a supposed fire in the upper stories of our house, and all the family in the house at that." Wisely, too, did he refer to the majority as "sowing in the fields of the Church the thorns of its errors, and cumbering its path with the *débris* and ruin of its own folly."

To these recent cases may be added the expulsion of Prof. Toy from teaching under ecclesiastical control at Louisville, and his election to a far more influential chair at Harvard University; the driving out from the American College at Beyrout of the young professors who accepted evolution as probable, and the rise of one of them, Mr. Nimr, to a far more commanding position than that which he left —the control of three leading journals at Cairo; the driving out of Robertson Smith from his position at Edinburgh, and his reception into the far more important and influential professorship at the English University of Cambridge; and multitudes of similar cases. From the days when Henry Dunster, the first President of Harvard College, was driven from his presidency, as Cotton Mather said, for "falling into the briers of Antipedobaptism" until now, the same spirit is shown in all such attempts. In each we have generally, on one side, a body of older theologians, who since their youth have learned nothing and forgotten nothing, sundry professors who do not wish to rewrite their lectures, and a mass of unthinking ecclesiastical persons of little or no importance save in making up a retrograde majority in an ecclesiastical tribunal; on the other side we have as generally the thinking, open-minded, devoted men who have listened to the revelation of their own time as well as of times past, and who are evidently thinking the future thought of the world.

Here we have survivals of that same oppression of thought by theology which has cost the modern world so dear; the system which forced great numbers of professors, under penalty of deprivation, to teach that the sun and planets revolve about the earth; that comets are fire-balls flung by an angry God at a wicked world; that insanity is diabolic possession; that anatomical investigation of the human frame is sin against the Holy Ghost; that chemistry leads to sorcery; that taking interest for money is forbidden by Scripture; that geology must conform to ancient Hebrew poetry. From the same source came in Austria the rule of the "Immaculate Oath," under which university professors, long before the dogma of the Immaculate Conception was defined by the Church, were obliged to swear to their belief in that dogma before they were permitted to teach even arithmetic or geometry; in England, the denunciation of inoculation against smallpox; in Scotland, the protests against using chloroform in childbirth as "vitiating the primal curse against woman"; in France, the use in clerical schools of a historical text-book from which Napoleon was left out; and, in America, the use of Catholic manuals in which the Inquisition is declared to have been a purely civil tribunal, or Protestant manuals in which the Puritans are shown to have been all that we could now wish they had been.

So, too, among multitudes of similar efforts abroad, we have during centuries the fettering of professors at English and Scotch universities by test oaths, subscriptions to articles, and catechisms without number. In our own country we have had in a vast multitude of denominational colleges, as the first qualification for a professorship, not ability in the subject to be taught, but fidelity to the particular shibboleth of the denomination controlling the college or university.

Happily, in these days such attempts generally defeat

themselves. The supposed victim is generally made a man of mark by persecution, and advanced to a higher and wider sphere of usefulness. In withstanding the march of scientific truth, any Conference, Synod, Board of Commissioners, Board of Trustees, or Faculty, is but as a nest of field-mice in the path of a steam plough.

The harm done to religion in these attempts is far greater than that done to science; for thereby suspicions are widely spread, especially among open-minded young men, that the accepted Christian system demands a concealment of truth, with the persecution of honest investigators, and therefore must be false. Well was it said in substance by President McCosh, of Princeton, that no more sure way of making unbelievers in Christianity among young men could be devised than preaching to them that the doctrines arrived at by the great scientific thinkers of this period are opposed to religion.

Yet it is but justice here to say that more and more there is evolving out of this past history of oppression a better spirit, which is making itself manifest with power in the leading religious bodies of the world. In the Church of Rome we have to-day such utterances as those of St. George Mivart, declaring that the Church must not attempt to interfere with science; that the Almighty in the Galileo case gave her a distinct warning that the priesthood of science must remain with the men of science. In the Anglican Church and its American daughter we have the acts and utterances of such men as Archbishop Tait, Bishop Temple, Dean Stanley, Dean Farrar, and many others, proving that the deepest religious thought is more and more tending to peace rather than warfare with science; and in the other churches, especially in America, while there is yet much to be desired, the welcome extended in many of them to Alexander Winchell, and the freedom given to views like his, augur well for a better state of things in the future.

From the science of Anthropology, when rightly viewed as a whole, has come the greatest aid to those who work to advance religion rather than to promote any particular system of theology; for Anthropology and its subsidiary sciences show more and more that man, since coming upon the earth, has risen, from the period when he had little, if any, idea of a great power above him, through successive stages of fetichism, shamanism, and idolatry, toward better forms of belief, making him more and more accessible to nobler forms of religion. The same sciences show, too, within the historic period, the same tendency, and especially within the events covered by our sacred books, a progress from fetichism, of which so many evidences crop out in the early Jewish worship as shown in the Old Testament Scriptures, through polytheism, when Jehovah was but "a god above all gods," through the period when he was "a jealous God," capricious and cruel, until he is revealed in such inspired utterances as those of the nobler Psalms, the great passages in Isaiah, the sublime preaching of Micah, and, above all, through the ideal given to the world by Jesus of Nazareth.

Well indeed has an eminent divine of the Church of England in our own time called on Christians to rejoice over this evolution, "between the God of Samuel, who ordered infants to be slaughtered, and the God of the Psalmist, whose tender mercies are over all his works; between the God of the Patriarchs, who was always repenting, and the God of the Apostles, who is the same yesterday, to-day, and forever, with whom there is no variableness nor shadow of turning; between the God of the Old Testament, who walked in the garden in the cool of the day, and the God of the New Testament, whom no man hath seen nor can see; between the God of Leviticus, who was so particular about the sacrificial furniture and utensils, and the God of the Acts, who dwelleth not in temples made with hands; between the God who hardened Pharaoh's heart, and the God who will have all men to be saved; between the God of Exodus, who is merciful only to those who love him, and the God of Christ—the heavenly Father—who is kind unto the unthankful and the evil."

However overwhelming, then, the facts may be which Anthropology, History, and their kindred sciences may, in the interest of simple truth, establish against the theological doctrine of "the Fall"; however completely they may fossilize various dogmas, catechisms, creeds, confessions, "plans of salvation" and "schemes of redemption," which have been evolved from the great minds of the theological period: science, so far from making inroads on religion, or even upon our Christian development of it, will strengthen all that is essential in it, giving new and nobler paths to man's highest aspirations. For the one great, legitimate, scientific conclusion of anthropology is, that, more and more, a better civilization of the world, despite all its survivals of savagery and barbarism, is developing men and women on whom the declarations of the nobler Psalms, of Isaiah, of Micah, the Sermon on the Mount, the first great commandment, and the second, which is like unto it, St. Paul's praise of charity and St. James's definition of "pure religion and undefiled," can take stronger hold for the more effective and more rapid uplifting of our race.*

---

* For the resolution of the Presbyterian Synod of Mississippi in 1857, see Prof. Woodrow's speech before the Synod of South Carolina, October 27 and 28, 1884, p. 6. As to the action of the Board of Directors of the Theological Seminary of Columbia, see ibid. As to the minority report in the Synod of South Carolina, see ibid., p. 24. For the pithy sentences regarding the conduct of the majority in the synods toward Dr. Woodrow, see the Rev. Mr. Flinn's article in the *Southern Presbyterian Review* for April, 1885, p. 272, and elsewhere. For the restrictions regarding the teaching of the Copernican theory and the true doctrine of comets in German universities, see various histories of astronomy, especially Mädler. For the immaculate oath (*Immaculaten-Eid*) as enforced upon the Austrian professors, see Luftkandl, *Die Josephinischen Ideen.* For the effort of the Church in France, after the restoration of the Bourbons, to teach a history of that country from which the name of Napoleon should be left out, see Father Loriquet's famous *Histoire de France à l'Usage de la Jeunesse*, Lyon, 1820, vol. ii; see especially table of contents at the end. The book bears on its title-page the well-known initials of the Jesuit motto, A. M. D. G. (Ad Majorem Dei Gloriam). For examples in England and Scotland, see various English histories, and especially Buckle's chapters on Scotland. For a longer collection of examples showing the suppression of anything like unfettered thought upon scientific subjects in American colleges, see *Inaugural Address at the Opening of Cornell University*, by the author of these chapters. For the citation regarding the evolution of better and nobler ideas of God, see *Church and Creed*: Sermons preached in the Chapel of the Foundling Hospital, London, by A. W. Momerie, M. A., LL. D., Professor of Logic and Metaphysics in King's College, London, London, 1890. For a very vigorous utterance on the other side, see a recent *charge* of the Bishop of Gloucester.

CHAPTER XI.

FROM "THE PRINCE OF THE POWER OF THE AIR" TO METEOROLOGY.

I. GROWTH OF A THEOLOGICAL THEORY.

THE popular beliefs of classic antiquity regarding storms, thunder, and lightning, took shape in myths representing Vulcan as forging thunderbolts, Jupiter as flinging them at his enemies, Æolus intrusting the winds in a bag to Æneas, and the like. An attempt at their further theological development is seen in the Pythagorean statement that lightnings are intended to terrify the damned in Tartarus.

But at a very early period we see the beginning of a scientific view. In Greece, the Ionic philosophers held that such phenomena are obedient to law. Plato, Aristotle, and many lesser lights, attempted to account for them on natural grounds; and their explanations, though crude, were based upon observation and thought. In Rome, Lucretius, Seneca, Pliny, and others, inadequate as their statements were, implanted at least the germs of a science. But, as the Christian Church rose to power, this evolution was checked; the new leaders of thought found, in the Scriptures recognized by them as sacred, the basis for a new view, or rather for a modification of the old view.

This ending of a scientific evolution based upon observation and reason, and this beginning of a sacred science based upon the letter of Scripture and on theology, are seen in the utterances of various fathers in the early Church. As to the general features of this new development, Tertullian held that sundry passages of Scripture prove lightning identical with hell-fire; and this idea was transmitted from generation to generation of later churchmen, who found an especial support of Tertullian's view in the sulphurous smell experienced during thunderstorms. St. Hilary thought the firmament very much lower than the heavens, and that it was created not only for the support of the upper waters, but also for the tempering of our atmosphere.* St. Ambrose held that thunder is caused by the winds breaking through the solid firmament, and cited from the prophet Amos the sublime passage regarding "Him that establisheth the thunders." † He shows, indeed, some conception of the true source of rain; but his whole reasoning is limited by various scriptural texts. He lays great stress upon the firmament as a solid outer shell of the universe: the heavens he holds to be not far outside this outer shell, and argues regarding their character from St. Paul's Epistle to the Corinthians and from the one hundred and forty-eighth Psalm. As to "the waters which are above the firmament," he takes up the objection of those who hold that, this outside of the universe being spherical, the waters must slide off it, especially if the firmament revolves; and he points out that it is by no means certain that the *outside* of the firmament *is* spherical, and insists that, if it does revolve, the water is just what is needed to lubricate and cool its axis.

St. Jerome held that God at the Creation, having spread out the firmament between heaven and earth, and having separated the upper waters from the lower, caused the upper waters to be frozen into ice, in order to keep all in place. A proof of this view Jerome found in the words of Ezekiel regarding "the crystal stretched above the cherubim." ‡

The germinal principle in accordance with which all these theories were evolved was most clearly proclaimed

---

* For Tertullian, see the *Apol. contra gentes*, c. 47; also Augustin de Angelis, *Lectiones Meteorologicæ*, p. 64. For Hilary, see *In Psalm. CXXXV* (Migne, *Patr. Lat.*, vol. ix, p. 773).

† "Firmans tonitrua" (Amos iv, 13); the phrase does not appear in our version.

‡ For Ambrose, see the *Hexæmeron*, lib. ii, cap. 3, 4; lib. iii, cap. 5 (Migne, *Patr. Lat.*, vol. xiv, pp. 148–150, 153, 165). The passage as to lubrication of the heavenly axis is as follows: "Deinde cum ipsi dicant volvi orbem cœli stellis ardentibus refulgentem, nonne divina providentia necessario prospexit, ut intra orbem cœli, et supra orbem redundaret aqua, quæ illa ferventis axis incendia temperaret?" For Jerome, see his *Epistola*, lxix, cap. 6 (Migne, *Patr. Lat.*, vol. xxii, p. 659).

to the world by St. Augustine in his famous utterance: "Nothing is to be accepted save on the authority of Scripture, since greater is that authority than all the powers of the human mind." * No treatise was safe thereafter which did not breathe the spirit and conform to the letter of this maxim. Unfortunately, what was generally understood by the "authority of Scripture" was the tyranny of sacred books imperfectly transcribed, viewed through distorting superstitions, and frequently interpreted by party spirit.

Following this precept of St. Augustine there were developed, in every field, theological views of science which have never led to a single truth—which, without exception, have forced mankind away from the truth, and have caused Christendom to stumble for centuries into abysses of error and sorrow. In meteorology, as in every other science with which he dealt, Augustine based everything upon the letter of the sacred text; and it is characteristic of the result that this man, so great when untrammelled, thought it his duty to guard especially the whole theory of the "waters above the heavens."

In the sixth century this theological reasoning was still further developed, as we have seen, by Cosmas Indicopleustes. Finding a sanction for the old Egyptian theory of the universe in the ninth chapter of Hebrews, he insisted that the earth is a flat parallelogram, and that from its outer edges rise immense walls supporting the firmament; then, throwing together the reference to the firmament in Genesis and the outburst of poetry in the Psalms regarding the "waters that be above the heavens," he insisted that over the terrestrial universe are solid arches bearing a vault supporting a vast cistern "containing the waters"; finally, taking from Genesis the expression regarding the "windows of heaven," he insisted that these windows are opened and closed by the angels whenever the Almighty wishes to send rain upon the earth or to withhold it.

* "Major est quippe Scripturæ hujus auctoritas, quam omnis humani ingenii capacitas."—Augustine, *De Genesi ad Lit.*, lib. ii, cap. 5 (Migne, *Patr. Lat.*, vol. xxxiv, pp. 266, 267). Or, as he is cited by Vincent of Beauvais (*Spec. Nat.*, lib. iv, 98): "Non est aliquid temere diffiniendum, sed quantum Scriptura dicit accipiendum, cujus major est auctoritas quam omnis humani ingenii capacitas."

This was accepted by the universal Church as a vast contribution to thought; for several centuries it was the orthodox doctrine, and various leaders in theology devoted themselves to developing and supplementing it.

About the beginning of the seventh century, Isidore, Bishop of Seville, was the ablest prelate in Christendom, and was showing those great qualities which led to his enrolment among the saints of the Church. His theological view of science marks an epoch. As to the "waters above the firmament," Isidore contends that they must be lower than the uppermost heaven, though higher than the lower heaven, because in the one hundred and forty-eighth Psalm they are mentioned *after* the heavenly bodies and the "heaven of heavens," but *before* the terrestrial elements. As to their purpose, he hesitates between those who held that they were stored up there by the prescience of God for the destruction of the world at the Flood, as the words of Scripture that "the windows of heaven were opened" seemed to indicate, and those who held that they were kept there to moderate the heat of the heavenly bodies. As to the firmament, he is in doubt whether it envelops the earth "like an eggshell," or is merely spread over it "like a curtain"; for he holds that the passage in the one hundred and fourth Psalm may be used to support either view.

Having laid these scriptural foundations, Isidore shows considerable power of thought; indeed, at times, when he discusses the rainbow, rain, hail, snow, and frost, his theories are rational, and give evidence that, if he could have broken away from his adhesion to the letter of Scripture, he might have given a strong impulse to the evolution of a true science.*

About a century later appeared, at the other extremity of Europe, the second in the trio of theological men of science in the early Middle Ages—Bede the Venerable. The nucleus of his theory also is to be found in the accepted view

---

* For Cosmas, see his *Topographia Christiana* (in Montfaucon, *Collectio nova patrum*, vol. ii), and the more complete account of his theory given in the chapter on *Geography* in this work. For Isidore, see the *Etymologiæ*, lib. xiii, cap. 7–9, *De ordine creaturarum*, cap. 3, 4, and *De natura rerum*, cap. 29, 30 (Migne, *Patr. Lat.*, vol. lxxxii, pp. 476, 477, vol. lxxxiii, pp. 920–922, 1001–1003).

of the "firmament" and of the "waters above the heavens," derived from Genesis. The firmament he holds to be spherical, and of a nature subtile and fiery; the upper heavens, he says, which contain the angels, God has tempered with ice, lest they inflame the lower elements. As to the waters placed above the firmament, lower than the spiritual heavens, but higher than all corporeal creatures, he says, "Some declare that they were stored there for the Deluge, but others, more correctly, that they are intended to temper the fire of the stars." He goes on with long discussions as to various elements and forces in Nature, and dwells at length upon the air, of which he says that the upper, serene air is over the heavens; while the lower, which is coarse, with humid exhalations, is sent off from the earth, and that in this are lightning, hail, snow, ice, and tempests, finding proof of this in the one hundred and forty-eighth Psalm, where these are commanded to "praise the Lord from the earth." *

So great was Bede's authority, that nearly all the anonymous speculations of the next following centuries upon these subjects were eventually ascribed to him. In one of these spurious treatises an attempt is made to get new light upon the sources of the waters above the heavens, the main reliance being the sheet containing the animals let down from heaven, in the vision of St. Peter. Another of these treatises is still more curious, for it endeavours to account for earthquakes and tides by means of the leviathan mentioned in Scripture. This characteristic passage runs as follows: "Some say that the earth contains the animal leviathan, and that he holds his tail after a fashion of his own, so that it is sometimes scorched by the sun, whereupon he strives to get hold of the sun, and so the earth is shaken by the motion of his indignation; he drinks in also, at times, such huge masses of the waves that when he belches them forth all the seas feel their effect." And this theological theory of the tides, as caused by the alternate suction and belching of leviathan, went far and wide.†

---

* See Bede, *De natura rerum* (Migne, *Patr. Lat.*, vol. xc).

† See the treatise *De mundi constitutione*, in Bede's *Opera* (Migne, *Patr. Lat.*, vol. xc, p. 884).

In the writings thus covered with the name of Bede there is much showing a scientific spirit, which might have come to something of permanent value had it not been hampered by the supposed necessity of conforming to the letter of Scripture. It is as startling as it is refreshing to hear one of these mediæval theorists burst out as follows against those who are content to explain everything by the power of God: "What is more pitiable than to say that a thing *is*, because God is able to do it, and not to show any reason why it is so, nor any purpose for which it is so; just as if God did everything that he is able to do! You talk like one who says that God is able to make a calf out of a log. But *did* he ever do it? Either, then, show a reason why a thing is so, or a purpose wherefore it is so, or else cease to declare it so." *

The most permanent contribution of Bede to scientific thought in this field was his revival of the view that the firmament is made of ice; and he supported this from the words in the twenty-sixth chapter of Job, "He bindeth up the waters in his thick cloud, and the cloud is not rent under them."

About the beginning of the ninth century appeared the third in that triumvirate of churchmen who were the oracles of sacred science throughout the early Middle Ages—Rabanus Maurus, Abbot of Fulda and Archbishop of Mayence. Starting, like all his predecessors, from the first chapter of Genesis, borrowing here and there from the ancient philosophers, and excluding everything that could conflict with the letter of Scripture, he follows, in his work upon the universe, his two predecessors, Isidore and Bede, developing especially St. Jerome's theory, drawn from Ezekiel, that the firmament is strong enough to hold up the "waters above the heavens," because it is made of ice.

For centuries the authority of these three great teachers was unquestioned, and in countless manuals and catechisms

* For this remonstrance, see the *Elementa philosophiæ*, in Bede's *Opera* (Migne, *Patr. Lat.*, vol. xc, p. 1139). This treatise, which has also been printed, under the title of *De philosophia mundi*, among the works of Honorius of Autun, is believed by modern scholars (Hauréau, Werner, Poole) to be the production of William of Conches.

their doctrine was translated and diluted for the common mind. But about the second quarter of the twelfth century a priest, Honorius of Autun, produced several treatises which show that thought on this subject had made some little progress. He explained the rain rationally, and mainly in the modern manner; with the thunder he is less successful, but insists that the thunderbolt "is not stone, as some assert." His thinking is vigorous and independent. Had theorists such as he been many, a new science could have been rapidly evolved, but the theological current was too strong.*

The strength of this current which overwhelmed the thought of Honorius is seen again in the work of the Dominican monk, John of San Geminiano, who in the thirteenth century gave forth his *Summa de Exemplis* for the use of preachers in his order. Of its thousand pages, over two hundred are devoted to illustrations drawn from the heavens and the elements. A characteristic specimen is his explanation of the Psalmist's phrase, "The arrows of the thunder." These, he tells us, are forged out of a dry vapour rising from the earth and kindled by the heat of the upper air, which then, coming into contact with a cloud just turning into rain, "is conglutinated like flour into dough," but, being too hot to be extinguished, its particles become merely sharpened at the lower end, and so blazing arrows, cleaving and burning everything they touch.†

But far more important, in the thirteenth century, was the fact that the most eminent scientific authority of that age, Albert the Great, Bishop of Ratisbon, attempted to reconcile the speculations of Aristotle with theological views derived from the fathers. In one very important respect he im-

---

* For Rabanus Maurus, see the *Comment. in Genesim* and *De Universo* (Migne, *Patr. Lat*, vol. cvii, cxi. For a charmingly naïve example of the primers referred to, see the little Anglo-Saxon manual of astronomy, sometimes attributed to Ælfric it is in the vernacular, but is translated in Wright's *Popular Treatises on Science during the Middle Ages*. Bede is, of course, its chief source. For Honorius, see the *De imagine mundi* and *Hexæmeron* (Migne, *Patr. Lat.*, vol. clxxii). The *De philosophia mundi*, the most rational of all, is, however, believed by modern scholars to be unjustly ascribed to him. See note above.

† See Joannes à S. Geminiano, *Summa*, c. 75.

proved upon the meteorological views of his great master. The thunderbolt, he says, is no mere fire, but the product of black clouds containing much mud, which, when it is baked by the intense heat, forms a fiery black or red stone that falls from the sky, tearing beams and crushing walls in its course: such he has seen with his own eyes.*

The monkish encyclopedists of the later Middle Ages added little to these theories. As we glance over the pages of Vincent of Beauvais, the monk Bartholomew, and William of Conches, we note only a growing deference to the authority of Aristotle as supplementing that of Isidore and Bede and explaining sacred Scripture. Aristotle is treated like a Church father, but extreme care is taken not to go beyond the great maxim of St. Augustine; then, little by little, Bede and Isidore fall into the background, Aristotle fills the whole horizon, and his utterances are second in sacredness only to the text of Holy Writ.

A curious illustration of the difficulties these mediæval scholars had to meet in reconciling the scientific theories of Aristotle with the letter of the Bible is seen in the case of the rainbow. It is to the honour of Aristotle that his conclusions regarding the rainbow, though slightly erroneous, were based upon careful observation and evolved by reasoning alone; but his Christian commentators, while anxious to follow him, had to bear in mind the scriptural statement that God had created the rainbow as a sign to Noah that there should never again be a Flood on the earth. Even so bold a thinker as Cardinal d'Ailly, whose speculations as to the geography of the earth did so much afterward in stimulating Columbus, faltered before this statement, acknowledging that God alone could explain it; but suggested that possibly never before the Deluge had a cloud been suffered to take such a position toward the sun as to cause a rainbow.

The learned cardinal was also constrained to believe that certain stars and constellations have something to do in causing the rain, since these would best explain Noah's fore-

* See Albertus Magnus, *II Sent.*, *Op.*, vol. xv, p. 137, a. (cited by Heller, *Gesch. d. Physik*, vol. i, p. 184) and his *Liber Methaurorum*, III, iv, 18 (of which I have used the edition of Venice, 1488).

knowledge of the Deluge. In connection with this scriptural doctrine of winds came a scriptural doctrine of earthquakes: they were believed to be caused by winds issuing from the earth, and this view was based upon the passage in the one hundred and thirty-fifth Psalm, "He bringeth the wind out of his treasuries." *

Such were the main typical attempts during nearly fourteen centuries to build up under theological guidance and within scriptural limitations a sacred science of meteorology. But these theories were mainly evolved in the effort to establish a basis and general theory of phenomena: it still remained to account for special manifestations, and here came a twofold development of theological thought.

On one hand, these phenomena were attributed to the Almighty, and, on the other, to Satan. As to the first of these theories, we constantly find the Divine wrath mentioned by the earlier fathers as the cause of lightning, hailstorms, hurricanes, and the like.

In the early days of Christianity we see a curious struggle between pagan and Christian belief upon this point. Near the close of the second century the Emperor Marcus Aurelius, in his effort to save the empire, fought a hotly contested battle with the Quadi, in what is now Hungary. While the issue of this great battle was yet doubtful there came suddenly a blinding storm beating into the faces of the Quadi, and this gave the Roman troops the advantage, enabling Marcus Aurelius to win a decisive victory. Votaries of each of the great religions claimed that this storm was caused by the object of their own adoration. The pagans insisted that Jupiter had sent the storm in obedience to their prayers, and on the Antonine Column at Rome we may still see the figure of Olympian Jove casting his thunderbolts and pouring a storm of rain from the open heavens against the Quadi. On the other hand, the Christians insisted that the storm had been sent by Jehovah in obedience to *their*

---

* For D'Ailly, see his *Concordia astronomicæ veritatis cum theologia* (Paris, 1483—in the *Imago mundi*—and Venice, 1490); also Eck's commentary on Aristotle's *Meteorologica* (Augsburg, 1519), lib. ii, *nota* 2; also Reisch, *Margarita philosophica*, lib. ix, c. 18.

prayers; and Tertullian, Eusebius, St. Gregory of Nyssa, and St. Jerome were among those who insisted upon this meteorological miracle; the first two, indeed, in the fervour of their arguments for its reality, allowing themselves to be carried considerably beyond exact historical truth.*

As time went on, the fathers developed this view more and more from various texts in the Jewish and Christian sacred books, substituting for Jupiter flinging his thunderbolts the Almighty wrapped in thunder and sending forth his lightnings. Through the Middle Ages this was fostered until it came to be accepted as a mere truism, entering into all mediæval thinking, and was still further developed by an attempt to specify the particular sins which were thus punished. Thus even the rational Florentine historian Villani ascribed floods and fires to the "too great pride of the city of Florence and the ingratitude of the citizens toward God," which, "of course," says a recent historian, "meant their insufficient attention to the ceremonies of religion." †

In the thirteenth century the Cistercian monk, Cæsarius of Heisterbach, popularized the doctrine in central Europe. His rich collection of anecdotes for the illustration of religious truths was the favourite recreative reading in the convents for three centuries, and exercised great influence over the thought of the later Middle Ages. In this work he relates several instances of the Divine use of lightning, both for rescue and for punishment. Thus he tells us how the steward (*cellerarius*) of his own monastery was saved from the clutch of a robber by a clap of thunder which, in answer to his prayer, burst suddenly from the sky and frightened the bandit from his purpose how, in a Saxon theatre, twenty men were struck down, while a priest escaped, not because he was not a greater sinner than the rest, but because the thunderbolt had respect for his profession! It is Cæsarius, too, who tells us the story of the priest of Treves, struck by lightning in his own church, whither he had gone to ring

* For the authorities, pagan and Christian, see the note of Merivale, in his *History of the Romans under the Empire*, chap. lxviii. He refers for still fuller citations to Fynes Clinton's *Fasti Rom.*, p. 24.

† See Trollope, *History of Florence*, vol. i, p. 64.

the bell against the storm, and whose sins were revealed by the course of the lightning, for it tore his clothes from him and consumed certain parts of his body, showing that the sins for which he was punished were vanity and unchastity.*

This mode of explaining the Divine interference more minutely is developed century after century, and we find both Catholics and Protestants assigning as causes of unpleasant meteorological phenomena whatever appears to them wicked or even unorthodox. Among the English Reformers, Tyndale quotes in this kind of argument the thirteenth chapter of I. Samuel, showing that, when God gave Israel a king, it thundered and rained. Archbishop Whitgift, Bishop Bale, and Bishop Pilkington insisted on the same view. In Protestant Germany, about the same period, Plieninger took a dislike to the new Gregorian calendar and published a volume of *Brief Reflections*, in which he insisted that the elements had given utterance to God's anger against it, calling attention to the fact that violent storms raged over almost all Germany during the very ten days which the Pope had taken out for the correction of the year, and that great floods began with the first days of the corrected year.†

Early in the seventeenth century, Majoli, Bishop of Voltoraria, in southern Italy, produced his huge work *Dies Caniculari*, or *Dog Days*, which remained a favourite encyclopædia in Catholic lands for over a hundred years. Treating of thunder and lightning, he compares them to bombs against the wicked, and says that the thunderbolt is "an exhalation condensed and cooked into stone," and that "it is not to be doubted that, of all instruments of God's vengeance, the thunderbolt is the chief"; that by means of it Sennacherib and his army were consumed; that Luther was struck by lightning in his youth as a caution against departing from the Catholic faith; that blasphemy and Sabbath-breaking are the sins to which this punishment is especially

* See Cæsarius Heisterbacensis, *Dialogus miraculorum*, lib. x, c. 28–30.

† For Tyndale, see his *Doctrinal Treatises*, p. 194, and for Whitgift, see his *Works*, vol. ii, pp. 477–483; Bale, *Works*, pp. 244, 245; and Pilkington, *Works*, pp. 177, 536 (all in *Parker Society Publications*). Bishop Bale cites especially Job xxxviii, Ecclesiasticus xiii, and Revelation viii, as supporting the theory. For Plieninger's words, see Janssen, *Geschichte des deutschen Volkes*, vol. v, p. 350.

assigned, and he cites the case of Dathan and Abiram. Fifty years later the Jesuit Stengel developed this line of thought still further in four thick quarto volumes on the judgments of God, adding an elaborate schedule for the use of preachers in the sermons of an entire year. Three chapters were devoted to thunder, lightning, and storms. That the author teaches the agency in these of diabolical powers goes without saying; but this can only act, he declares, by Divine permission, and the thunderbolt is always the finger of God, which rarely strikes a man save for his sins, and the nature of the special sin thus punished may be inferred from the bodily organs smitten. A few years later, in Protestant Swabia, Pastor Georg Nuber issued a volume of "weather-sermons," in which he discusses nearly every sort of elemental disturbances—storms, floods, droughts, lightning, and hail. These, he says, come direct from God for human sins, yet no doubt with discrimination, for there are five sins which God especially punishes with lightning and hail—namely, impenitence, incredulity, neglect of the repair of churches, fraud in the payment of tithes to the clergy, and oppression of subordinates, each of which points he supports with a mass of scriptural texts.*

This doctrine having become especially precious both to Catholics and to Protestants, there were issued handbooks of prayers against bad weather: among these was the *Spiritual Thunder and Storm Booklet*, produced in 1731 by a Protestant scholar, Stöltzlin, whose three or four hundred pages of prayer and song, "sighs for use when it lightens fearfully," and "cries of anguish when the hailstorm is drawing on," show a wonderful adaptability to all possible meteorological emergencies. The preface of this volume is contributed by Prof. Dilherr, pastor of the great church of St. Sebald at Nuremberg, who, in discussing the Divine purposes of storms, adds to the three usually assigned —namely, God's wish to manifest his power, to display his anger, and to drive sinners to repentance—a fourth, which,

---

* For Majoli, see *Dies Can.*, I, i; for Stengel, see the *De judiciis divinis*, vol. ii, pp. 15–61, and especially the example of the *impurus et saltator sacerdos, fulmine castratus*, pp. 26, 27. For Nuber, see his *Conciones meteoricæ*, Ulm, 1661.

he says, is that God may show us "with what sort of a storm-bell he will one day ring in the last judgment."

About the end of the first quarter of the eighteenth century we find, in Switzerland, even the eminent and rational Professor of Mathematics, Scheuchzer, publishing his *Physica Sacra*, with the Bible as a basis, and forced to admit that the elements, in the most literal sense, utter the voice of God. The same pressure was felt in New England. Typical are the sermons of Increase Mather on *The Voice of God in Stormy Winds*. He especially lays stress on the voice of God speaking to Job out of the whirlwind, and upon the text, "Stormy wind fulfilling his word." He declares, "When there are great tempests, the angels oftentimes have a hand therein, . . . yea, and sometimes evil angels." He gives several cases of blasphemers struck by lightning, and says, "Nothing can be more dangerous for mortals than to contemn dreadful providences, and, in particular, dreadful tempests."

His distinguished son, Cotton Mather, disentangled himself somewhat from the old view, as he had done in the interpretation of comets. In his *Christian Philosopher*, his *Thoughts for the Day of Rain*, and his *Sermon preached at the Time of the Late Storm* (in 1723), he is evidently tending toward the modern view. Yet, from time to time, the older view has reasserted itself, and in France, as recently as the year 1870, we find the Bishop of Verdun ascribing the drought afflicting his diocese to the sin of Sabbath-breaking.*

This theory, which attributed injurious meteorological phenomena mainly to the purposes of God, was a natural development, and comparatively harmless; but at a very early period there was evolved another theory, which, having been ripened into a doctrine, cost the earth dear indeed. Never, perhaps, in the modern world has there been a dogma more prolific of physical, mental, and moral agony

---

* For Stöltzlin, see his *Geistliches Donner- und Wetter-Büchlein* (Zürich, 1731). For Increase Mather, see his *The Voice of God*, etc. (Boston, 1704). This rare volume is in the rich collection of the American Antiquarian Society at Worcester. For Cotton Mather's view, see the chapter *From Signs and Wonders to Law*, in this work. For the Bishop of Verdun, see the *Semaine relig. de Lorraine*, 1870, p. 445 (cited by "Paul Parfait," in his *Dossier des Pèlerinages*, pp. 141–143).

throughout whole nations and during whole centuries. This theory, its development by theology, its fearful results to mankind, and its destruction by scientific observation and thought, will next be considered.

II. DIABOLIC AGENCY IN STORMS.

While the fathers and schoolmen were labouring to deduce a science of meteorology from our sacred books, there oozed up in European society a mass of traditions and observances which had been lurking since the days of paganism; and, although here and there appeared a churchman to oppose them, the theologians and ecclesiastics ere long began to adopt them and to clothe them with the authority of religion.

Both among the pagans of the Roman Empire and among the barbarians of the North the Christian missionaries had found it easier to prove the new God supreme than to prove the old gods powerless. Faith in the miracles of the new religion seemed to increase rather than to diminish faith in the miracles of the old; and the Church at last began admitting the latter as facts, but ascribing them to the devil. Jupiter and Odin sank into the category of ministers of Satan, and transferred to that master all their former powers. A renewed study of Scripture by theologians elicited overwhelming proofs of the truth of this doctrine. Stress was especially laid on the declaration of Scripture, "The gods of the heathen are devils." * Supported by this and other texts, it soon became a dogma. So strong was the hold it took, under the influence of the Church, that not until late in the seventeenth century did its substantial truth begin to be questioned.

With no field of action had the sway of the ancient deities been more identified than with that of atmospheric phenomena. The Roman heard Jupiter, and the Teuton heard Thor, in the thunder. Could it be doubted that these powerful beings would now take occasion, unless hindered by the command of the Almighty, to vent their spite against those

* For so the Vulgate and all the early versions rendered Ps. xcvi, 5.

who had deserted their altars? Might not the Almighty himself be willing to employ the malice of these powers of the air against those who had offended him?

It was, indeed, no great step, for those whose simple faith accepted rain or sunshine as an answer to their prayers, to suspect that the untimely storms or droughts, which baffled their most earnest petitions, were the work of the archenemy, "the prince of the power of the air."

The great fathers of the Church had easily found warrant for this doctrine in Scripture. St. Jerome declared the air to be full of devils, basing this belief upon various statements in the prophecies of Isaiah and in the Epistle to the Ephesians. St. Augustine held the same view as beyond controversy.*

During the Middle Ages this doctrine of the diabolical origin of storms went on gathering strength. Bede had full faith in it, and narrates various anecdotes in support of it. St. Thomas Aquinas gave it his sanction, saying in his all-authoritative *Summa*, "Rains and winds, and whatsoever occurs by local impulse alone, can be caused by demons." "It is," he says, "a dogma of faith that the demons can produce wind, storms, and rain of fire from heaven."

Albert the Great taught the same doctrine, and showed how a certain salve thrown into a spring produced whirlwinds. The great Franciscan—the "seraphic doctor"—St. Bonaventura, whose services to theology earned him one of the highest places in the Church, and to whom Dante gave special honour in paradise, set upon this belief his high authority. The lives of the saints, and the chronicles of the Middlė Ages, were filled with it. Poetry and painting accepted the idea and developed it. Dante wedded it to verse, and at Venice this thought may still be seen embodied in

---

* For St. Jerome, see his *Com. in Ep. ad Ephesios* (lib. iii, cap. 6); commenting on the text, "Our battle is not with flesh and blood," he explains this as meaning the devils in the air, and adds: "Nam et in alio loco de dæmonibus quod in aere isto vagentur, Apostolus ait: *In quibus ambulastis aliquando juxta sæculum mundi istius, secundum principem potestatis aeris spiritus, qui nunc operatur in filios diffidentiæ* (Eph. ii, 2). Hæc autem omnium doctorum opinio est, quod aer iste qui cœlum et terram medius dividens, inane appellatur, plenus sit contrariis fortitudinibus." See also his *Com. in Isaiam*, lib. xiii, cap. 50 (Migne, *Patr. Lat.*, vol. xxiv, p. 477). For Augustine, see the *De Civitate Dei, passim*.

one of the grand pictures of Bordone: a shipload of demons is seen approaching Venice in a storm, threatening destruction to the city, but St. Mark, St. George, and St. Nicholas attack the vessel, and disperse the hellish crew.*

The popes again and again sanctioned this doctrine, and it was amalgamated with various local superstitions, pious imaginations, and interesting arguments, to strike the fancy of the people at large. A strong argument in favour of a diabolical origin of the thunderbolt was afforded by the eccentricities of its operation. These attracted especial attention in the Middle Ages, and the popular love of marvel generalized isolated phenomena into rules. Thus it was said that the lightning strikes the sword in the sheath, gold in the purse, the foot in the shoe, leaving sheath and purse and shoe unharmed; that it consumes a human being internally without injuring the skin; that it destroys nets in the water, but not on the land; that it kills one man, and leaves untouched another standing beside him; that it can tear through a house and enter the earth without moving a stone from its place; that it injures the heart of a tree, but not the bark; that wine is poisoned by it, while poisons struck by it lose their venom; that a man's hair may be consumed by it and the man be unhurt.†

These peculiar phenomena, made much of by the allegorizing sermonizers of the day, were used in moral lessons from every pulpit. Thus the Carmelite, Matthias Farinator, of Vienna, who at the Pope's own instance compiled early in the fifteenth century that curious handbook of illustrative examples for preachers, the *Lumen Animæ*, finds a spiritual analogue for each of these anomalies.‡

This doctrine grew, robust and noxious, until, in the

---

* For Bede, see the *Hist. Eccles.*, vol. i, p. 17; *Vita Cuthberti*, c. 17 (Migne, tome xliv). For Thomas Aquinas, see the *Summa*, pars I, qu. lxxx, art. 2. The second citation I owe to Rydberg, *Magic of the Middle Ages*, p. 73, where the whole interesting passage is given at length. For Albertus Magnus, see the *De Potentia Dæmonum* (cited by Maury, *Légendes Pieuses*). For Bonaventura, see the *Comp. Theol. Veritat.*, ii, 26. For Dante, see *Purgatorio*, c. 5. On Bordone's picture, see Maury, *Légendes Pieuses*, p. 18, note.

† See, for lists of such *admiranda*, any of the early writers—e.g., Vincent of Beauvais, Reisch's *Margarita*, or Eck's *Aristotle*.

‡ See the *Lumen Animæ*, Eichstadt, 1479.

fifteenth, sixteenth, and seventeenth centuries, we find its bloom in a multitude of treatises by the most learned of the Catholic and Protestant divines, and its fruitage in the torture chambers and on the scaffolds throughout Christendom. At the Reformation period, and for nearly two hundred years afterward, Catholics and Protestants vied with each other in promoting this growth. John Eck, the great opponent of Luther, gave to the world an annotated edition of Aristotle's *Physics*, which was long authoritative in the German universities; and, though the text is free from this doctrine, the woodcut illustrating the earth's atmosphere shows most vividly, among the clouds of mid-air, the devils who there reign supreme.*

Luther, in the other religious camp, supported the superstition even more zealously, asserting at times his belief that the winds themselves are only good or evil spirits, and declaring that a stone thrown into a certain pond in his native region would cause a dreadful storm because of the devils kept prisoners there.†

Just at the close of the same century, Catholics and Protestants welcomed alike the great work of Delrio. In this, the power of devils over the elements is proved first from the Holy Scriptures, since, he declares, "they show that Satan brought fire down from heaven to consume the servants and flocks of Job, and that he stirred up a violent wind, which overwhelmed in ruin the sons and daughters of Job at their feasting." Next, Delrio insists on the agreement of all the orthodox fathers, that it was the devil himself who did this, and attention is called to the fact that the hail with which the Egyptians were punished is expressly declared in Holy Scripture to have been brought by the evil angels. Citing from the Apocalypse, he points to the four angels standing at the four corners of the earth, holding back the winds and preventing their doing great damage to mortals; and he dwells especially upon the fact that the devil is called by the apostle a "prince of the power of the air." He then

* See Eck, *Aristotelis Meteorologica*, Augsburg, 1519.

† For Luther, see the *Table Talk*; also Michelet, *Life of Luther* (translated by Hazlitt, p. 321).

goes on to cite the great fathers of the Church—Clement, Jerome, Augustine, and Thomas Aquinas.*

This doctrine was spread not only in ponderous treatises, but in light literature and by popular illustrations. In the *Compendium Maleficarum* of the Italian monk Guacci, perhaps the most amusing book in the whole literature of witchcraft, we may see the witch, *in propria persona*, riding the diabolic goat through the clouds while the storm rages around and beneath her; and we may read a rich collection of anecdotes, largely contemporary, which establish the required doctrine beyond question.

The first and most natural means taken against this work of Satan in the air was prayer; and various petitions are to be found scattered through the Christian liturgies—some very beautiful and touching. This means of escape has been relied upon, with greater or less faith, from those days to these. Various mediæval saints and reformers, and devoted men in all centuries, from St. Giles to John Wesley, have used it with results claimed to be miraculous. Whatever theory any thinking man may hold in the matter, he will certainly not venture a reproachful word: such prayers have been in all ages a natural outcome of the mind of man in trouble.†

But against the "power of the air" were used other means of a very different character and tendency, and foremost among these was exorcism. In an exorcism widely used and ascribed to Pope Gregory XIII, the formula is given: "I, a priest of Christ, . . . do command ye, most foul spirits, who do stir up these clouds, . . . that ye depart from them, and disperse yourselves into wild and untilled places, that

---

* For Delrio, see his *Disquisitiones Magicæ*, first printed at Liége in 1599–1600, but reprinted again and again throughout the seventeenth century. His interpretation of Psalm lxxviii, 47–49, was apparently shared by the translators of our own authorized version. For citations by him, see Revelation vii, 1; Ephesians ii, 2. Even according to modern commentators (e. g., Alford), the word here translated "power" denotes not *might*, but *government*, *court*, *hierarchy*; and in this sense it was always used by the ecclesiastical writers, whose conception is best rendered by our plural—"powers." See Delrio, *Disquisitiones Magicæ*, lib. ii, c. 11.

† For Guacci, see his *Compendium Maleficarum* (Milan, 1608). For the cases of St. Giles, John Wesley, and others stilling the tempests, see Brewer, *Dictionary of Miracles*, s. v. *Prayer*.

ye may be no longer able to harm men or animals or fruits or herbs, or whatsoever is designed for human use." But this is mild, indeed, compared to some later exorcisms, as when the ritual runs: "All the people shall rise, and the priest, turning toward the clouds, shall pronounce these words: 'I exorcise ye, accursed demons, who have dared to use, for the accomplishment of your iniquity, those powers of Nature by which God in divers ways worketh good to mortals; who stir up winds, gather vapours, form clouds, and condense them into hail. . . . I exorcise ye, . . . that ye relinquish the work ye have begun, dissolve the hail, scatter the clouds, disperse the vapours, and restrain the winds.'" The rubric goes on to order that then there shall be a great fire kindled in an open place, and that over it the sign of the cross shall be made, and the one hundred and fourteenth Psalm chanted, while malodorous substances, among them sulphur and asafœtida, shall be cast into the flames. The purpose seems to have been literally to "smoke out" Satan.*

Manuals of exorcisms became important—some bulky quartos, others handbooks. Noteworthy among the latter is one by the Italian priest Locatelli, entitled *Exorcisms most Powerful and Efficacious for the Dispelling of Aërial Tempests, whether raised by Demons at their own Instance or at the Beck of some Servant of the Devil.* †

The Jesuit Gretser, in his famous book on *Benedictions and Maledictions*, devotes a chapter to this subject, dismissing summarily the scepticism that questions the power of devils over the elements, and adducing the story of Job as conclusive. ‡

Nor was this theory of exorcism by any means confined to the elder Church. Luther vehemently upheld it, and

---

* See Polidorus Valerius, *Practica exorcistarum*; also the *Thesaurus exorcismorum* (Cologne, 1626), pp. 158–162.

† That is, *Exorcismi*, etc. A "corrected" second edition was printed at Laybach, 1680, in 24mo, to which is appended another manual of *Preces et conjurationes contra aëreas tempestates, omnibus sacerdotibus utiles et necessaria*, printed at the monastery of Kempten (in Bavaria) in 1667. The latter bears as epigraph the passage from the gospels describing Christ's stilling of the winds

‡ See Gretser, *De benedictionibus et maledictionibus*, lib. ii, c. 48.

prescribed especially the first chapter of St. John's gospel as of unfailing efficacy against thunder and lightning, declaring that he had often found the mere sign of the cross, with the text, "The word was made flesh," sufficient to put storms to flight.*

From the beginning of the Middle Ages until long after the Reformation the chronicles give ample illustration of the successful use of such exorcisms. So strong was the belief in them that it forced itself into minds comparatively rational, and found utterance in treatises of much importance.

But, since exorcisms were found at times ineffectual, other means were sought, and especially fetiches of various sorts. One of the earliest of these appeared when Pope Alexander I, according to tradition, ordained that holy water should be kept in churches and bedchambers to drive away devils.† Another safeguard was found in relics, and of similar efficacy were the so-called "conception billets" sold by the Carmelite monks. They contained a formula upon consecrated paper, at which the devil might well turn pale. Buried in the corner of a field, one of these was thought to give protection against bad weather and destructive insects.‡

But highest in repute during centuries was the *Agnus Dei*—a piece of wax blessed by the Pope's own hand, and stamped with the well-known device representing the "Lamb of God." Its powers were so marvellous that Pope Urban V thought three of these cakes a fitting gift from himself to the Greek Emperor. In the Latin doggerel recounting their virtues, their meteorological efficacy stands first, for especial stress is laid on their power of dispelling the thunder. The stress thus laid by Pope Urban, as the infallible guide of Christendom, on the efficacy of this fetich, gave it great value throughout Europe, and the doggerel verses reciting

---

* So, at least, says Gretser (in his *De ben. et mal.*, as above).

† "Instituit ut aqua quam sanctam appellamus sale admixta interpositis sacris orationibus et in templis et in cubiculis ad fugandos dæmones retineretur."—Platina, *Vitæ Pontif.* But the story is from the False Decretals.

‡ See Rydberg, *The Magic of the Middle Ages*, translated by Edgren, pp. 63–66.

its virtues sank deep into the popular mind. It was considered a most potent means of dispelling hail, pestilence, storms, conflagrations, and enchantments; and this feeling was deepened by the rules and rites for its consecration. So solemn was the matter, that the manufacture and sale of this particular fetich was, by a papal bull of 1471, reserved for the Pope himself, and he only performed the required ceremony in the first and seventh years of his pontificate. Standing unmitred, he prayed: "O God, . . . we humbly beseech thee that thou wilt bless these waxen forms, figured with the image of an innocent lamb, . . . that, at the touch and sight of them, the faithful may break forth into praises, and that the crash of hailstorms, the blast of hurricanes, the violence of tempests, the fury of winds, and the malice of thunderbolts may be tempered, and evil spirits flee and tremble before the standard of thy holy cross, which is graven upon them." *

Another favourite means with the clergy of the older Church for bringing to naught the "power of the air," was found in great processions bearing statues, relics, and holy

---

* These pious charms are still in use in the Church, and may be found described in any ecclesiastical cyclopædia. The doggerel verses run as follows:

"Tonitrua magna terret,
Et peccata nostra delet;
Ab incendio præservat,
A submersione servat,
A morte cita liberat,
Et Cacodæmones fugat,
Inimicos nostros domat,
Prægnantem cum partu salvat,
Dona dignis multa confert,
Utque malis mala defert.
Portio, quamvis parva sit,
Ut magna tamen proficit."

See these verses cited in full faith, so late as 1743, in Father Vincent of Berg's *Enchiridium*, pp. 23, 24, where is an ample statement of the virtues of the *Agnus Dei*, and instructions for its use. A full account of the rites used in consecrating this fetich, with the prayers and benedictions which gave colour to this theory of the powers of the *Agnus Dei*, may be found in the ritual of the Church. I have used the edition entitled *Sacrarum ceremoniarum sive rituum Sanctæ Romanæ Ecclesiæ libri tres*, Rome, 1560, in folio. The form of the papal prayer is as follows: "Deus, . . . te suppliciter deprecamur, ut . . . has cereas formas, innocentissimi agni imagine figuratas, benędicere . . . digneris, ut per ejus tactum et visum fideles invitentur ad laudes, fragor grandinum, procella turbinum, impetus tempestatum, ventorum rabies, infesta tonitrua temperentur, fugiant atque tremiscant maligni spiritus ante Sanctæ Crucis vexillum, quod in illis exsculptum est. . . ." (*Sacr. Cer. Rom. Eccl.*, as above). If any are curious as to the extent to which this consecrated wax was a specific for all spiritual and most temporal ills during the sixteenth and seventeenth centuries, let them consult the Jesuit *Litteræ annuæ, passim*.

emblems through the streets. Yet even these were not always immediately effective. One at Liége, in the thirteenth century, thrice proved unsuccessful in bringing rain, when at last it was found that the image of the Virgin had been forgotten! A new procession was at once formed, the *Salve Regina* sung, and the rain came down in such torrents as to drive the devotees to shelter.*

In Catholic lands this custom remains to this day, and very important features in these processions are the statues and the reliquaries of patron saints. Some of these excel in bringing sunshine, others in bringing rain. The Cathedral of Chartres is so fortunate as to possess sundry relics of St. Taurin, especially potent against dry weather, and some of St. Piat, very nearly as infallible against wet weather. In certain regions a single saint gives protection alternately against wet and dry weather—as, for example, St. Godeberte at Noyon. Against storms St. Barbara is very generally considered the most powerful protectress; but, in the French diocese of Limoges, Notre Dame de Crocq has proved a most powerful rival, for when, a few years since, all the neighbouring parishes were ravaged by storms, not a hailstone fell in the canton which she protected. In the diocese of Tarbès, St. Exupère is especially invoked against hail, peasants flocking from all the surrounding country to his shrine.†

But the means of baffling the powers of the air which came to be most widely used was the ringing of consecrated church bells.

This usage had begun in the time of Charlemagne, and there is extant a prohibition of his against the custom of baptizing bells and of hanging certain tags‡ on their tongues as a protection against hailstorms; but even Charlemagne

---

* John of Winterthur describes many such processions in Switzerland in the thirteenth century, and all the monkish chronicles speak of them. See also Rydberg, *Magic of the Middle Ages*, p. 74.

† As to protection by special saints as stated, see the *Guide du touriste et du pèlerin à Chartres*, 1867 (cited by "Paul Parfait," in his *Dossier des Pèlerinages*); also pp. 139–145 of the *Dossier*.

‡ *Perticæ*. See Montanus, *Hist. Nachricht von den Glocken* (Chemnitz, 1726), p. 121; and Meyer, *Der Aberglaube des Mittelalters*, p. 186.

was powerless against this current of mediæval superstition. Theological reasons were soon poured into it, and in the year 968 Pope John XIII gave it the highest ecclesiastical sanction by himself baptizing the great bell of his cathedral church, the Lateran, and christening it with his own name.*

This idea was rapidly developed, and we soon find it supported in ponderous treatises, spread widely in sermons, and popularized in multitudes of inscriptions cast upon the bells themselves. This branch of theological literature may still be studied in multitudes of church towers throughout Europe. A bell at Basel bears the inscription, "Ad fugandos demones." Another, in Lugano, declares "The sound of this bell vanquishes tempests, repels demons, and summons men." Another, at the Cathedral of Erfurt, declares that it can "ward off lightning and malignant demons." A peal in the Jesuit church at the university town of Pont-à-Mousson bore the words, "They praise God, put to flight the clouds, affright the demons, and call the people." This is dated 1634. Another bell in that part of France declares, "It is I who dissipate the thunders" (*Ego sum qui dissipo tonitrua*).†

Another, in one of the forest cantons of Switzerland, bears a doggerel couplet, which may be thus translated:

> "On the devil my spite I'll vent,
> And, God helping, bad weather prevent."‡

Very common were inscriptions embodying this doctrine in sonorous Latin.

Naturally, then, there grew up a ritual for the consecration of bells. Knollys, in his quaint translation of the old

---

* For statements regarding Pope John and bell superstitions, see Higgins's *Anacalypsis*, vol. ii, p. 70. See also Platina, *Vitæ Pontif.*, s. v. John XIII, and Paronius, *Annales Ecclesiastici*, sub anno 968. The conjecture of Baronius that the bell was named after St. John the Baptist, is even more startling than the accepted tradition of the Pope's sponsorship,

† For these illustrations, with others equally striking, see Meyer, *Der Aberglaube des Mittelalters*, pp. 185, 186. For the later examples, see Germain, *Anciennes cloches lorraines* (Nancy, 1885), pp. 23, 27.

‡ "An dem Tüfel will ich mich rächen,
Mit der hilf gotz alle bösen wetter zerbrechen."
(See Meyer, as above.)

chronicler Sleidan, gives us the usage in the simple English of the middle of the sixteenth century:

"In lyke sorte [as churches] are the belles used. And first, forsouth, they must hange so, as the Byshop may goe round about them. Whiche after he hath sayde certen Psalmes, he consecrateth water and salte, and mingleth them together, wherwith he washeth the belle diligently both within and without, after wypeth it drie, and with holy oyle draweth in it the signe of the crosse, and prayeth God, that whan they shall rynge or sounde that bell, all the disceiptes of the devyll may vanyshe away, hayle, thondryng, lightening, wyndes, and tempestes, and all untemperate weathers may be aswaged. Whan he hath wipte out the crosse of oyle wyth a linen cloth, he maketh seven other crosses in the same, and within one only. After saying certen Psalmes, he taketh a payre of sensours and senseth the bel within, and prayeth God to sende it good lucke. In many places they make a great dyner, and kepe a feast as it were at a solemne wedding." *

These bell baptisms became matters of great importance. Popes, kings, and prelates were proud to stand as sponsors. Four of the bells at the Cathedral of Versailles having been destroyed during the French Revolution, four new ones were baptized, on the 6th of January, 1824, the Voltairean King, Louis XVIII, and the pious Duchess d'Angoulême standing as sponsors.

In some of these ceremonies zeal appears to have outrun knowledge, and one of Luther's stories, at the expense of the older Church, was that certain authorities thus christened a bell "Hosanna," supposing that to be the name of a woman.

To add to the efficacy of such baptisms, water was sometimes brought from the river Jordan. †

The prayers used at bell baptisms fully recognise this doctrine. The ritual of Paris embraces the petition that, "whensoever this bell shall sound, it shall drive away the

* Sleidan's *Commentaries*, English translation, as above, fol. 334 (lib. xxi, sub anno 1549).

† See Montanus, as above, who cites Beck, *Lutherthum vor Luthero*, p. 294, for the statement that many bells were carried to the Jordan by pilgrims for this purpose.

malign influences of the assailing spirits, the horror of their apparitions, the rush of whirlwinds, the stroke of lightning, the harm of thunder, the disasters of storms, and all the spirits of the tempest." Another prayer begs that "the sound of this bell may put to flight the fiery darts of the enemy of men"; and others vary the form but not the substance of this petition. The great Jesuit theologian, Bellarmin, did indeed try to deny the reality of this baptism; but this can only be regarded as a piece of casuistry suited to Protestant hardness of heart, or as strategy in the warfare against heretics.*

Forms of baptism were laid down in various manuals sanctioned directly by papal authority, and sacramental efficacy was everywhere taken for granted.† The development of this idea in the older Church was too strong to be resisted;‡ but, as a rule, the Protestant theologians of the Reformation, while admitting that storms were caused by Satan and his legions, opposed the baptism of bells, and denied the theory of their influence in dispersing storms. Luther, while never doubting that troublesome meteorological phenomena were caused by devils, regarded with contempt the idea that

---

* For prayers at bell baptisms, see Arago, *Œuvres*, Paris, 1854, vol. iv, p. 322.

† As has often been pointed out, the ceremony was in all its details—even to the sponsors, the wrapping a garment about the baptized, the baptismal fee, the feast—precisely the same as when a child was baptized. Magius, who is no sceptic, relates from his own experience an instance of this sort, where a certain bishop stood sponsor for two bells, giving them both his own name—William. (See his *De Tintinnabulis*, vol. xiv.)

‡ And no wonder, when the oracle of the Church, Thomas Aquinas, expressly pronounced church bells, "provided they have been duly consecrated and baptized," the foremost means of "frustrating the atmospheric mischiefs of the devil," and likened steeples in which bells are ringing to a hen brooding her chickens, "for the tones of the consecrated metal repel the demons and avert storm and lightning"; when pre-Reformation preachers of such universal currency as Joannes Herolt declared, "Bells, as all agree, are baptized with the result that they are secure from the power of Satan, terrify the demons, compel the powers"; when Geiler of Kaisersberg especially commended bell-ringing as a means of beating off the devil in storms; and when a canonist like Durandus explained the purpose of the rite to be, that "the demons hearing the trumpets of the Eternal King, to wit, the bells, may flee in terror, and may cease from the stirring up of tempests." See Herolt, *Sermones Discipuli*, vol. xvii, and Durandus, *De ritibus ecclesiæ*, vol. ii, p. 12. I owe the first of these citations to Rydberg, and the others to Montanus. For Geiler, see Dacheux, *Geiler de Kaisersberg*, pp. 280, 281.

the demons were so childish as to be scared by the clang of bells; his theory made them altogether too powerful to be affected by means so trivial. The great English Reformers, while also accepting very generally the theory of diabolic interference in storms, reproved strongly the baptizing of bells, as the perversion of a sacrament and involving blasphemy. Bishop Hooper declared reliance upon bells to drive away tempests, futile. Bishop Pilkington, while arguing that tempests are direct instruments of God's wrath, is very severe against using "unlawful means," and among these he names "the hallowed bell"; and these opinions were very generally shared by the leading English clergy.*

Toward the end of the sixteenth century the Elector of Saxony strictly forbade the ringing of bells against storms, urging penance and prayer instead; but the custom was not so easily driven out of the Protestant Church, and in some quarters was developed a Protestant theory of a rationalistic sort, ascribing the good effects of bell-ringing in storms to the calling together of the devout for prayer or to the suggestion of prayers during storms at night. As late as the end of the seventeenth century we find the bells of Protestant churches in northern Germany rung for the dispelling of tempests. In Catholic Austria this bell-ringing seems to have become a nuisance in the last century, for the Emperor Joseph II found it necessary to issue an edict against it; but this doctrine had gained too large headway to be arrested by argument or edict, and the bells may be heard ringing during storms to this day in various remote districts in Europe.† For this was no mere superficial view.

---

* The baptism of bells was, indeed, one of the express complaints of the German Protestant princes at the Reformation. See their *Gravam. Cent. German. Grav.*, p. 51. For Hooper, see his *Early Writings*, p. 197 (in *Parker Society Publications*). For Pilkington, see his *Works*, p. 177 (in same). Among others sharing these opinions were Tyndale, Bishop Ridley, Archbishop Sandys, Becon, Calfhill, and Rogers. It is to be noted that all these speak of the rite as "baptism."

† For Elector of Saxony, see Peuchen, *Disp. circa tempestates*, Jena, 1697. For the Protestant theory of bells, see, e. g., the *Conciones Selectæ* of Superintendent Conrad Dieterich (cited by Peuchen, *Disp. circa tempestates*). For Protestant ringing of bells to dispel tempests, see Schwimmer, *Physicalische Luftfragen*, 1692 (cited by Peuchen, as above). He pictures the whole population of a Thuringian district flocking to the churches on the approach of a storm.

It was really part of a deep theological current steadily developed through the Middle Ages, the fundamental idea of the whole being the direct influence of the bells upon the "Power of the Air"; and it is perhaps worth our while to go back a little and glance over the coming of this current into the modern world. Having grown steadily through the Middle Ages, it appeared in full strength at the Reformation period; and in the sixteenth century Olaus Magnus, Archbishop of Upsala and Primate of Sweden, in his great work on the northern nations, declares it a well-established fact that cities and harvests may be saved from lightning by the ringing of bells and the burning of consecrated incense, accompanied by prayers; and he cautions his readers that the workings of the thunderbolt are rather to be marvelled at than inquired into. Even as late as 1673 the Franciscan professor Lealus, in Italy, in a schoolbook which was received with great applause in his region, taught unhesitatingly the agency of demons in storms, and the power of bells over them, as well as the portentousness of comets and the movement of the heavens by angels. He dwells especially, too, upon the perfect protection afforded by the waxen *Agnus Dei*. How strong this current was, and how difficult even for philosophical minds to oppose, is shown by the fact that both Descartes and Francis Bacon speak of it with respect, admitting the fact, and suggesting very mildly that the bells may accomplish this purpose by the concussion of the air.*

But no such moderate doctrine sufficed, and the renowned Bishop Binsfeld, of Treves, in his noted treatise on the credibility of the confessions of witches, gave an entire chapter to the effect of bells in calming atmospheric disturbances. Basing his general doctrine upon the first chapter of Job and the second chapter of Ephesians, he insisted on the reality of diabolic agency in storms; and then, by theological reasoning, corroborated by the statements extorted in the torture chamber, he showed the efficacy of bells

* For Olaus Magnus, see the *De gentibus septentrionalibus* (Rome, 1555), lib. i, c. 12, 13. For Descartes, see his *De meteor.*, c. 7. For Bacon, see his *Natural History*, cent. 2, 127. In his *Historia Ventorum* he again alludes to the belief, and without comment.

in putting the hellish legions to flight.* This continued, therefore, an accepted tenet, developed in every nation, and coming to its climax near the end of the seventeenth century. At that period—the period of Isaac Newton—Father Augustine de Angelis, rector of the Clementine College at Rome, published under the highest Church authority his lectures upon meteorology. Coming from the centre of Catholic Christendom, at so late a period, they are very important as indicating what had been developed under the influence of theology during nearly seventeen hundred years. This learned head of a great college at the heart of Christendom taught that "the surest remedy against thunder is that which our Holy Mother the Church practises, namely, the ringing of bells when a thunderbolt impends: thence follows a twofold effect, physical and moral—a physical, because the sound variously disturbs and agitates the air, and by agitation disperses the hot exhalations and dispels the thunder; but the moral effect is the more certain, because by the sound the faithful are stirred to pour forth their prayers, by which they win from God the turning away of the thunderbolt." Here we see in this branch of thought, as in so many others, at the close of the seventeenth century, the dawn of rationalism. Father De Angelis now keeps demoniacal influence in the background. Little, indeed, is said of the efficiency of bells in putting to flight the legions of Satan: the wise professor is evidently preparing for that inevitable compromise which we see in the history of every science when it is clear that it can no longer be suppressed by ecclesiastical fulminations.†

III. THE AGENCY OF WITCHES.

But, while this comparatively harmless doctrine of thwarting the powers of the air by fetiches and bell-ringing was developed, there were evolved another theory, and a series of practices sanctioned by the Church, which must forever be considered as among the most fearful calamities in human

* See Binsfeld, *De Confessionbus Malef.*, pp. 308–314, edition of 1623.
† For De Angelis, see his *Lectiones Meteorol.*, p. 75.

history. Indeed, few errors have ever cost so much shedding of innocent blood over such wide territory and during so many generations. Out of the old doctrine—pagan and Christian—of evil agency in atmospheric phenomena was evolved the belief that certain men, women, and children may secure infernal aid to produce whirlwinds, hail, frosts, floods, and the like.

As early as the ninth century one great churchman, Agobard, Archbishop of Lyons, struck a heavy blow at this superstition. His work, *Against the Absurd Opinion of the Vulgar touching Hail and Thunder*, shows him to have been one of the most devoted apostles of right reason whom human history has known. By argument and ridicule, and at times by a lofty eloquence, he attempted to breast this tide. One passage is of historical significance. He declares: "The wretched world lies now under the tyranny of foolishness; things are believed by Christians of such absurdity as no one ever could aforetime induce the heathen to believe." *

All in vain; the tide of superstition continued to roll on; great theologians developed it and ecclesiastics favoured it; until as we near the end of the mediæval period the infallible voice of Rome is heard accepting it, and clinching this belief into the mind of Christianity. For, in 1437, Pope Eugene IV, by virtue of the teaching power conferred on him by the Almighty, and under the divine guarantee against any possible error in the exercise of it, issued a bull exhorting the inquisitors of heresy and witchcraft to use greater diligence against the human agents of the Prince of Darkness, and especially against those who have the power to produce bad weather. In 1445 Pope Eugene returned again to the charge, and again issued instructions and commands infallibly committing the Church to the doctrine. But a greater than Eugene followed, and stamped the idea yet more deeply into the mind of the Church. On the 7th of December, 1484, Pope Innocent VIII sent forth his bull *Summis Desiderantes*. Of all documents ever issued from Rome, imperial

* For a very interesting statement of Agobard's position and work, with citations from his *Liber contra insulsam vulgi opinionem de grandine et tonitruis*, see Poole, *Illustrations of the History of Mediæval Thought*, pp. 40 *et seq.* The works of Agobard are in vol. civ of Migne's *Patrol. Lat.*

or papal, this has doubtless, first and last, cost the greatest shedding of innocent blood. Yet no document was ever more clearly dictated by conscience. Inspired by the scriptural command, "Thou shalt not suffer a witch to live," Pope Innocent exhorted the clergy of Germany to leave no means untried to detect sorcerers, and especially those who by evil weather destroy vineyards, gardens, meadows, and growing crops. These precepts were based upon various texts of Scripture, especially upon the famous statement in the book of Job; and, to carry them out, witch-finding inquisitors were authorized by the Pope to scour Europe, especially Germany, and a manual was prepared for their use—the Witch-Hammer, *Malleus Maleficarum*. In this manual, which was revered for centuries, both in Catholic and Protestant countries, as almost divinely inspired, the doctrine of Satanic agency in atmospheric phenomena was further developed, and various means of detecting and punishing it were dwelt upon.*

With the application of torture to thousands of women, in accordance with the precepts laid down in the *Malleus*, it was not difficult to extract masses of proof for this sacred theory of meteorology. The poor creatures, writhing on the rack, held in horror by those who had been nearest and dearest to them, anxious only for death to relieve their sufferings, confessed to anything and everything that would satisfy the inquisitors and judges. All that was needed was that the inquisitors should ask leading questions † and sug-

---

* For the bull of Pope Eugene, see Raynaldus, *Annales Eccl.*, pp. 1437, 1445. The Latin text of the bull *Summis Desiderantes* may be found in the *Malleus Maleficarum*, in Binsfeld's *De Confessionibus* cited below, or in Roskoff's *Geschichte des Teufels* (Leipsic, 1869), vol. i, pp. 222–225. There is, so far as I know, no good analysis, in any English book, of the contents of the *Witch-Hammer*; but such may be found in Roskoff's *Geschichte des Teufels*, or in Soldan's *Geschichte der Hexenprozesse*. Its first dated edition is that of 1489; but Prof. Burr has shown that it was printed as early as 1486. It was, happily, never translated into any modern tongue.

† For still extant lists of such questions, see the *Zeitschrift für deutsche Culturgeschichte* for 1858, pp. 522–528, or Diefenbach, *Der Hexenwahn in Deutschland*, pp. 15–17. Father Vincent of Berg (in his *Enchiridium*) gives a similar list for use by priests in the confession of the accused. Manuscript lists of this sort which have actually done service in the courts of Baden and Bavaria may be seen in the library of Cornell University.

gest satisfactory answers: the prisoners, to shorten the torture, were sure sooner or later to give the answer required, even though they knew that this would send them to the stake or scaffold. Under the doctrine of "excepted cases," there was no limit to torture for persons accused of heresy or witchcraft; even the safeguards which the old pagan world had imposed upon torture were thus thrown down, and the prisoner *must* confess.

The theological literature of the Middle Ages was thus enriched with numberless statements regarding modes of Satanic influence on the weather. Pathetic, indeed, are the records; and none more so than the confessions of these poor creatures, chiefly women and children, during hundreds of years, as to their manner of raising hailstorms and tempests. Such confessions, by tens of thousands, are still to be found in the judicial records of Germany, and indeed of all Europe. Typical among these is one on which great stress was laid during ages, and for which the world was first indebted to one of these poor women. Crazed by the agony of torture, she declared that, returning with a demon through the air from the witches' sabbath, she was dropped upon the earth in the confusion which resulted among the hellish legions when they heard the bells sounding the *Ave Maria*. It is sad to note that, after a contribution so valuable to sacred science, the poor woman was condemned to the flames. This revelation speedily ripened the belief that, whatever might be going on at the witches' sabbath—no matter how triumphant Satan might be—at the moment of sounding the consecrated bells the Satanic power was paralyzed. This theory once started, proofs came in to support it, during a hundred years, from the torture chambers in all parts of Europe.

Throughout the later Middle Ages the Dominicans had been the main agents in extorting and promulgating these revelations, but in the centuries following the Reformation the Jesuits devoted themselves with even more keenness and vigour to the same task. Some curious questions incidentally arose. It was mooted among the orthodox authorities whether the damage done by storms should or should not be assessed upon the property of convicted witches. The theologians inclined decidedly to the affirmative; the jurists, on the whole, to the negative.*

In spite of these tortures, lightning and tempests continued, and great men arose in the Church throughout Europe in every generation to point out new cruelties for the discovery of "weather-makers," and new methods for bringing their machinations to naught.

But here and there, as early as the sixteenth century, we begin to see thinkers endeavouring to modify or oppose these methods. At that time Paracelsus called attention to the reverberation of cannon as explaining the rolling of thunder, but he was confronted by one of his greatest contemporaries. Jean Bodin, as superstitious in natural as he was rational in political science, made sport of the scientific theory, and declared thunder to be "a flaming exhalation set in motion by evil spirits, and hurled downward with a great crash and a horrible smell of sulphur." In support of this view, he dwelt upon the confessions of tortured witches, upon the acknowledged agency of demons in the Will-o'-the-wisp, and specially upon the passage in the one hundred and fourth Psalm, "Who maketh his angels spirits, his ministers a flaming fire."

To resist such powerful arguments by such powerful men was dangerous indeed. In 1513, Pomponatius, professor at Padua, published a volume of *Doubts as to the Fourth Book of Aristotle's Meteorologica*, and also dared to question this power of devils; but he soon found it advisable to explain that, while as a *philosopher* he might doubt, yet as a *Christian* he of course believed everything taught by Mother Church—devils and all—and so escaped the fate of several others who dared to question the agency of witches in atmospheric and other disturbances.

A few years later Agrippa of Nettesheim made a somewhat similar effort to breast this theological tide in northern Europe. He had won a great reputation in various fields, but especially in natural science, as science was then understood. Seeing the folly and cruelty of the prevailing theory,

---

* For proofs of the vigour of the Jesuits in this persecution, see not only the histories of witchcraft, but also the *Annuæ litteræ* of the Jesuits themselves, *passim*.

he attempted to modify it, and in 1518, as Syndic of Metz, endeavoured to save a poor woman on trial for witchcraft. But the chief inquisitor, backed by the sacred Scriptures, the papal bulls, the theological faculties, and the monks, was too strong for him; he was not only forced to give up his office, but for this and other offences of a similar sort was imprisoned, driven from city to city and from country to country, and after his death his clerical enemies, especially the Dominicans, pursued his memory with calumny, and placed over his grave probably the most malignant epitaph ever written.

As to argument, these efforts were met especially by Jean Bodin in his famous book, the *Démonomanie des Sorciers*, published in 1580. It was a work of great power by a man justly considered the leading thinker in France, and perhaps in Europe. All the learning of the time, divine and human, he marshalled in support of the prevailing theory. With inexorable logic he showed that both the veracity of sacred Scripture and the infallibility of a long line of popes and councils of the Church were pledged to it, and in an eloquent passage this great publicist warned rulers and judges against any mercy to witches—citing the example of King Ahab condemned by the prophet to die for having pardoned a man worthy of death, and pointing significantly to King Charles IX of France, who, having pardoned a sorcerer, died soon afterward.*

In the last years of the sixteenth century the persecutions for witchcraft and magic were therefore especially cruel; and in the western districts of Germany the main instrument in them was Binsfeld, Suffragan Bishop of Treves.

* To the argument cited above, Bodin adds: "Id certissimam dæmonis præsentiam significat: nam ubicunque dæmones cum hominibus nefaria societatis fide copulantur, fœdissimum semper relinquunt sulphuris odorem, quod sortilegi sæpissime experiuntur et confitentur." See Bodin's *Universæ Naturæ Theatrum*, Frankfort, 1597, pp. 208–211. The first edition of the book by Pomponatius, which was the earliest of his writings, is excessively rare, but it was reprinted at Venice just a half-century later. It is in his *De incantationibus*, however, that he speaks especially of devils. As to Pomponatius, see, besides these, Creighton's *History of the Papacy during the Reformation*, and an excellent essay in Franck's *Moralistes et Philosophes*. For Agrippa, see his biography by Prof. Henry Morley, London, 1856. For Bodin, see a statement of his general line of argument in Lecky, *Rationalism in Europe*, vol. i, chap. i.

At that time Cornelius Loos was a professor at the university of that city. He was a devoted churchman, and one of the most brilliant opponents of Protestantism, but he finally saw through the prevailing belief regarding occult powers, and in an evil hour for himself embodied his idea in a book entitled *True and False Magic.* The book, though earnest, was temperate, but this helped him and his cause not at all. The texts of Scripture clearly sanctioning belief in sorcery and magic stood against him, and these had been confirmed by the infallible teachings of the Church and the popes from time immemorial; the book was stopped in the press, the manuscript confiscated, and Loos thrown into a dungeon.

The inquisitors having wrought their will upon him, in the spring of 1593 he was brought out of prison, forced to recant on his knees before the assembled dignitaries of the Church, and thenceforward kept constantly under surveillance and at times in prison. Even this was considered too light a punishment, and his arch-enemy, the Jesuit Delrio, declared that, but for his death by the plague, he would have been finally sent to the stake.*

That this threat was not unmeaning had been seen a few years earlier in a case even more noted, and in the same city. During the last decades of the sixteenth century, Dietrich Flade, an eminent jurist, was rector of the University of Treves, and chief judge of the Electoral Court, and in the latter capacity he had to pass judgment upon persons tried on the capital charge of magic and witchcraft. For a time he yielded to the long line of authorities, ecclesiastical and judicial, supporting the reality of this crime; but he at last seems to have realized that it was unreal, and that the confessions in his torture chamber, of compacts with Satan, riding on broomsticks to the witch-sabbath,

* What remains of the manuscript of Loos, which until recently was supposed to be lost, was found, hidden away on the shelves of the old Jesuit library at Treves, by Mr. George Lincoln Burr, now a professor at Cornell University; and Prof. Burr's copy of the manuscript is now in the library of that institution. For a full account of the discovery and its significance, see the *New York Nation* for November 11, 1886. The facts regarding the after-life of Loos were discovered by Prof. Burr in manuscript records at Brussels.

raising tempests, producing diseases, and the like, were either the results of madness or of willingness to confess anything and everything, and even to die, in order to shorten the fearful tortures to which the accused were in all cases subjected until a satisfactory confession was obtained.

On this conviction of the unreality of many at least of the charges Flade seems to have acted, and he at once received his reward. He was arrested by the authority of the archbishop and charged with having sold himself to Satan—the fact of his hesitation in the persecution being perhaps what suggested his guilt. He was now, in his turn, brought into the torture chamber over which he had once presided, was racked until he confessed everything which his torturers suggested, and finally, in 1589, was strangled and burnt.

Of that trial a record exists in the library of Cornell University in the shape of the original minutes of the case, and among them the depositions of Flade when under torture, taken down from his own lips in the torture chamber. In these depositions this revered and venerable scholar and jurist acknowledged the truth of every absurd charge brought against him—anything, everything, which would end the fearful torture: compared with that, death was nothing.*

Nor was even a priest secure who ventured to reveal the unreality of magic. When Friedrich Spee, the Jesuit poet of western Germany, found, in taking the confessions of those about to be executed for magic, that without exception, just when about to enter eternity and utterly beyond hope of pardon, they all retracted their confessions made under torture, his sympathies as a man rose above his loyalty to his order, and he published his *Cautio Criminalis* as a warning, stating with entire moderation the facts he had observed and the necessity of care. But he did not dare publish it under his own name, nor did he even dare publish it in a Catholic town; he gave it to the world anonymously, and, in order to prevent any tracing of the work to him through the confessional, he secretly caused it to be published in the Protestant town of Rinteln.

---

* For the case of Flade, see the careful study by Prof. Burr, *The Fate of Dietrich Flade*, in the *Papers of the American Historical Association*, 1891.

Nor was this all. Nothing shows so thoroughly the hold that this belief in magic had obtained as the conduct of Spee's powerful friend and contemporary, John Philip von Schönborn, later the Elector and Prince Archbishop of Mayence.

As a youth, Schönborn had loved and admired Spee, and had especially noted his persistent melancholy and his hair whitened even in his young manhood. On Schönborn's pressing him for the cause, Spee at last confessed that his sadness, whitened hair, and premature old age were due to his recollections of the scores of men and women and children whom he had been obliged to see tortured and sent to the scaffold and stake for magic and witchcraft, when he as their father confessor positively knew them to be innocent. The result was that, when Schönborn became Elector and Archbishop of Mayence, he stopped the witch persecutions in that province, and prevented them as long as he lived. But here was shown the strength of theological and ecclesiastical traditions and precedents. Even a man so strong by family connections, and enjoying such great temporal and spiritual power as Schönborn, dared not openly give his reasons for this change of policy. So far as is known, he never uttered a word publicly against the reality of magic, and under his successor in the electorate witch trials were resumed.

The great upholders of the orthodox view retained full possession of the field. The victorious Bishop Binsfeld, of Treves, wrote a book to prove that everything confessed by the witches under torture, especially the raising of storms and the general controlling of the weather, was worthy of belief; and this book became throughout Europe a standard authority, both among Catholics and Protestants. Even more inflexible was Remigius, criminal judge in Lorraine. On the title-page of his manual he boasts that within fifteen years he had sent nine hundred persons to death for this imaginary crime.*

* For Spee and Schönborn, see Soldan and other German authorities. There are copies of the first editions of the *Cautio Criminalis* in the library of Cornell University. Binsfeld's book bore the title of *Tractatus de confessionibus maleficorum et sagarum*. First published at Treves in 1589, it appeared subsequently four times in the original Latin, as well as in two distinct German translations, and

Protestantism fell into the superstition as fully as Catholicism. In the same century John Wier, a disciple of Agrippa, tried to frame a pious theory which, while satisfying orthodoxy, should do something to check the frightful cruelties around him. In his book *De Præstigiis Dæmonum*, published in 1563, he proclaimed his belief in witchcraft, but suggested that the compacts with Satan, journeys through the air on broomsticks, bearing children to Satan, raising storms and producing diseases—to which so many women and children confessed under torture—were delusions suggested and propagated by Satan himself, and that the persons charged with witchcraft were therefore to be considered "as possessed"—that is, rather as sinned against than sinning.*

But neither Catholics nor Protestants would listen for a moment to any such suggestion. Wier was bitterly denounced and persecuted. Nor did Bekker, a Protestant divine in Holland, fare any better in the following century. For his *World Bewitched*, in which he ventured not only to question the devil's power over the weather, but to deny his bodily existence altogether, he was solemnly tried by the synod of his Church and expelled from his pulpit, while his views were condemned as heresy, and overwhelmed with a flood of refutations whose mere catalogue would fill pages; and these cases were typical of many.

The Reformation had, indeed, at first deepened the superstition; the new Church being anxious to show itself equally orthodox and zealous with the old. During the century following the first great movement, the eminent Lutheran jurist and theologian Benedict Carpzov, whose boast was that he had read the Bible fifty-three times, especially distinguished himself by his skill in demonstrating the reality of witchcraft, and by his cruelty in detecting and punishing it. The torture chambers were set at work more vigorously than ever, and a long line of theological jurists followed to maintain the system and to extend it.

---

in a French one. Remigius's manual was entitled *Dæmonolatreia*, and was first printed at Lyons in 1595.

* For Wier, or Weyer, see, beside his own works, the excellent biography by Prof. Binz, of Bonn.

To argue against it, or even doubt it, was exceedingly dangerous. Even as late as the beginning of the eighteenth century, when Christian Thomasius, the greatest and bravest German between Luther and Lessing, began the efforts which put an end to it in Protestant Germany, he did not dare at first, bold as he was, to attack it in his own name, but presented his views as the university thesis of an irresponsible student.*

The same stubborn resistance to the gradual encroachment of the scientific spirit upon the orthodox doctrine of witchcraft was seen in Great Britain. Typical as to the attitude both of Scotch and English Protestants were the theory and practice of King James I, himself the author of a book on *Demonology*, and nothing if not a theologian. As to theory, his treatise on *Demonology* supported the worst features of the superstition; as to practice, he ordered the learned and acute work of Reginald Scot, *The Discoverie of Witchcraft*, one of the best treatises ever written on the subject, to be burned by the hangman, and he applied his own knowledge to investigating the causes of the tempests which beset his bride on her voyage from Denmark. Skilful use of unlimited torture soon brought these causes to light. A Dr. Fian, while his legs were crushed in the "boots" and wedges were driven under his finger nails, confessed that several hundred witches had gone to sea in a sieve from the port of Leith, and had raised storms and tempests to drive back the princess.

With the coming in of the Puritans the persecution was even more largely, systematically, and cruelly developed. The great witch-finder, Matthew Hopkins, having gone through the county of Suffolk and tested multitudes of poor old women by piercing them with pins and needles, declared that county to be infested with witches. Thereupon Parliament issued a commission, and sent two eminent Presbyterian divines to accompany it, with the result that in

---

* For Thomasius, see his various biographies by Luden and others; also the treatises on witchcraft of Soldan and others. Manuscript notes of his lectures, and copies of his earliest books on witchcraft as well as on other forms of folly, are to be found in the library of Cornell University.

that county alone sixty persons were hanged for witchcraft in a single year. In Scotland matters were even worse. The *auto da fé* of Spain was celebrated in Scotland under another name, and with Presbyterian ministers instead of Roman Catholic priests as the main attendants. At Leith, in 1664, nine women were burned together. Condemnations and punishments of women in batches were not uncommon. Torture was used far more freely than in England, both in detecting witches and in punishing them. The natural argument developed in hundreds of pulpits was this: If the All-wise God punishes his creatures with tortures infinite in cruelty and duration, why should not his ministers, as far as they can, imitate him?

The strongest minds in both branches of the Protestant Church in Great Britain devoted themselves to maintaining the superstition. The newer scientific modes of thought, and especially the new ideas regarding the heavens, revealed first by Copernicus and Galileo and later by Newton, Huygens, and Halley, were gradually dissipating the whole domain of the Prince of the Power of the Air; but from first to last a long line of eminent divines, Anglican and Calvinistic, strove to resist the new thought. On the Anglican side, in the seventeenth century, Meric Casaubon, Doctor of Divinity and a high dignitary of Canterbury,—Henry More, in many respects the most eminent scholar in the Church,—Cudworth, by far the most eminent philosopher, and Dr. Joseph Glanvil, the most cogent of all writers in favour of witchcraft, supported the orthodox superstition in treatises of great power; and Sir Matthew Hale, the greatest jurist of the period, condemning two women to be burned for witchcraft, declared that he based his judgment on the direct testimony of Holy Scripture. On the Calvinistic side were the great names of Richard Baxter, who applauded some of the worst cruelties in England, and of Increase and Cotton Mather, who stimulated the worst in America; and these marshalled in behalf of this cruel superstition a long line of eminent divines, the most earnest of all, perhaps, being John Wesley.

Nor was the Lutheran Church in Sweden and the other Scandinavian countries behind its sister churches, either in persecuting witchcraft or in repressing doubts regarding the doctrine which supported it.

But in spite of all these great authorities in every land, in spite of such summary punishments as those of Flade, Loos, and Bekker, and in spite of the virtual exclusion from church preferment of all who doubted the old doctrine, the new scientific view of the heavens was developed more and more; the physical sciences were more and more cultivated; the new scientific atmosphere in general more and more prevailed; and at the end of the seventeenth century this vast growth of superstition began to wither and droop. Montaigne, Bayle, and Voltaire in France, Thomasius in Germany, Calef in New England, and Beccaria in Italy, did much also to create an intellectual and moral atmosphere fatal to it.

And here it should be stated, to the honour of the Church of England, that several of her divines showed great courage in opposing the dominant doctrine. Such men as Harsnet, Archbishop of York, and Morton, Bishop of Lichfield, who threw all their influence against witch-finding cruelties even early in the seventeenth century, deserve lasting gratitude. But especially should honour be paid to the younger men in the Church, who wrote at length against the whole system: such men as Wagstaffe and Webster and Hutchinson, who in the humbler ranks of the clergy stood manfully for truth, with the certainty that by so doing they were making their own promotion impossible.

By the beginning of the eighteenth century the doctrine was evidently dying out. Where torture had been abolished, or even made milder, "weather-makers" no longer confessed, and the fundamental proofs in which the system was rooted were evidently slipping away. Even the great theologian Fromundus, at the University of Louvain, the oracle of his age, who had demonstrated the futility of the Copernican theory, had foreseen this and made the inevitable attempt at compromise, declaring that devils, though *often*, are not *always* or even for the most part the causes of thunder. The learned Jesuit Caspar Schott, whose *Physica Curiosa* was one of the most popular books of the seventeenth century, also ventured to make the same mild statement. But even such concessions by such great champions of orthodoxy did not prevent frantic efforts in various quarters to bring the world back under the old dogma: as late as 1743 there was published in Catholic Germany a manual by Father Vincent of Berg, in which the superstition was taught to its fullest extent, with the declaration that it was issued for the use of priests under the express sanction of the theological professors of the University of Cologne; and twenty-five years later, in 1768, we find in Protestant England John Wesley standing firmly for witchcraft, and uttering his famous declaration, "The giving up of witchcraft is in effect the giving up of the Bible." The latest notable demonstration in Scotland was made as late as 1773, when "the divines of the Associated Presbytery" passed a resolution declaring their belief in witchcraft, and deploring the general scepticism regarding it.*

---

* For Carpzov and his successors, see authorities already given. The best account of James's share in the extortion of confessions may be found in the collection of *Curious Tracts* published at Edinburgh in 1820. See also King James's own *Demonologie*, and Pitcairn's *Criminal Trials of Scotland*, vol. i, part ii, pp. 213–223. For Casaubon, see his *Credulity and Incredulity in Things Natural*, pp. 66, 67. For Glanvil, More, Casaubon, Baxter, Wesley, and others named, see Lecky, as above. As to Increase Mather, in his sermons, already cited, on *The Voice of God in Stormy Winds*, Boston, 1704, he says: "When there are great tempests, the Angels oftentimes have a Hand therein. . . . Yea, and sometimes, by Divine Permission, Evil Angels have a Hand in such Storms and Tempests as are very hurtful to Men on the Earth." Yet "for the most part, such Storms are sent by the Providence of God as a Sign of His Displeasure for the Sins of Men," and sometimes "as Prognosticks and terrible Warnings of Great Judgements not far off." From the height of his erudition Mather thus rebukes the timid voice of scientific scepticism: "There are some who would be esteemed the Wits of the World, that ridicule those as Superstitious and Weak Persons, which look upon Dreadful Tempests as Prodromous of other Judgements. Nevertheless, the most Learned and Judicious Writers, not only of the Gentiles, but amongst Christians, have Embraced such a Persuasion; their Sentiments therein being Confirmed by the Experience of many Ages." For another curious turn given to this theory, with reference to sanitary science, see Deodat Lawson's famous sermon at Salem, in 1692, on *Christ's Fidelity a Shield against Satan's Malignity*, p. 21 of the second edition. For Cotton Mather, see his biography by Barrett Wendell, pp. 91, 92; also the chapter on *Diabolism and Hysteria* in this work. For Fromundus, see his *Meteorologica* (London, 1656), lib. iii, c. 9, and lib. ii, c. 3. For Schott, see his *Physica Curiosa* (edition of Würzburg, 1667), p. 1249. For Father Vincent of Berg, see his *Enchiridium quadripartitum* (Cologne, 1743). Besides benedictions and exorcisms for all emergencies, it contains full directions for the

IV. FRANKLIN'S LIGHTNING-ROD.

But in the midst of these efforts by Catholics like Father Vincent and by Protestants like John Wesley to save the old sacred theory, it received its death-blow. In 1752 Franklin made his experiments with the kite on the banks of the Schuylkill; and, at the moment when he drew the electric spark from the cloud, the whole tremendous fabric of theological meteorology reared by the fathers, the popes, the mediæval doctors, and the long line of great theologians, Catholic and Protestant, collapsed; the "Prince of the Power of the Air" tumbled from his seat; the great doctrine which had so long afflicted the earth was prostrated forever.

The experiment of Franklin was repeated in various parts of Europe, but, at first, the Church seemed careful to take no notice of it. The old church formulas against the Prince of the Power of the Air were still used, but the theological theory, especially in the Protestant Church, began to grow milder. Four years after Franklin's discovery Pastor Karl Koken, member of the Consistory and official preacher to the City Council of Hildesheim, was moved by a great hailstorm to preach and publish a sermon on *The Revelation of God in Weather*. Of "the Prince of the Power of the Air" he says nothing; the theory of diabolical agency he throws overboard altogether; his whole attempt is to save the older and more harmless theory, that the storm is the voice of God. He insists that, since Christ told Nicodemus that men "know not whence the wind cometh," it can not be of mere natural origin, but is sent directly by God himself, as David intimates in the Psalm, "out of His secret places." As to the hailstorm, he lays great stress upon the plague of hail sent by the Almighty upon Egypt, and clinches all by insisting that God showed at Mount Sinai his purpose to startle the body before impressing the conscience.

manufacture of the *Agnus Dei*, and of another sacred panacea called "*Heiligthum*," not less effective against evil powers,—gives formulæ to be worn for protection against the devil,—suggests a list of signs by which diabolical possession may be recognised, and prescribes the questions to be asked by priests in the examination of witches. For Wesley, see his *Journal* for 1768. The whole citation is given in Lecky.

While the theory of diabolical agency in storms was thus drooping and dying, very shrewd efforts were made at compromise. The first of these attempts we have already noted, in the effort to explain the efficacy of bells in storms by their simple use in stirring the faithful to prayer, and in the concession made by sundry theologians, and even by the great Lord Bacon himself, that church bells might, under the sanction of Providence, disperse storms by agitating the air. This gained ground somewhat, though it was resisted by one eminent Church authority, who answered shrewdly that, in that case, cannon would be even more pious instruments. Still another argument used in trying to save this part of the theological theory was that the bells were consecrated instruments for this purpose, "like the horns at whose blowing the walls of Jericho fell." *

But these compromises were of little avail. In 1766 Father Sterzinger attacked the very groundwork of the whole diabolic theory. He was, of course, bitterly assailed, insulted, and hated; but the Church thought it best not to condemn him. More and more the "Prince of the Power of the Air" retreated before the lightning-rod of Franklin. The older Church, while clinging to the old theory, was finally obliged to confess the supremacy of Franklin's theory practically; for his lightning-rod did what exorcisms, and holy water, and processions, and the *Agnus Dei*, and the ringing of church bells, and the rack, and the burning of witches, had failed to do. This was clearly seen, even by the poorest peasants in eastern France, when they observed that the grand spire of Strasburg Cathedral, which neither the sacredness of the place, nor the bells within it, nor the holy water and relics beneath it, could protect from frequent injuries by lightning, was once and for all protected by Franklin's rod. Then came into the minds of multitudes the answer to the question which had so long exercised the leading theologians of Europe and America, namely, "Why should the Almighty strike his own consecrated temples, or suffer Satan to strike them?"

---

* For Koken, see his *Offenbarung Gottes in Wetter*, Hildesheim, 1756; and for the answer to Bacon, see Gretser's *De benedictionibus*, lib. ii, cap. 46.

Yet even this practical solution of the question was not received without opposition.

In America the earthquake of 1755 was widely ascribed, especially in Massachusetts, to Franklin's rod. The Rev. Thomas Prince, pastor of the Old South Church, published a sermon on the subject, and in the appendix expressed the opinion that the frequency of earthquakes may be due to the erection of "iron points invented by the sagacious Mr. Franklin." He goes on to argue that "in Boston are more erected than anywhere else in New England, and Boston seems to be more dreadfully shaken. Oh! there is no getting out of the mighty hand of God."

Three years later, John Adams, speaking of a conversation with Arbuthnot, a Boston physician, says: "He began to prate upon the presumption of philosophy in erecting iron rods to draw the lightning from the clouds. He railed and foamed against the points and the presumption that erected them. He talked of presuming upon God, as Peter attempted to walk upon the water, and of attempting to control the artillery of heaven."

As late as 1770 religious scruples regarding lightning-rods were still felt, the theory being that, as thunder and lightning were tokens of the Divine displeasure, it was impiety to prevent their doing their full work. Fortunately, Prof. John Winthrop, of Harvard, showed himself wise in this, as in so many other things: in a lecture on earthquakes he opposed the dominant theology; and as to arguments against Franklin's rods, he declared, "It is as much our duty to secure ourselves against the effects of lightning as against those of rain, snow, and wind by the means God has put into our hands."

Still, for some years theological sentiment had to be regarded carefully. In Philadelphia, a popular lecturer on science for some time after Franklin's discovery thought it best in advertising his lectures to explain that "the erection of lightning-rods is not chargeable with presumption nor inconsistent with any of the principles either of natural or revealed religion." *

* Regarding opposition to Franklin's rods in America, see Prince's *Sermon*,

In England, the first lightning conductor upon a church was not put up until 1762, ten years after Franklin's discovery. The spire of St. Bride's Church in London was greatly injured by lightning in 1750, and in 1764 a storm so wrecked its masonry that it had to be mainly rebuilt; yet for years after this the authorities refused to attach a lightning-rod. The Protestant Cathedral of St. Paul's, in London, was not protected until sixteen years after Franklin's discovery, and the tower of the great Protestant church at Hamburg not until a year later still. As late as 1783 it was declared in Germany, on excellent authority, that within a space of thirty-three years nearly four hundred towers had been damaged and one hundred and twenty bell-ringers killed.

In Roman Catholic countries a similar prejudice was shown, and its cost at times was heavy. In Austria, the church of Rosenberg, in the mountains of Carinthia, was struck so frequently and with such loss of life that the peasants feared at last to attend service. Three times was the spire rebuilt, and it was not until 1778—twenty-six years after Franklin's discovery—that the authorities permitted a rod to be attached. Then all trouble ceased.

A typical case in Italy was that of the tower of St. Mark's, at Venice. In spite of the angel at its summit and the bells consecrated to ward off the powers of the air, and the relics in the cathedral hard by, and the processions in the adjacent square, the tower was frequently injured and even ruined by lightning. In 1388 it was badly shattered; in 1417, and again in 1489, the wooden spire surmounting it was utterly consumed; it was again greatly injured in 1548, 1565, 1653, and in 1745 was struck so powerfully that the whole tower, which had been rebuilt of stone and brick, was shattered in thirty-seven places. Although the invention of Franklin had been introduced into Italy by the physicist Beccaria, the tower of St. Mark's still went unprotected, and was again badly struck in 1761 and 1762; and not until 1766—fourteen

---

especially p. 23; also Quincy, *History of Harvard University*, vol. ii, p. 219; also *Works of John Adams*, vol. ii, pp. 51, 52; also Parton's *Life of Franklin*, vol. i, p. 294.

years after Franklin's discovery—was a lightning-rod placed upon it; and it has never been struck since.*

So, too, though the beautiful tower of the Cathedral of Siena, protected by all possible theological means, had been struck again and again, much opposition was shown to placing upon it what was generally known as "the heretical rod"; but the tower was at last protected by Franklin's invention, and in 1777, though a very heavy bolt passed down the rod, the church received not the slightest injury. This served to reconcile theology and science, so far as that city was concerned; but the case which did most to convert the Italian theologians to the scientific view was that of the church of San Nazaro, at Brescia. The Republic of Venice had stored in the vaults of this church over two hundred thousand pounds of powder. In 1767, seventeen years after Franklin's discovery, no rod having been placed upon it, it was struck by lightning, the powder in the vaults was exploded, one sixth of the entire city destroyed, and over three thousand lives were lost.†

Such examples as these, in all parts of Europe, had their effect. The formulas for conjuring off storms, for consecrating bells to ward off lightning and tempests, and for putting to flight the powers of the air, were still allowed to stand in the liturgies; but the lightning-rod, the barometer, and the thermometer, carried the day. A vigorous line of investigators succeeding Franklin completed his victory. The traveller in remote districts of Europe still hears the church bells ringing during tempests; the Polish or Italian peasant is still persuaded to pay fees for sounding bells to keep off hailstorms; but the universal tendency favours more and more the use of the lightning-rod, and of the insurance offices where men can be relieved of the ruinous results of meteorological disturbances in accordance with the scientific laws of average, based upon the ascertained recurrence of storms. So, too, though many a poor seaman trusts to his charm that has been bathed in holy water, or

---

* For reluctance in England to protect churches with Franklin's rods, see Priestley, *History of Electricity*, London, 1775, vol. i, pp. 407, 465 *et seq.*

† See article on *Lightning* in the *Edinburgh Review* for October, 1844.

that has touched some relic, the tendency among mariners is to value more and more those warnings which are sent far and wide each day over the earth and under the sea by the electric wires in accordance with laws ascertained by observation.

Yet, even in our own time, attempts to revive the old theological doctrine of meteorology have not been wanting. Two of these, one in a Roman Catholic and another in a Protestant country, will serve as types of many, to show how completely scientific truth has saturated and permeated minds supposed to be entirely surrendered to the theological view.

The Island of St. Honorat, just off the southern coast of France, is deservedly one of the places most venerated in Christendom. The monastery of Lérins, founded there in the fourth century, became a mother of similar institutions in western Europe, and a centre of religious teaching for the Christian world. In its atmosphere, legends and myths grew in beauty and luxuriance. Here, as the chroniclers tell us, at the touch of St. Honorat, burst forth a stream of living water, which a recent historian of the monastery declares a greater miracle than that of Moses; here he destroyed, with a touch of his staff, the reptiles which infested the island, and then forced the sea to wash away their foul remains. Here, to please his sister, Sainte-Marguerite, a cherry tree burst into full bloom every month; here he threw his cloak upon the waters and it became a raft, which bore him safely to visit the neighbouring island; here St. Patrick received from St. Just the staff with which he imitated St. Honorat by driving all reptiles from Ireland.

Pillaged by Saracens and pirates, the island was made all the more precious by the blood of Christian martyrs. Popes and kings made pilgrimages to it; saints, confessors, and bishops went forth from it into all Europe; in one of its cells St. Vincent of Lérins wrote that famous definition of pure religion which, for nearly fifteen hundred years, has virtually superseded that of St. James. Naturally, the monastery became most illustrious, and its seat "the Mediterranean Isle of Saints."

But toward the close of the last century, its inmates having become slothful and corrupt, it was dismantled, all save a small portion torn down, and the island became the property first of impiety, embodied in a French actress, and finally of heresy, embodied in an English clergyman.

Bought back for the Church by the Bishop of Fréjus in 1859, there was little revival of life for twelve years. Then came the reaction, religious and political, after the humiliation of France and the Vatican by Germany; and of this reaction the monastery of St. Honorat was made one of the most striking outward and visible signs. Pius IX interested himself directly in it, called into it a body of Cistercian monks, and it became the chief seat of their order in France. To restore its sacredness the strict system of La Trappe was established—labour, silence, meditation on death. The word thus given from Rome was seconded in France by cardinals, archbishops, and all churchmen especially anxious for promotion in this world or salvation in the next. Worn-out dukes and duchesses of the Faubourg Saint-Germain united in this enterprise of pious reaction with the frivolous youngsters, the *petits crevés*, who haunt the purlieus of Notre Dame de Lorette. The great church of the monastery was handsomely rebuilt and a multitude of altars erected; and beautiful frescoes and stained windows came from the leaders of the reaction. The whole effect was, perhaps, somewhat theatrical and thin, but it showed none the less earnestness in making the old "Isle of Saints" a protest against the hated modern world.

As if to bid defiance still further to modern liberalism, great store of relics was sent in; among these, pieces of the true cross, of the white and purple robes, of the crown of thorns, sponge, lance, and winding-sheet of Christ,—the hair, robe, veil, and girdle of the Blessed Virgin; relics of St. John the Baptist, St. Joseph, St. Mary Magdalene, St. Paul, St. Barnabas, the four evangelists, and a multitude of other saints: so many that the bare mention of these treasures requires twenty-four distinct heads in the official catalogue recently published at the monastery. Besides all this—what was considered even more powerful in warding off harm from the revived monastery—the bones of Christian martyrs were brought from the Roman catacombs and laid beneath the altars.*

All was thus conformed to the mediæval view; nothing was to be left which could remind one of the nineteenth century; the "ages of faith" were to be restored in their simplicity. Pope Leo XIII commended to the brethren the writings of St. Thomas Aquinas as their one great object of study, and works published at the monastery dwelt upon the miracles of St. Honorat as the most precious refutation of modern science.

High in the cupola, above the altars and relics, were placed the bells. Sent by pious donors, they were solemnly baptized and consecrated in 1871, four bishops officiating, a multitude of the faithful being present from all parts of Europe, and the sponsors of the great tenor bell being the Bourbon claimant to the ducal throne of Parma and his duchess. The good bishop who baptized the bells consecrated them with a formula announcing their efficacy in driving away the "Prince of the Power of the Air," and the lightning and tempests he provokes.

And then, above all, at the summit of the central spire, high above relics, altars, and bells, was placed—*a lightning-rod!* †

The account of the monastery, published under the direction of the present worthy abbot, more than hints at the saving, by its bells, of a ship which was wrecked a few years since on that coast; and yet, to protect the bells and church and monks and relics from the very foe whom, in the mediæval faith, all these were thought most powerful to drive away, recourse was had to the scientific discovery of that "arch-infidel," Benjamin Franklin!

Perhaps the most striking recent example in Protestant lands of this change from the old to the new occurred not

---

* See the *Guide des Visiteurs à Lérins*, published at the monastery in 1880, p. 204; also the *Histoire de Lérins*, mentioned below.

† See *Guide*, as above, p. 84. *Les Isles de Lérins*, by the Abbé Alliez (Paris, 1860), and the *Histoire de Lérins*, by the same author, are the authorities for the general history of the abbey, and are especially strong in presenting the miracles of St Honorat, etc. The *Cartulaire* of the monastery, recently published, is also valuable. But these do not cover the recent revival, for an account of which recourse must be had to the very interesting and naïve *Guide* already cited.

long since in one of the great Pacific dependencies of the British crown. At a time of severe drought an appeal was made to the bishop, Dr. Moorhouse, to order public prayers for rain. The bishop refused, advising the petitioners for the future to take better care of their water supply, virtually telling them, "Heaven helps those who help themselves." But most noteworthy in this matter was it that the English Government, not long after, scanning the horizon to find some man to take up the good work laid down by the lamented Bishop Fraser, of Manchester, chose Dr. Moorhouse; and his utterance upon meteorology, which a few generations since would have been regarded by the whole Church as blasphemy, was universally alluded to as an example of strong good sense, proving him especially fit for one of the most important bishoprics in England.

Throughout Christendom, the prevalence of the conviction that meteorology is obedient to laws is more and more evident. In cities especially, where men are accustomed each day to see posted in public places charts which show the storms moving over various parts of the country, and to read in the morning papers scientific prophecies as to the weather, the old view can hardly be very influential.

Significant of this was the feeling of the American people during the fearful droughts a few years since in the States west of the Missouri. No days were appointed for fasting and prayer to bring rain; there was no attribution of the calamity to the wrath of God or the malice of Satan; but much was said regarding the folly of our people in allowing the upper regions of their vast rivers to be denuded of forests, thus subjecting the States below to alternations of drought and deluge. Partly as a result of this, a beginning has been made of teaching forest culture in many schools, tree-planting societies have been formed, and "Arbor Day" is recognised in several of the States. A true and noble theology can hardly fail to recognise, in the love of Nature and care for our fellow-men thus promoted, something far better, both from a religious and a moral point of view, than any efforts to win the Divine favour by flattery, or to avert Satanic malice by fetichism.

CHAPTER XII.

FROM MAGIC TO CHEMISTRY AND PHYSICS.

I.

IN all the earliest developments of human thought we find a strong tendency to ascribe mysterious powers over Nature to men and women especially gifted or skilled. Survivals of this view are found to this day among savages and barbarians left behind in the evolution of civilization, and especially is this the case among the tribes of Australia, Africa, and the Pacific coast of America. Even in the most enlightened nations still appear popular beliefs, observances, or sayings, drawn from this earlier phase of thought.

Between the prehistoric savage developing this theory, and therefore endeavouring to deal with the powers of Nature by magic, and the modern man who has outgrown it, appears a long line of nations struggling upward through it. As the hieroglyphs, cuneiform inscriptions, and various other records of antiquity are read, the development of this belief can be studied in Egypt, India, Babylonia, Assyria, Persia, and Phœnicia. From these civilizations it came into the early thought of Greece and Rome, but especially into the Jewish and Christian sacred books. Both in the Old Testament and in the New we find magic, witchcraft, and soothsaying constantly referred to as realities.*

* For magic in prehistoric times and survivals of it since, with abundant citation of authorities, see Tylor, *Primitive Culture*, chap. iv; also *The Early History of Mankind*, by the same author, third edition, pp. 115 *et seq.*, also p. 380; also Andrew Lang, *Myth, Ritual, and Religion*, vol. i, chap. iv. For magic in Egypt, see Lenormant, *Chaldean Magic*, chaps. vi–viii; also Maspero, *Histoire Ancienne des Peuples de l'Orient*; also Maspero and Sayce, *The Dawn of Civilization*, p. 282, and for the threat of the magicians to wreck heaven, see ibid., p. 17, note, and especially the citations from Chabas, *Le Papyrus Magique Harris*, in chap.

The first distinct impulse toward a higher view of research into natural laws was given by the philosophers of Greece. It is true that philosophical opposition to physical research was at times strong, and that even a great thinker like Socrates considered certain physical investigations as an impious intrusion into the work of the gods. It is also true that Plato and Aristotle, while bringing their thoughts to bear upon the world with great beauty and force, did much to draw mankind away from those methods which in modern times have produced the best results.

Plato developed a world in which the physical sciences had little if any real reason for existing; Aristotle, a world in which the same sciences were developed largely indeed by observation of what is, but still more by speculation on what ought to be. From the former of these two great men came into Christian theology many germs of mediæval magic, and from the latter sundry modes of reasoning which aided in the evolution of these; yet the impulse to human thought given by these great masters was of inestimable value to our race, and one legacy from them was especially precious—the idea that a science of Nature is possible, and that the highest occupation of man is the discovery of its laws. Still another gift from them was greatest of all, for they gave scientific freedom. They laid no interdict upon new paths; they interposed no barriers to the extension of knowledge; they threatened no doom in this life or in the

---

vii; also Maury, *La Magie et l'Astrologie dans l'Antiquité et au Moyen Age*. For magic in Chaldea, see Lenormant as above; also Maspero and Sayce, pp. 780 *et seq.* For examples of magical powers in India, see Max Müller's *Sacred Books of the East*, vol. xvii, pp. 121 *et seq.* For a legendary view of magic in Media, see the *Zend Avesta*, part i, p. 14, translated by Darmsteter; and for a more highly developed view, see the *Zend Avesta*, part iii, p. 239, translated by Mill. For magic in Greece and Rome, and especially in the Neoplatonic school, as well as in the Middle Ages, see especially Maury, *La Magie et l'Astrologie*, chaps. iii–v. For various sorts of magic recognised and condemned in our sacred books, see Deuteronomy xviii, 10, 11; and for the burning of magical books at Ephesus under the influence of St. Paul, see Acts xix, 14. See also Ewald, *History of Israel*, Martineau's translation, fourth edition, vol. ii, pp. 55–63; vol. iii, pp. 45–51. For a very elaborate summing up of the passages in our sacred books recognizing magic as a fact, see De Haen, *De Magia*, Leipsic, 1775, chaps. i, ii, and iii, of first part. For the general subject of magic, see Ennemoser, *History of Magic*, translated by Howitt, which, however, constantly mixes sorcery with magic proper.

next against investigators on new lines; they left the world free to seek any new methods and to follow any new paths which thinking men could find.

This legacy of belief in science, of respect for scientific pursuits, and of freedom in scientific research, was especially received by the school of Alexandria, and above all by Archimedes, who began, just before the Christian era, to open new paths through the great field of the inductive sciences by observation, comparison, and experiment.*

The establishment of Christianity, beginning a new evolution of theology, arrested the normal development of the physical sciences for over fifteen hundred years. The cause of this arrest was twofold: First, there was created an atmosphere in which the germs of physical science could hardly grow—an atmosphere in which all seeking in Nature for truth as truth was regarded as futile. The general belief derived from the New Testament Scriptures was, that the end of the world was at hand; that the last judgment was approaching; that all existing physical nature was soon to be destroyed: hence, the greatest thinkers in the Church generally poured contempt upon all investigators into a science of Nature, and insisted that everything except the saving of souls was folly.

This belief appears frequently through the entire period of the Middle Ages; but during the first thousand years it is clearly dominant. From Lactantius and Eusebius, in the third century, pouring contempt, as we have seen, over studies in astronomy, to Peter Damian, the noted chancellor of Pope Gregory VII, in the eleventh century, declaring all worldly sciences to be "absurdities" and "fooleries," it

* As to the beginnings of physical science in Greece, and of the theological opposition to physical science, also Socrates's view regarding certain branches as interdicted to human study, see Grote's *History of Greece*, vol. i, pp. 495 and 504, 505; also Jowett's introduction to his translation of the *Timæus*, and Whewell's *History of the Inductive Sciences*. For examples showing the incompatibility of Plato's methods in physical science with that pursued in modern times, see Zeller, *Plato and the Older Academy*, English translation by Alleyne and Goodwin, pp. 375 *et seq*. The supposed opposition to freedom of opinion in the *Laws* of Plato, toward the end of his life, can hardly make against the whole spirit of Greek thought.

becomes a very important element in the atmosphere of thought.*

Then, too, there was established a standard to which all science which did struggle up through this atmosphere must be made to conform—a standard which favoured magic rather than science, for it was a standard of rigid dogmatism obtained from literal readings in the Jewish and Christian Scriptures. The most careful inductions from ascertained facts were regarded as wretchedly fallible when compared with any view of nature whatever given or even hinted at in any poem, chronicle, code, apologue, myth, legend, allegory, letter, or discourse of any sort which had happened to be preserved in the literature which had come to be held as sacred.

For twelve centuries, then, the physical sciences were thus discouraged or perverted by the dominant orthodoxy. Whoever studied nature studied it either openly to find illustrations of the sacred text, useful in the "saving of souls," or secretly to gain the aid of occult powers, useful in securing personal advantage. Great men like Bede, Isidore of Seville, and Rabanus Maurus, accepted the scriptural standard of science and used it as a means of Christian edification. The views of Bede and Isidore on kindred subjects have been shown in former chapters; and typical of the view taken by Rabanus is the fact that in his great work on the *Universe* there are only two chapters which seem directly or indirectly to recognise even the beginnings of a real philosophy of nature. A multitude of less-known men found warrant in Scripture for magic applied to less worthy purposes.†

---

* For the view of Peter Damian and others through the Middle Ages as to the futility of scientific investigation, see citations in Eicken, *Geschichte und System der mittelalterlichen Weltanschauung*, chap. vi.

† As typical examples, see the utterances of Eusebius and Lactantius regarding astronomers given in the chapter on Astronomy. For a summary of Rabanus Maurus's doctrine of physics, see Heller, *Geschichte der Physik*, vol. i, pp. 172 *et seq*. For Bede and Isidore, see the earlier chapters of this work. For an excellent statement regarding the application of scriptural standards to scientific research in the Middle Ages, see Kretschmer, *Die physische Erdkunde im christlichen Mittelalter*, pp. 5 *et seq*. For the distinctions in magic recognised in the mediæval Church, see the long catalogue of various sorts given in the Abbé Migne's *Encyclopédie Théologique*, third series, article *Magie*.

But after the thousand years had passed to which various thinkers in the Church, upon supposed scriptural warrant, had lengthened out the term of the earth's existence, "the end of all things" seemed further off than ever; and in the twelfth and thirteenth centuries, owing to causes which need not be dwelt upon here, came a great revival of thought, so that the forces of theology and of science seemed arrayed for a contest. On one side came a revival of religious fervour, and to this day the works of the cathedral builders mark its depth and strength; on the other side came a new spirit of inquiry incarnate in a line of powerful thinkers.

First among these was Albert of Bollstadt, better known as Albert the Great, the most renowned scholar of his time. Fettered though he was by the methods sanctioned in the Church, dark as was all about him, he had conceived better methods and aims; his eye pierced the mists of scholasticism; he saw the light, and sought to draw the world toward it. He stands among the great pioneers of physical and natural science; he aided in giving foundations to botany and chemistry; he rose above his time, and struck a heavy blow at those who opposed the possibility of human life on opposite sides of the earth; he noted the influence of mountains, seas, and forests upon races and products, so that Humboldt justly finds in his works the germs of physical geography as a comprehensive science.

But the old system of deducing scientific truth from scriptural texts was renewed in the development of scholastic theology; and ecclesiastical power, acting through thousands of subtle channels, was made to aid this development. The old idea of the futility of physical science and of the vast superiority of theology was revived. Though Albert's main effort was to Christianize science, he was dealt with by the authorities of the Dominican order, subjected to suspicion and indignity, and only escaped persecution for sorcery by yielding to the ecclesiastical spirit of the time, and working finally in theological channels by scholastic methods.

It was a vast loss to the earth; and certainly, of all organizations that have reason to lament the pressure of ecclesiasticism which turned Albert the Great from natural philosophy to theology, foremost of all in regret should be the Christian Church, and especially the Roman branch of it. Had there been evolved in the Church during the thirteenth century a faith strong enough to accept the truths in natural science which Albert and his compeers could have given, and to have encouraged their growth, this faith and this encouragement would to this day have formed the greatest argument for proving the Church directly under Divine guidance; they would have been among the brightest jewels in her crown. The loss to the Church by this want of faith and courage has proved in the long run even greater than the loss to science.*

The next great man of that age whom the theological and ecclesiastical forces of the time turned from the right path was Vincent of Beauvais. During the first half of the twelfth century he devoted himself to the study of Nature in several of her most interesting fields. To astronomy, botany, and zoölogy he gave special attention, but in a larger way he made a general study of the universe, and in a series of treatises undertook to reveal the whole field of science. But his work simply became a vast commentary on the account of creation given in the book of Genesis. Beginning with the work of the Trinity at the creation, he goes on to detail the work of angels in all their fields, and makes excursions into every part of creation, visible and invisible, but always with the most complete subordination of his thought to the literal statements of Scripture. Could he have taken

* For a very careful discussion of Albert's strength in investigation and weakness in yielding to scholastic authority, see Kopp, *Ansichten über die Aufgabe der Chemie von Geber bis Stahl*, Braunschweig, 1875, pp. 64 *et seq.* For a very extended and enthusiastic biographical sketch, see Pouchet. For comparison of his work with that of Thomas Aquinas, see Milman, *History of Latin Christianity*, vol. vi, p. 461. "Il était aussi très-habile dans les arts mécaniques, ce que le fit soupçonner d'être sorcier" (Sprengel, *Histoire de la Médecine*, vol. ii, p. 389). For Albert's biography treated strictly in accordance with ecclesiastical methods, see *Albert the Great*, by Joachim Sighart, translated by the Rev. T. A. Dickson, of the Order of Preachers, published under the sanction of the Dominican censor and of the Cardinal Archbishop of Westminster, London, 1876. How an Englishman like Cardinal Manning could tolerate among Englishmen such an unctuous glossing over of historical truth is one of the wonders of contemporary history. For choice specimens, see chapters ii and iv. For one of the best and most recent summaries, see Heller, *Geschichte der Physik*, Stuttgart, 1882, vol. i, pp. 179 *et seq.*

the path of experimental research, the world would have been enriched with most precious discoveries; but the force which had given wrong direction to Albert of Bollstadt, backed as it was by the whole ecclesiastical power of his time, was too strong, and in all the life labour of Vincent nothing appears of any permanent value. He reared a structure which the adaptation of facts to literal interpretations of Scripture and the application of theological subtleties to nature combine to make one of the most striking monuments of human error.*

But the theological spirit of the thirteenth century gained its greatest victory in the work of St. Thomas Aquinas. In him was the theological spirit of his age incarnate. Although he yielded somewhat at one period to love of natural science, it was he who finally made that great treaty or compromise which for ages subjected science entirely to theology. He it was who reared the most enduring barrier against those who in that age and in succeeding ages laboured to open for science the path by its own methods toward its own ends.

He had been the pupil of Albert the Great, and had gained much from him. Through the earlier systems of philosophy, as they were then known, and through the earlier theologic thought, he had gone with great labour and vigour; and all his mighty powers, thus disciplined and cultured, he brought to bear in making a truce which was to give theology permanent supremacy over science.

The experimental method had already been practically initiated: Albert of Bollstadt and Roger Bacon had begun their work in accordance with its methods; but St. Thomas gave all his thoughts to bringing science again under the sway of theological methods and ecclesiastical control. In his commentary on Aristotle's treatise upon *Heaven and Earth* he gave to the world a striking example of what his method could produce, illustrating all the evils which arise in combining theological reasoning and literal interpretation of Scripture with scientific facts; and this work remains to

---

* For Vincent de Beauvais, see *Études sur Vincent de Beauvais*, par l'Abbé Bourgeat, chaps. xii, xiii, and xiv; also Pouchet, *Histoire des Sciences Naturelles au Moyen Age*, Paris, 1853, pp. 470 *et seq.*; also other histories cited hereafter.

this day a monument of scientific genius perverted by theology.*

The ecclesiastical power of the time hailed him as a deliverer; it was claimed that miracles were vouchsafed, proving that the blessing of Heaven rested upon his labours, and among the legends embodying this claim is that given by the Bollandists and immortalized by a renowned painter. The great philosopher and saint is represented in the habit of his order, with book and pen in hand, kneeling before the image of Christ crucified, and as he kneels the image thus addresses him: "Thomas, thou hast written well concerning me; what price wilt thou receive for thy labour?" The myth-making faculty of the people at large was also brought into play. According to a widespread and circumstantial legend, Albert, by magical means, created an android—an artificial man, living, speaking, and answering all questions with such subtlety that St. Thomas, unable to answer its reasoning, broke it to pieces with his staff.

Historians of the Roman Church like Rohrbacher, and historians of science like Pouchet, have found it convenient to propitiate the Church by dilating upon the glories of St. Thomas Aquinas in thus making an alliance between religious and scientific thought, and laying the foundations for a "sanctified science"; but the unprejudiced historian can not indulge in this enthusiastic view: the results both for the Church and for science have been most unfortunate. It was a wretched delay in the evolution of fruitful thought, for the first result of this great man's great compromise was to close for ages that path in science which above all others leads to discoveries of value—the experimental method—and to reopen that old path of mixed theology and science which, as Hallam declares, "after three or four hundred years had not untied a single knot or added one unequivocal truth to the domain of philosophy"—the path which, as all modern history proves, has ever since led only to delusion and evil.†

* For citations showing this subordination of science to theology, see Eicken, chap. vi.

† For the work of Aquinas, see his *Liber de Cœlo et Mundo*, section xx; also, *Life and Labours of St. Thomas of Aquin*, by Archbishop Vaughan, pp. 459 *et seq.*

The theological path thus opened by these strong men became the main path for science during ages, and it led the world ever further and further from any fruitful fact or useful method. Roger Bacon's investigations already begun were discredited: worthless mixtures of scriptural legends with imperfectly authenticated physical facts took their place. Thus it was that for twelve hundred years the minds in control of Europe regarded all real science as *futile*, and diverted the great current of earnest thought into theology.

The next stage in this evolution was the development of an idea which acted with great force throughout the Middle Ages—the idea that science is *dangerous*. This belief was also of very ancient origin. From the time when the Egyptian magicians made their tremendous threat that unless their demands were granted they would reach out to the four corners of the earth, pull down the pillars of heaven, wreck the abodes of the gods above and crush those of men below, fear of these representatives of science is evident in the ancient world.

But differences in the character of magic were recognised, some sorts being considered useful and some baleful. Of the former was magic used in curing diseases, in deter-

---

For his labours in natural science, see Hoefer, *Histoire de la Chimie*, Paris, 1843, vol. i, p. 381. For theological views of science in the Middle Ages, and rejoicing thereat, see Pouchet, *Hist. des Sci. Nat. au Moyen Age*, *ubi supra*. Pouchet says: "En général au milieu du moyen âge les sciences sont essentiellement chrétiennes, leur but est tout-à-fait religieux, et elles semblent beaucoup moins s'inquiéter de l'avancement intellectuel de l'homme que de son salut eternel." Pouchet calls this "conciliation" into a "harmonieux ensemble" "la plus glorieuse des conquêtes intellectuelles du moyen âge." Pouchet belongs to Rouen, and the shadow of Rouen Cathedral seems thrown over all his history. See, also, l'Abbé Rohrbacher, *Hist. de l'Église Catholique*, Paris, 1858, vol. xviii, pp. 421 *et seq*. The abbé dilates upon the fact that "the Church organizes the agreement of all the sciences by the labours of St. Thomas of Aquin and his contemporaries." For the complete subordination of science to theology by St. Thomas, see Eicken, chap. vi. For the theological character of science in the Middle Ages, recognised by a Protestant philosophic historian, see the well-known passage in Guizot, *History of Civilization in Europe*; and by a noted Protestant ecclesiastic, see Bishop Hampden's *Life of Thomas Aquinas*, chaps. xxxvi, xxxvii; see also Hallam, *Middle Ages*, chap. ix. For dealings of Pope John XXII, of the Kings of France and England, and of the Republic of Venice, see Figuier, *L'Alchimie et les Alchimistes*, pp. 140, 141, where, in a note, the text of the bull *Spondet pariter* is given. For popular legends regarding Albert and St. Thomas, see Éliphas Lévi, *Hist. de la Magie*, liv. iv, chap. iv.

mining times auspicious for enterprises, and even in contributing to amusement; of the latter was magic used to bring disease and death on men and animals or tempests upon the growing crops. Hence gradually arose a general distinction between white magic, which dealt openly with the more beneficent means of nature, and black magic, which dealt secretly with occult, malignant powers.

Down to the Christian era the fear of magic rarely led to any persecution very systematic or very cruel. While in Greece and Rome laws were at times enacted against magicians, they were only occasionally enforced with rigour, and finally, toward the end of the pagan empire, the feeling against them seemed dying out altogether. As to its more kindly phases, men like Marcus Aurelius and Julian did not hesitate to consult those who claimed to foretell the future. As to black magic, it seemed hardly worth while to enact severe laws, when charms, amulets, and even gestures could thwart its worst machinations.

Moreover, under the old empire a real science was coming in, and thought was progressing. Both the theory and practice of magic were more and more held up to ridicule. Even as early a writer as Ennius ridiculed the idea that magicians, who were generally poor and hungry themselves, could bestow wealth on others; Pliny, in his *Natural Philosophy*, showed at great length their absurdities and cheatery; others followed in the same line of thought, and the whole theory, except among the very lowest classes, seemed dying out.

But with the development of Christian theology came a change. The idea of the active interference of Satan in magic, which had come into the Hebrew mind with especial force from Persia during the captivity of Israel, had passed from the Hebrew Scriptures into Christianity, and had been made still stronger by various statements in the New Testament. Theologians laid stress especially upon the famous utterances of the Psalmist that "all the gods of the heathen are devils," and of St. Paul that "the things which the Gentiles sacrifice, they sacrifice to devils"; and it was widely held that these devils were naturally indignant at their dethronement and anxious to wreak vengeance upon Christianity. Magicians were held to be active agents of these dethroned gods, and this persuasion was strengthened by sundry old practitioners in the art of magic—impostors who pretended to supernatural powers, and who made use of old rites and phrases inherited from paganism.

Hence it was that as soon as Christianity came into power it more than renewed the old severities against the forbidden art, and one of the first acts of the Emperor Constantine after his conversion was to enact a most severe law against magic and magicians, under which the main offender might be burned alive. But here, too, it should be noted that a distinction between the two sorts of magic was recognised, for Constantine shortly afterward found it necessary to issue a proclamation stating that his intention was only to prohibit deadly and malignant magic; that he had no intention of prohibiting magic used to cure diseases and to protect the crops from hail and tempests. But as new emperors came to the throne who had not in them that old leaven of paganism which to the last influenced Constantine, and as theology obtained a firmer hold, severity against magic increased. Toleration of it, even in its milder forms, was more and more denied. Black magic and white were classed together.

This severity went on increasing and threatened the simplest efforts in physics and chemistry; even the science of mathematics was looked upon with dread. By the twelfth and thirteenth centuries, the older theology having arrived at the climax of its development in Europe, terror of magic and witchcraft took complete possession of the popular mind. In sculpture, painting, and literature it appeared in forms ever more and more striking. The lives of saints were filled with it. The cathedral sculpture embodied it in every part. The storied windows made it all the more impressive. The missal painters wrought it not only into prayer books, but, despite the fact that hardly a trace of the belief appears in the Psalms, they illustrated it in the great illuminated psalters from which the noblest part of the service was sung before the high altar. The service books showed every form of agonizing petition for delivery from this dire influence, and every form of exorcism for thwarting it.

All the great theologians of the Church entered into this belief and aided to develop it. The fathers of the early Church were full and explicit, and the mediæval doctors became more and more minute in describing the operations of the black art and in denouncing them. It was argued that, as the devil afflicted Job, so he and his minions continue to cause diseases; that, as Satan is the Prince of the power of the air, he and his minions cause tempests; that the cases of Nebuchadnezzar and Lot's wife prove that sorcerers can transform human beings into animals or even lifeless matter; that, as the devils of Gadara were cast into swine, all animals could be afflicted in the same manner; and that, as Christ himself had been transported through the air by the power of Satan, so any human being might be thus transported to "an exceeding high mountain."

Thus the horror of magic and witchcraft increased on every hand, and in 1317 Pope John XXII issued his bull *Spondent pariter*, levelled at the alchemists, but really dealing a terrible blow at the beginnings of chemical science. That many alchemists were knavish is no doubt true, but no infallibility in separating the evil from the good was shown by the papacy in this matter. In this and in sundry other bulls and briefs we find Pope John, by virtue of his infallibility as the world's instructor in all that pertains to faith and morals, condemning real science and pseudo-science alike. In two of these documents, supposed to be inspired by wisdom from on high, he complains that both he and his flock are in danger of their lives by the arts of the sorcerers; he declares that such sorcerers can send devils into mirrors and finger rings, and kill men and women by a magic word; that they had tried to kill him by piercing a waxen image of him with needles in the name of the devil. He therefore called on all rulers, secular and ecclesiastical, to hunt down the miscreants who thus afflicted the faithful, and he especially increased the powers of inquisitors in various parts of Europe for this purpose.

The impulse thus given to childish fear and hatred against the investigation of nature was felt for centuries; more and more chemistry came to be known as one of the "seven devilish arts."

Thus began a long series of demonstrations against magic from the centre of Christendom. In 1437, and again in 1445, Pope Eugene IV issued bulls exhorting inquisitors to be more diligent in searching out and delivering over to punishment magicians and witches who produced bad weather, the result being that persecution received a fearful impulse. But the worst came forty years later still, when, in 1484, there came the yet more terrible bull of Pope Innocent VIII, known as *Summis Desiderantes*, which let inquisitors loose upon Germany, with Sprenger at their head, armed with the *Witch-Hammer*, the fearful manual *Malleus Maleficarum*, to torture and destroy men and women by tens of thousands for sorcery and magic. Similar bulls were issued in 1504 by Julius II, and in 1523 by Adrian VI.

The system of repression thus begun lasted for hundreds of years. The Reformation did little to change it, and in Germany, where Catholics and Protestants vied with each other in proving their orthodoxy, it was at its worst. On German soil more than one hundred thousand victims are believed to have been sacrificed to it between the middle of the fifteenth and the middle of the sixteenth centuries.

Thus it was that from St. Augustine to St. Thomas Aquinas, from Aquinas to Luther, and from Luther to Wesley, theologians of both branches of the Church, with hardly an exception, enforced the belief in magic and witchcraft, and, as far as they had power, carried out the injunction, "Thou shalt not suffer a witch to live."

How this was ended by the progress of scientific modes of thought I shall endeavour to show elsewhere: here we are only concerned with the effect of this widespread terrorism on the germs and early growth of the physical sciences.

Of course, the atmosphere created by this persecution of magicians was deadly to any open beginnings of experimental science. The conscience of the time, acting in obedience to the highest authorities of the Church, and, as was supposed, in defence of religion, now brought out a missile which it hurled against scientific investigators with deadly effect. The mediæval battlefields of thought were strewn with various forms of it. This missile was the charge of unlawful compact with Satan, and it was most effective. We find it used against every great investigator of nature in those times and for ages after. The list of great men in those centuries charged with magic, as given by Naudé, is astounding; it includes every man of real mark, and in the midst of them stands one of the most thoughtful popes, Sylvester II (Gerbert), and the foremost of mediæval thinkers on natural science, Albert the Great. It came to be the accepted idea that, as soon as a man conceived a wish to study the works of God, his first step must be a league with the devil.

It was entirely natural, then, that in 1163 Pope Alexander III, in connection with the Council of Tours, forbade the study of physics to all ecclesiastics, which, of course, in that age meant prohibition of all such scientific studies to the only persons likely to make them. What the Pope then expressly forbade was, in the words of the papal bull, "the study of physics or the laws of the world," and it was added that any person violating this rule "shall be avoided by all and excommunicated." *

The first great thinker who, in spite of some stumbling into theologic pitfalls, persevered in a truly scientific path, was Roger Bacon. His life and works seem until recently to have been generally misunderstood: he was formerly ranked as a superstitious alchemist who happened upon some inventions, but more recent investigation has shown him to be one of the great masters in the evolution of human thought. The advance of sound historical judgment seems likely to bring the fame of the two who bear the name of Bacon nearly to equality. Bacon of the chancellorship and of the *Novum Organum* may not wane, but Bacon of the prison cell and the *Opus Majus* steadily approaches him in brightness.

More than three centuries before Francis Bacon advocated the experimental method, Roger Bacon practised it, and the results as now revealed are wonderful. He wrought

---

* For the charge of magic against scholars and others, see Naudé, *Apologie pour les Grands Hommes soupçonnés de Magie, passim*; also Maury, *Hist. de la Magie*, troisième édition, pp. 214, 215; also Cuvier, *Hist. des Sciences Naturelles*, vol. i, p. 396. For the prohibition by the Council of Tours and Alexander III, see the *Acta Conciliorum* (ed. Harduin), tom. vi, pars ii, p. 1598, Canon viii.

with power in many sciences, and his knowledge was sound and exact. By him, more than by any other man of the Middle Ages, was the world brought into the more fruitful paths of scientific thought—the paths which have led to the most precious inventions; and among these are clocks, lenses, and burning specula, which were given by him to the world, directly or indirectly. In his writings are found formulæ for extracting phosphorus, manganese, and bismuth. It is even claimed, with much appearance of justice, that he investigated the power of steam, and he seems to have very nearly reached some of the principal doctrines of modern chemistry. But it should be borne in mind that his *method* of investigation was even greater than its *results*. In an age when theological subtilizing was alone thought to give the title of scholar, he insisted on *real* reasoning and the aid of natural science by mathematics; in an age when experimenting was sure to cost a man his reputation, and was likely to cost him his life, he insisted on experimenting, and braved all its risks. Few greater men have lived. As we follow Bacon's process of reasoning regarding the refraction of light, we see that he was divinely inspired.

On this man came the brunt of the battle. The most conscientious men of his time thought it their duty to fight him, and they fought him steadily and bitterly. His sin was not disbelief in Christianity, not want of fidelity to the Church, not even dissent from the main lines of orthodoxy; on the contrary, he showed in all his writings a desire to strengthen Christianity, to build up the Church, and to develop orthodoxy. He was attacked and condemned mainly because he did not believe that philosophy had become complete, and that nothing more was to be learned; he was condemned, as his opponents expressly declared, "on account of certain suspicious novelties"—"*propter quasdam novitates suspectas*."

Upon his return to Oxford, about 1250, the forces of unreason beset him on all sides. Greatest of all his enemies was Bonaventura. This enemy was the theologic idol of the period: the learned world knew him as the "seraphic Doctor"; Dante gave him an honoured place in the great poem of the Middle Ages; the Church finally enrolled him among the saints. By force of great ability in theology he had become, in the middle of the thirteenth century, general of the Franciscan order: thus, as Bacon's master, his hands were laid heavily on the new teaching, so that in 1257 the troublesome monk was forbidden to lecture; all men were solemnly warned not to listen to his teaching, and he was ordered to Paris, to be kept under surveillance by the monastic authorities. Herein was exhibited another of the myriad examples showing the care exercised over scientific teaching by the Church. The reasons for thus dealing with Bacon were evident: First, he had dared attempt scientific explanations of natural phenomena, which under the mystic theology of the Middle Ages had been referred simply to supernatural causes. Typical was his explanation of the causes and character of the rainbow. It was clear, cogent, a great step in the right direction as regards physical science: but there, in the book of Genesis, stood the legend regarding the origin of the rainbow, supposed to have been dictated immediately by the Holy Spirit; and, according to that, the "bow in the cloud" was not the result of natural laws, but a "sign" arbitrarily placed in the heavens for the simple purpose of assuring mankind that there was not to be another universal deluge.

But this was not the worst: another theological idea was arrayed against him—the idea of Satanic intervention in science; hence he was attacked with that goodly missile which with the epithets "infidel" and "atheist" has decided the fate of so many battles—the charge of magic and compact with Satan.

He defended himself with a most unfortunate weapon—a weapon which exploded in his hands and injured him more than the enemy; for he argued against the idea of compacts with Satan, and showed that much which is ascribed to demons results from natural means. This added fuel to the flame. To limit the power of Satan was deemed hardly less impious than to limit the power of God.

The most powerful protectors availed him little. His friend Guy of Foulques, having in 1265 been made Pope under the name of Clement IV, shielded him for a time; but the fury of the enemy was too strong, and when he made ready to perform a few experiments before a small audience, we are told that all Oxford was in an uproar. It was believed that Satan was about to be let loose. Everywhere priests, monks, fellows, and students rushed about, their garments streaming in the wind, and everywhere rose the cry, "Down with the magician!" and this cry, "Down with the magician!" resounded from cell to cell and from hall to hall.

Another weapon was also used upon the battlefields of science in that time with much effect. The Arabs had made many noble discoveries in science, and Averroes had, in the opinion of many, divided the honours with St. Thomas Aquinas; these facts gave the new missile—it was the epithet "Mohammedan"; this, too, was flung with effect at Bacon.

The attack now began to take its final shape. The two great religious orders, Franciscan and Dominican, then in all the vigour of their youth, vied with each other in fighting the new thought in chemistry and physics. St. Dominic solemnly condemned research by experiment and observation; the general of the Franciscan order took similar ground. In 1243 the Dominicans interdicted every member of their order from the study of medicine and natural philosophy, and in 1287 this interdiction was extended to the study of chemistry.

In 1278 the authorities of the Franciscan order assembled at Paris, solemnly condemned Bacon's teaching, and the general of the Franciscans, Jerome of Ascoli, afterward Pope, threw him into prison, where he remained for fourteen years. Though Pope Clement IV had protected him, Popes Nicholas III and IV, by virtue of their infallibility, decided that he was too dangerous to be at large, and he was only released at the age of eighty—but a year or two before death placed him beyond the reach of his enemies. How deeply the struggle had racked his mind may be gathered from that last affecting declaration of his, "Would that I had not given myself so much trouble for the love of science!"

The attempt has been made by sundry champions of the Church to show that some of Bacon's utterances against ecclesiastical and other corruptions in his time were the main cause of the severity which the Church authorities exercised against him. This helps the Church but little, even if it be well based; but it is not well based. That some of his utterances of this sort made him enemies is doubtless true, but the charges on which St. Bonaventura silenced him, and Jerome of Ascoli imprisoned him, and successive popes kept him in prison for fourteen years, were "dangerous novelties" and suspected sorcery.

Sad is it to think of what this great man might have given to the world had ecclesiasticism allowed the gift. He held the key of treasures which would have freed mankind from ages of error and misery. With his discoveries as a basis, with his method as a guide, what might not the world have gained! Nor was the wrong done to that age alone; it was done to this age also. The nineteenth century was robbed at the same time with the thirteenth. But for that interference with science the nineteenth century would be enjoying discoveries which will not be reached before the twentieth century, and even later. Thousands of precious lives shall be lost, tens of thousands shall suffer discomfort, privation, sickness, poverty, ignorance, for lack of discoveries and methods which, but for this mistaken dealing with Roger Bacon and his compeers, would now be blessing the earth.

In two recent years sixty thousand children died in England and in Wales of scarlet fever; probably quite as many died in the United States. Had not Bacon been hindered, we should have had in our hands, by this time, the means to save two thirds of these victims; and the same is true of typhoid, typhus, cholera, and that great class of diseases of whose physical causes science is just beginning to get an inkling. Put together all the efforts of all the atheists who have ever lived, and they have not done so much harm to Christianity and the world as has been done by the narrow-minded, conscientious men who persecuted Roger Bacon, and closed the path which he gave his life to open.

But despite the persecution of Bacon and the defection of those who ought to have followed him, champions of the experimental method rose from time to time during the succeeding centuries. We know little of them personally; our main knowledge of their efforts is derived from the endeavours of their persecutors.

Under such guidance the secular rulers were naturally vigorous. In France Charles V forbade, in 1380, the possession of furnaces and apparatus necessary for chemical processes; under this law the chemist John Barrillon was thrown into prison, and it was only by the greatest effort that his life was saved. In England Henry IV, in 1404, issued a similar decree. In Italy the Republic of Venice, in 1418, followed these examples. The judicial torture and murder of Antonio de Dominis were not simply for heresy; his investigations in the phenomena of light were an additional crime. In Spain everything like scientific research was crushed out among Christians. Some earnest efforts were afterward made by Jews and Moors, but these were finally ended by persecution; and to this hour the Spanish race, in some respects the most gifted in Europe, which began its career with everything in its favour and with every form of noble achievement, remains in intellectual development behind every other in Christendom.

To question the theological view of physical science was, even long after the close of the Middle Ages, exceedingly perilous. We have seen how one of Roger Bacon's unpardonable offences was his argument against the efficacy of magic, and how, centuries afterward, Cornelius Agrippa, Weyer, Flade, Loos, Bekker, and a multitude of other investigators and thinkers, suffered confiscation of property, loss of position, and even torture and death, for similar views.*

---

* For an account of Bacon's treatise, *De Nullitate Magiæ*, see Hoefer. For the uproar caused by Bacon's teaching at Oxford, see Kopp, *Geschichte der Chemie*, Braunschweig, 1869, vol. i, p. 63; and for a somewhat reactionary discussion of Bacon's relation to the progress of chemistry, see a recent work by the same author, *Ansichten über die Aufgabe der Chemie*, Braunschweig, 1874, pp. 85 *et seq.*; also, for an excellent summary, see Hoefer, *Hist. de la Chimie*, vol. i, pp. 368 *et seq.* For probably the most thorough study of Bacon's general works in science, and for his views of the universe, see Prof. Werner, *Die Kosmologie und allgemeine Naturlehre des Roger Baco*, Wien, 1879. For summaries of his work in other fields, see Whewell, vol. i, pp. 367, 368; Draper, p. 438; Saisset, *Descartes et ses Précurseurs*, deuxième édition, pp. 397 *et seq.*; Nourrisson, *Progrès de la Pensée humaine*, pp. 271, 272; Sprengel, *Histoire de la Médecine*, Paris, 1865, vol. ii, p. 397; Cuvier, *Histoire des Sciences Naturelles*, vol. i, p. 417. As to Bacon's orthodoxy, see Saisset, pp. 53, 55. For special examination of causes of Bacon's condemnation, see

The theological atmosphere, which in consequence settled down about the great universities and colleges, seemed likely to stifle all scientific effort in every part of Europe, and it is one of the great wonders in human history that in spite of this deadly atmosphere a considerable body of thinking men, under such protection as they could secure, still persisted in devoting themselves to the physical sciences.

In Italy, in the latter half of the sixteenth century, came a striking example of the difficulties which science still encountered even after the Renaissance had undermined the old beliefs. At that time John Baptist Porta was conducting his investigations, and, despite a considerable mixture of pseudo-science, they were fruitful. His was not "black magic," claiming the aid of Satan, but "white magic," bringing into service the laws of nature—the precursor of applied

---

Waddington, cited by Saisset, p. 14. For a brief but admirable statement of Roger Bacon's relation to the world in his time, and of what he might have done had he not been thwarted by theology, see Döllinger, *Studies in European History*, English translation, London, 1890, pp. 178, 179. For a good example of the danger of denying the full power of Satan, even in much more recent times and in a Protestant country, see account of treatment of Bekker's *Monde Enchanté* by the theologians of Holland, in Nisard, *Histoire des Livres Populaires*, vol. i, pp. 172, 173. Kopp, in his *Ansichten*, pushes criticism even to some scepticism as to Roger Bacon being the *discoverer* of many of the things generally attributed to him; but, after all deductions are carefully made, enough remains to make Bacon the greatest benefactor to humanity during the Middle Ages. For Roger Bacon's deep devotion to religion and the Church, see citation and remarks in Schneider, *Roger Bacon*, Augsburg, 1873, p. 112; also, citation from the *Opus Majus* in Eicken, chap. vi. On Bacon as a "Mohammedan," see Saisset, p. 17. For the interdiction of studies in physical science by the Dominicans and Franciscans, see Henri Martin, *Histoire de France*, vol. iv, p. 283. For the suppression of chemical teaching by the Parliament of Paris, see ibid., vol. xii, pp. 14, 15. For proofs that the world is steadily working toward great discoveries as to the cause and prevention of zymotic diseases and of their propagation, see Beale's *Disease Germs*, Baldwin Latham's *Sanitary Engineering*, Michel Lévy's *Traité d'Hygiène Publique et Privée*. For a summary of the bull *Spondent pariter*, and for an example of injury done by it, see Schneider, *Geschichte der Alchemie*, p. 160; and for a studiously moderate statement, Milman, *Latin Christianity*, book xii, chap. vi. For character and general efforts of John XXII, see Lea, *Inquisition*, vol. iii, p. 436, also pp. 452 *et seq.* For the character of the two papal briefs, see Rydberg, p. 177. For the bull *Summis Desiderantes*, see previous chapters of this work. For Antonio de Dominis, see Montucla, *Hist. des Mathématiques*, vol. i, p. 705; Humboldt, *Cosmos*; Libri, vol. iv, pp. 145 *et seq.* For Weyer, Flade, Bekker, Loos, and others, see the chapters of this work on Meteorology, Demoniacal Possession and Insanity, and Diabolism and Hysteria.

science. His book on meteorology was the first in which sound ideas were broached on this subject; his researches in optics gave the world the camera obscura, and possibly the telescope; in chemistry he seems to have been the first to show how to reduce the metallic oxides, and thus to have laid the foundation of several important industries. He did much to change natural philosophy from a black art to a vigorous open science. He encountered the old ecclesiastical policy. The society founded by him for physical research, "I Secreti," was broken up, and he was summoned to Rome by Pope Paul III and forbidden to continue his investigations.

So, too, in France. In 1624, some young chemists at Paris having taught the experimental method and cut loose from Aristotle, the faculty of theology beset the Parliament of Paris, and the Parliament prohibited these new chemical researches under the severest penalties.

The same war continued in Italy. Even after the belief in magic had been seriously weakened, the old theological fear and dislike of physical science continued. In 1657 occurred the first sitting of the Accademia del Cimento at Florence, under the presidency of Prince Leopold de' Medici. This academy promised great things for science; it was open to all talent; its only fundamental law was "the repudiation of any favourite system or sect of philosophy, and the obligation to investigate Nature by the pure light of experiment"; it entered into scientific investigations with energy. Borelli in mathematics, Redi in natural history, and many others, enlarged the boundaries of knowledge. Heat, light, magnetism, electricity, projectiles, digestion, and the incompressibility of water were studied by the right method and with results that enriched the world.

The academy was a fortress of science, and siege was soon laid to it. The votaries of scholastic learning denounced it as irreligious, quarrels were fomented, Leopold was bribed with a cardinal's hat and drawn away to Rome, and, after ten years of beleaguering, the fortress fell: Borelli was left a beggar; Oliva killed himself in despair.

So, too, the noted Academy of the Lincei at times incurred the ill will of the papacy by the very fact that it included thoughtful investigators. It was "patronized" by Pope Urban VIII in such manner as to paralyze it, and it was afterward vexed by Pope Gregory XVI. Even in our own time sessions of scientific associations were discouraged and thwarted by as kindly a pontiff as Pius IX.*

A hostility similar in kind, though less in degree, was shown in Protestant countries.

Even after Thomasius in Germany and Voltaire in France and Beccaria in Italy had given final blows to the belief in magic and witchcraft throughout Christendom, the traditional orthodox distrust of the physical sciences continued for a long time.

In England a marked dislike was shown among various leading ecclesiastics and theologians towards the Royal Society, and later toward the Association for the Advancement of Science; and this dislike, as will hereafter be seen, sometimes took shape in serious opposition.

As a rule, both in Protestant and Catholic countries instruction in chemistry and physics was for a long time discouraged by Church authorities; and, when its suppression was no longer possible, great pains were taken to subordinate it to instruction supposed to be more fully in accordance with the older methods of theological reasoning.

---

* For Porta, see the English translation of his main summary, *Natural Magick*, London, 1658. The first chapters are especially interesting, as showing what the word "magic" had come to mean in the mind of a man in whom mediæval and modern ideas were curiously mixed; see also Hoefer, *Histoire de la Chimie*, vol. ii, pp. 102–106; also Kopp; also Sprengel, *Histoire de la Médecine*, vol. iii, p. 239; also Musset-Pathay. For the Accademia del Cimento, see Napier, *Florentine History*, vol. v, p. 485; Tiraboschi, *Storia della Litteratura*; Henri Martin, *Histoire de France*; Jevons, *Principles of Science*, vol. ii, pp. 36–40. For value attached to Borelli's investigations by Newton and Huygens, see Brewster's *Life of Sir Isaac Newton*, London, 1875, pp. 128, 129. Libri, in his *Essai sur Galilée*, p. 37, says that Oliva was summoned to Rome and so tortured by the Inquisition that, to escape further cruelty, he ended his life by throwing himself from a window. For interference by Pope Gregory XVI with the Academy of the Lincei, and with public instruction generally, see Carutti, *Storia della Accademia dei Lincei*, p. 126. Pius IX, with all his geniality, seems to have allowed his hostility to voluntary associations to carry him very far at times. For his answer to an application made through Lord Odo Russell regarding a society for the prevention of cruelty to animals and his answer that "such an association could not be sanctioned by the Holy See, being founded on a theological error, to wit, that Christians owed any duties to animals," see Frances Power Cobbe, *Hopes of the Human Race*, p. 207.

I have now presented in outline the more direct and open struggle of the physical sciences with theology, mainly as an exterior foe. We will next consider their warfare with the same foe in its more subtle form, mainly as a vitiating and sterilizing principle in science itself.

We have seen thus far, first, how such men as Eusebius, Lactantius, and their compeers, opposed scientific investigation as futile; next, how such men as Albert the Great, St. Thomas Aquinas, and the multitude who followed them, turned the main current of mediæval thought from science to theology; and, finally, how a long line of Church authorities from Popes John XXII and Innocent VIII, and the heads of the great religious orders, down to various theologians and ecclesiastics, Catholic and Protestant, of a very recent period, endeavoured first to crush and afterward to discourage scientific research as dangerous.

Yet, injurious as all this was to the evolution of science, there was developed something in many respects more destructive; and this was the influence of mystic theology, penetrating, permeating, vitiating, sterilizing nearly every branch of science for hundreds of years. Among the forms taken by this development in the earlier Middle Ages we find a mixture of physical science with a pseudo-science obtained from texts of Scripture. In compounding this mixture, Jews and Christians vied with each other. In this process the sacred books were used as a fetich; every word, every letter, being considered to have a divine and hidden meaning. By combining various scriptural letters in various abstruse ways, new words of prodigious significance in magic were obtained, and among them the great word embracing the seventy-two mystical names of God—the mighty word "*Schemhamphoras.*" Why should men seek knowledge by observation and experiment in the book of Nature, when the book of Revelation, interpreted by the Kabbalah, opened such treasures to the ingenious believer?

So, too, we have ancient mystical theories of number which the theological spirit had made Christian, usurping an enormous place in mediæval science. The sacred power of the number three was seen in the Trinity; in the three main divisions of the universe—the empyrean, the heavens, and the earth; in the three angelic hierarchies; in the three choirs of seraphim, cherubim, and thrones; in the three of dominions, virtues, and powers; in the three of principalities, archangels, and angels; in the three orders in the Church—bishops, priests, and deacons; in the three classes—the baptized, the communicants, and the monks; in the three degrees of attainment—light, purity, and knowledge; in the three theological virtues—faith, hope, and charity—and in much else. All this was brought into a theologico-scientific relation, then and afterward, with the three dimensions of space; with the three divisions of time—past, present, and future; with the three realms of the visible world—sky, earth, and sea; with the three constituents of man—body, soul, and spirit; with the threefold enemies of man—the world, the flesh, and the devil; with the three kingdoms in nature—mineral, vegetable, and animal; with "the three colours"—red, yellow, and blue; with "the three eyes of the honey-bee"—and with a multitude of other analogues equally precious. The sacred power of the number seven was seen in the seven golden candlesticks and the seven churches in the Apocalypse; in the seven cardinal virtues and the seven deadly sins; in the seven liberal arts and the seven devilish arts, and, above all, in the seven sacraments. And as this proved in astrology that there could be only seven planets, so it proved in alchemy that there must be exactly seven metals. The twelve apostles were connected with the twelve signs in the zodiac, and with much in physical science. The seventy-two disciples, the seventy-two interpreters of the Old Testament, the seventy-two mystical names of God, were connected with the alleged fact in anatomy that there were seventy-two joints in the human frame.

Then, also, there were revived such theologic and metaphysical substitutes for scientific thought as the declaration that the perfect line is a circle, and hence that the planets must move in absolute circles—a statement which led astronomy astray even when the great truths of the Copernican theory were well in sight; also, the declaration that nature abhors a vacuum—a statement which led physics astray until Torricelli made his experiments; also, the declaration that we see the lightning before we hear the thunder because "sight is nobler than hearing."

In chemistry we have the same theologic tendency to magic, and, as a result, a muddle of science and theology, which from one point of view seems blasphemous and from another idiotic, but which none the less sterilized physical investigation for ages. That debased Platonism which had been such an important factor in the evolution of Christian theology from the earliest days of the Church continued its work. As everything in inorganic nature was supposed to have spiritual significance, the doctrines of the Trinity and Incarnation were turned into an argument in behalf of the philosopher's stone; arguments for the scheme of redemption and for transubstantiation suggested others of similar construction to prove the transmutation of metals; the doctrine of the resurrection of the human body was by similar mystic jugglery connected with the processes of distillation and sublimation. Even after the Middle Ages were past, strong men seemed unable to break away from such reasoning as this—among them such leaders as Basil Valentine in the fifteenth century, Agricola in the sixteenth, and Van Helmont in the seventeenth.

The greatest theologians contributed to the welter of unreason from which this pseudo-science was developed. One question largely discussed was, whether at the Redemption it was necessary for God to take the human form. Thomas Aquinas answered that it was necessary, but William Occam and Duns Scotus answered that it was not; that God might have taken the form of a stone, or of a log, or of a beast. The possibilities opened to wild substitutes for science by this sort of reasoning were infinite. Men have often asked how it was that the Arabians accomplished so much in scientific discovery as compared with Christian investigators; but the answer is easy: the Arabians were comparatively free from these theologic allurements which in Christian Europe flickered in the air on all sides, luring men into paths which led no-whither.

Strong investigators, like Arnold of Villanova, Raymond Lully, Basil Valentine, Paracelsus, and their compeers, were thus drawn far out of the only paths which led to fruitful truths. In a work generally ascribed to the first of these, the student is told that in mixing his chemicals he must repeat the psalm *Exsurge Domine*, and that on certain chemical vessels must be placed the last words of Jesus on the cross. Vincent of Beauvais insisted that, as the Bible declares that Noah, when five hundred years old, had children born to him, he must have possessed alchemical means of preserving life; and much later Dickinson insisted that the patriarchs generally must have owed their long lives to such means. It was loudly declared that the reality of the philosopher's stone was proved by the words of St. John in the Revelation. "To him that overcometh I will give a white stone." The reasonableness of seeking to develop gold out of the baser metals was for many generations based upon the doctrine of the resurrection of the physical body, which, though explicitly denied by St. Paul, had become a part of the creed of the Church. Martin Luther was especially drawn to believe in the alchemistic doctrine of transmutation by this analogy. The Bible was everywhere used, both among Protestants and Catholics, in support of these mystic adulterations of science, and one writer, as late as 1751, based his alchemistic arguments on more than a hundred passages of Scripture. As an example of this sort of reasoning, we have a proof that the elect will preserve the philosopher's stone until the last judgment, drawn from a passage in St. Paul's Epistle to the Corinthians, "We have this treasure in earthen vessels."

The greatest thinkers devoted themselves to adding new ingredients to this strange mixture of scientific and theologic thought. The Catholic philosophy of Thomas Aquinas, the Protestant mysticism of Jacob Boehme, and the alchemistic reveries of Basil Valentine were all cast into this seething mass.

And when alchemy in its old form had been discredited, we find scriptural arguments no less perverse, and even comical, used on the other side. As an example of this, just before the great discoveries by Stahl, we find the valuable scientific efforts of Becher opposed with the following syllogism: "King Solomon, according to the Scriptures, possessed the united wisdom of heaven and earth; but King

Solomon knew nothing about alchemy [or chemistry in the form it then took], and sent his vessels to Ophir to seek gold, and levied taxes upon his subjects; *ergo* alchemy [or chemistry] has no reality or truth." And we find that Becher is absolutely turned away from his labours, and obliged to devote himself to proving that Solomon used more money than he possibly could have obtained from Ophir or his subjects, and therefore that he must have possessed a knowledge of chemical methods and the philosopher's stone as the result of them.*

Of the general reasoning enforced by theology regarding physical science, every age has shown examples; yet out of them all I will select but two, and these are given because

---

* For an extract from Agrippa's *Occulta Philosophia* giving examples of the way in which mystical names were obtained from the Bible, see Rydberg, *Magic of the Middle Ages*, pp. 143 *et seq*. For the germs of many mystic beliefs regarding number and the like, which were incorporated into mediæval theology, see Zeller, *Plato and the Older Academy*, English translation, pp. 254 and 572, and elsewhere. As to the connection of spiritual things with inorganic nature in relation to chemistry, see Eicken, p. 634. On the injury to science wrought by Platonism acting through mediæval theology, see Hoefer, *Histoire de la Chimie*, vol. i, p. 90. As to the influence of mysticism upon strong men in science, see Hoefer; also Kopp, *Geschichte der Alchemie*, vol. i, p. 211. For a very curious Catholic treatise on sacred numbers, see the Abbé Auber, *Symbolisme Religieux*, Paris, 1870; also Detzel, *Christliche Ikonographie*, pp. 44 *et seq*.; and for an equally important Protestant work, see Samuell, *Seven the Sacred Number*, London, 1887. It is interesting to note that the latter writer, having been forced to give up the seven planets, consoles himself with the statement that "the earth is the seventh planet, counting from Neptune and calling the asteroids one" (see p. 426). For the *electrum magicum*, the seven metals composing it, and its wonderful qualities, see extracts from Paracelsus's writings in Hartmann's *Life of Paracelsus*, London, 1887, pp. 169 *et seq*. As to the more rapid transmission of light than sound, the following expresses the scholastic method well: "What is the cause why we see sooner the lightning than we heare the thunder clappe? That is because our sight is both nobler and sooner perceptive of its object than our eare; as being the more active part, and priore to our hearing: besides, the visible species are more subtile and less corporeal than the audible species."—Person's *Varieties*, Meteors, p. 82. For Basil Valentine's view, see Hoefer, vol. i, pp. 453–465; Schmieder, *Geschichte der Alchemie*, pp. 197–209; *Allgemeine deutsche Biographie*, article *Basilius*. For the discussions referred to on possibilities of God assuming forms of stone, or log, or beast, see Lippert, *Christenthum, Volksglaube, und Volksbrauch*, pp. 372, 373, where citations are given, etc. For the syllogism regarding Solomon, see Figuier, *L'Alchimie et les Alchimistes*, pp. 106, 107. For careful appreciation of Becher's position in the history of chemistry, see Kopp, *Ansichten über die Aufgabe der Chemie*, etc., *von Geber bis Stahl*, Braunschweig, 1875, pp. 201 *et seq*. For the text proving the existence of the philosopher's stone from the book of Revelation, see Figuier, p. 22.

they show how this mixture of theological with scientific ideas took hold upon the strongest supporters of better reasoning even after the power of mediæval theology seemed broken.

The first of these examples is Melanchthon. He was the scholar of the Reformation, and justly won the title "Preceptor of Germany." His mind was singularly open, his sympathies broad, and his usual freedom from bigotry drew down upon him that wrath of Protestant heresy-hunters which embittered the last years of his life and tortured him upon his deathbed. During his career at the University of Wittenberg he gave a course of lectures on physics, and in these he dwelt upon scriptural texts as affording scientific proofs, accepted the interference of the devil in physical phenomena as in other things, and applied the mediæval method throughout his whole work.*

Yet far more remarkable was the example, a century later, of the man who more than any other led the world out of the path opened by Aquinas, and into that through which modern thought has advanced to its greatest conquests. Strange as it may at first seem, Francis Bacon, whose keenness of sight revealed the delusions of the old path and the promises of the new, and whose boldness did so much to turn the world from the old path into the new, presents in his own writings one of the most striking examples of the evil he did so much to destroy.

The *Novum Organon*, considering the time when it came from his pen, is doubtless one of the greatest exhibitions of genius in the history of human thought. It showed the modern world the way out of the scholastic method and reverence for dogma into the experimental method and reverence for fact. In it occur many passages which show that the great philosopher was fully alive to the danger both to religion and to science arising from their mixture. He declares that the "corruption of philosophy from superstition and theology introduced the greatest amount of evil both into whole systems of philosophy and into their parts." He denounces those who "have endeavoured to found a

---

* For Melanchthon's ideas on physics, see his *Initia Doctrinæ Physicæ*, Wittenberg, 1557, especially pp. 243 and 274; also in vol. xiii of Bretschneider's edition of the collected works, and especially pp. 339–343.

natural philosophy on the books of Genesis and Job and other sacred Scriptures, so 'seeking the dead among the living.'" He speaks of the result as "an unwholesome mixture of things human and divine; not merely fantastic philosophy, but heretical religion." He refers to the opposition of the fathers to the doctrine of the rotundity of the earth, and says that, "thanks to some of them, you may find the approach to any kind of philosophy, however improved, entirely closed up." He charges that some of these divines are "afraid lest perhaps a deeper inquiry into nature should penetrate beyond the allowed limits of sobriety"; and finally speaks of theologians as sometimes craftily conjecturing that, if science be little understood, "each single thing can be referred more easily to the hand and rod of God," and says, "*This is nothing more or less than wishing to please God by a lie.*"

No man who has reflected much upon the annals of his race can, without a feeling of awe, come into the presence of such clearness of insight and boldness of utterance, and the first thought of the reader is that, of all men, Francis Bacon is the most free from the unfortunate bias he condemns; that he, certainly, can not be deluded into the old path. But as we go on through his main work we are surprised to find that the strong arm of Aquinas has been stretched over the intervening ages, and has laid hold upon this master-thinker of the seventeenth century; for only a few chapters beyond those containing the citations already made we find Bacon alluding to the recent voyage of Columbus, and speaking of the prophecy of Daniel regarding the latter days, that "many shall run to and fro, and knowledge be increased," as clearly signifying "that . . . the circumnavigation of the world and the increase of science should happen in the same age." *

In his great work on the *Advancement of Learning* the firm grasp which the methods he condemned held upon him is shown yet more clearly. In the first book of it he asserts that "that excellent book of Job, if it be revolved with dili-

---

* See the *Novum Organon*, translated by the Rev. G. W. Kitchin, Oxford, 1855, chaps. lxv and lxxxix.

gence, will be found pregnant and swelling with natural philosophy," and he endeavours to show that in it the "roundness of the earth," the "fixing of the stars, ever standing at equal distances," the "depression of the southern pole," the "matter of generation," and "matter of minerals" are "with great elegancy noted." But, curiously enough, he uses to support some of these truths the very texts which the fathers of the Church used to destroy them, and those for which he finds Scripture warrant most clearly are such as science has since disproved. So, too, he says that Solomon was enabled in his Proverbs, "by donation of God, to compile a natural history of all verdure." *

Such was the struggle of the physical sciences in general. Let us now look briefly at one special example out of many, which reveals, as well as any, one of the main theories which prompted theological interference with them.

It will doubtless seem amazing to many that for ages the weight of theological thought in Christendom was thrown against the idea of the suffocating properties of certain gases, and especially of carbonic acid. Although in antiquity we see men forming a right theory of gases in mines, we find that, early in the history of the Church, St. Clement of Alexandria put forth the theory that these gases are manifestations of diabolic action, and that, throughout Christendom, suffocation in caverns, wells, and cellars was attributed to the direct action of evil spirits. Evidences of this view abound through the mediæval period, and during the Reformation period a great authority, Agricola, one of the most earnest and truthful of investigators, still adhered to the belief that these gases in mines were manifestations of devils, and he specified two classes—one of malignant imps, who blow out the miners' lamps, and the other of friendly imps, who

---

* See Bacon, *Advancement of Learning*, edited by W. Aldis Wright, London, 1873, pp. 47, 48. Certainly no more striking examples of the strength of the evil which he had all along been denouncing could be exhibited than these in his own writings. Nothing better illustrates the sway of the mediæval theology, or better explains his blindness to the discoveries of Copernicus and to the experiments of Gilbert. For a very contemptuous statement of Lord Bacon's claim to his position as a philosopher, see Lange, *Geschichte des Materialismus*, Leipsic, 1874, vol. i, p. 219. For a more just statement, see Brewster, *Life of Sir Isaac Newton*. See also Jevons, *Principles of Science*, London, 1874, vol. ii, p. 298.

simply tease the workmen in various ways. He went so far as to say that one of these spirits in the Saxon mine of Annaberg destroyed twelve workmen at once by the power of his breath.

At the end of the sixteenth century we find a writer on mineralogy complaining that the mines in France and Germany had been in large part abandoned on account of the "evil spirits of metals which had taken possession of them."

Even as late as the seventeenth century, Van Helmont, after he had broken away from alchemy and opened one of the great paths to chemistry—even after he had announced to the world the existence of various gases and the mode of their generation—was not strong enough to free himself from theologic bias; he still inclined to believe that the gases he had discovered, were in some sense living spirits, beneficent or diabolical.

But at various periods glimpses of the truth had been gained. The ancient view had not been entirely forgotten; and as far back as the first part of the thirteenth century Albert the Great suggested a natural cause in the possibility of exhalations from minerals causing a "corruption of the air"; but he, as we have seen, was driven or dragged off into theological studies, and the world relapsed into the theological view.

Toward the end of the fifteenth century there had come a great genius laden with important truths in chemistry, but for whom the world was not ready—Basil Valentine. His discoveries anticipated much that has brought fame and fortune to chemists since, yet so fearful of danger was he that his work was carefully concealed. Not until after his death was his treatise on alchemy found, and even then it was for a long time not known where and when he lived. The papal bull, *Spondent pariter*, and the various prohibitions it bred, forcing other alchemists to conceal their laboratories, led him to let himself be known during his life at Erfurt simply as an apothecary, and to wait until after his death to make a revelation of truth which during his lifetime might have cost him dear. Among the legacies of this greatest of the alchemists was the doctrine that the air which asphyxiates workers in mines is similar to that which is produced by fermentation of malt, and a recommendation that, in order to drive away the evil and to prevent serious accidents, fires be lighted and jets of steam used to ventilate the mines—stress being especially laid upon the idea that the danger in the mines is produced by "exhalations of metals."

Thanks to men like Valentine, this idea of the interference of Satan and his minions with the mining industry was gradually weakened, and the working of the deserted mines was resumed; yet even at a comparatively recent period we find it still lingering, and among leading divines in the very heart of Protestant Germany. In 1715 a cellar-digger having been stifled at Jena, the medical faculty of the university decided that the cause was not the direct action of the devil, but a deadly gas. Thereupon Prof. Loescher, of the University of Wittenberg, entered a solemn protest, declaring that the decision of the medical faculty was "only a proof of the lamentable license which has so taken possession of us, and which, if we are not earnestly on our guard, will finally turn away from us the blessing of God."* But denunciations of this kind could not hold back the little army of science; in spite of adverse influences, the evolution of physics and chemistry went on. More and more there rose men bold enough to break away from theological methods and strong enough to resist ecclesiastical bribes and threats. As alchemy in its first form, seeking for the philosopher's stone and the transmutation of metals, had given way to alchemy in its second form, seeking for the elixir of life and remedies more or less magical for disease, so now the latter yielded to the search for truth as truth. More and more the "solemnly constituted impostors" were resisted in every field. A great line of physicists and chemists began to appear.†

---

* For Loescher's protest, see Julian Schmidt, *Geschichte des geistigen Lebens*, etc., vol. i, p. 319.

† For the general view of noxious gases as imps of Satan, see Hoefer, *Histoire de la Chimie*, vol. i, p. 350; vol. ii, p. 48. For the work of Black, Priestley, Bergmann, and others, see main authorities already cited, and especially the admirable paper of Dr. R. G. Eccles on *The Evolution of Chemistry*, New York, D. Appleton & Co., 1891. For the treatment of Priestley, see Spence's *Essays*, London, 1892; also Rutt, *Life and Correspondence of Priestley*, vol. ii, pp. 115 *et seq.*

II.

Just at the middle of the seventeenth century, and at the very centre of opposition to physical science, Robert Boyle began the new epoch in chemistry. Strongly influenced by the writings of Bacon and the discoveries of Galileo, he devoted himself to scientific research, establishing at Oxford a laboratory and putting into it a chemist from Strasburg. For this he was at once bitterly attacked. In spite of his high position, his blameless life, his liberal gifts to charity and learning, the Oxford pulpit was especially severe against him, declaring that his researches were destroying religion and his experiments undermining the university. Public orators denounced him, the wits ridiculed him, and his associates in the peerage were indignant that he should condescend to pursuits so unworthy. But Boyle pressed on. His discoveries opened new paths in various directions and gave an impulse to a succession of vigorous investigators. Thus began the long series of discoveries culminating in those of Black, Bergmann, Cavendish, Priestley, and Lavoisier, who ushered in the chemical science of the nineteenth century.

Yet not even then without a sore struggle against unreason. And it must here be noticed that this unreason was not all theological. The unreasoning heterodox when intrusted with irresponsible power can be as short-sighted and cruel as the unreasoning orthodox. Lavoisier, one of the best of our race, not only a great chemist but a true man, was sent to the scaffold by the Parisian mob, led by bigoted "liberals" and atheists, with the sneer that the republic had no need of *savants*. As to Priestley, who had devoted his life to science and to every good work among his fellow-men, the Birmingham mob, favoured by the Anglican clergymen who harangued them as "fellow-churchmen," wrecked his house, destroyed his library, philosophical instruments, and papers containing the results of long years of scientific research, drove him into exile, and would have murdered him if they could have laid their hands upon him.

Nor was it entirely his devotion to rational liberty, nor

even his disbelief in the doctrine of the Trinity, which brought on this catastrophe. That there was a deep distrust of his scientific pursuits, was evident when the leaders of the mob took pains to use his electrical apparatus to set fire to his papers.

Still, though theological modes of thought continued to sterilize much effort in chemistry, the old influence was more and more thrown off, and truth sought more and more for truth's sake. "Black magic" with its Satanic machinery vanished, only reappearing occasionally among marvel-mongers and belated theologians. "White magic" became legerdemain.

In the early years of the nineteenth century, physical research, though it went on with ever-increasing vigour, felt in various ways the reaction which followed the French Revolution. It was not merely under the Bourbons and Hapsburgs that resistance was offered; even in England the old spirit lingered long. As late as 1832, when the British Association for the Advancement of Science first visited Oxford, no less amiable a man than John Keble—at that time a power in the university—condemned indignantly the conferring of honorary degrees upon the leading men thus brought together. In a letter of that date to Dr. Pusey he complained bitterly, to use his own words, that "the Oxford doctors have truckled sadly to the spirit of the times in receiving the hotchpotch of philosophers as they did." It is interesting to know that among the men thus contemptuously characterized were Brewster, Faraday, and Dalton.

Nor was this a mere isolated exhibition of feeling; it lasted many years, and was especially shown on both sides of the Atlantic in all higher institutions of learning where theology was dominant. Down to a period within the memory of men still in active life, students in the sciences, not only at Oxford and Cambridge but at Harvard and Yale, were considered a doubtful if not a distinctly inferior class, intellectually and socially—to be relegated to different instructors and buildings, and to receive their degrees on a different occasion and with different ceremonies from those appointed for students in literature. To the State University of Michigan, among the greater American institutions of learning which have never possessed or been possessed by a theological seminary, belongs the honour of first breaking down this wall of separation.

But from the middle years of the century chemical science progressed with ever-accelerating force, and the work of Bunsen, Kirchhoff, Dalton, and Faraday has, in the last years of the century, led up to the establishment of Mendeleef's law, by which chemistry has become predictive, as astronomy had become predictive by the calculations of Newton, and biology by the discoveries of Darwin.

While one succession of strong men were thus developing chemistry out of one form of magic, another succession were developing physics out of another form.

First in this latter succession may be mentioned that line of thinkers who divined and reasoned out great physical laws—a line extending from Galileo and Kepler and Newton to Ohm and Faraday and Joule and Helmholtz. These, by revealing more and more clearly the reign of law, steadily undermined the older theological view of arbitrary influence in nature. Next should be mentioned the line of profound observers, from Galileo and Torricelli to Kelvin. These have as thoroughly undermined the old theologic substitution of phrases for facts. When Galileo dropped the differing weights from the Leaning Tower of Pisa, he began the end of Aristotelian authority in physics. When Torricelli balanced a column of mercury against a column of water and each of these against a column of air, he ended the theologic phrase that "nature abhors a vacuum." When Newton approximately determined the velocity of sound, he ended the theologic argument that we see the flash before we hear the roar because "sight is nobler than hearing." When Franklin showed that lightning is caused by electricity, and Ohm and Faraday proved that electricity obeys ascertained laws, they ended the theological idea of a divinity seated above the clouds and casting thunderbolts.

Resulting from the labour of both these branches of physical science, we have the establishment of the great laws of the indestructibility of matter, the correlation of forces, and chemical affinity. Thereby is ended, with various other sacred traditions, the theological theory of a visible universe created out of nothing, so firmly imbedded in the theological thought of the Middle Ages and in the Westminster Catechism.*

In our own time some attempt has been made to renew this war against the physical sciences. Joseph de Maistre, uttering his hatred of them, declaring that mankind has paid too dearly for them, asserting that they must be subjected to theology, likening them to fire—good when confined and dangerous when scattered about—has been one of the main leaders among those who can not relinquish the idea that our body of sacred literature should be kept a controlling text-book of science. The only effect of such teachings has been to weaken the legitimate hold of religion upon men.

In Catholic countries exertion has of late years been mainly confined to excluding science or diluting it in university teachings. Early in the present century a great effort was made by Ferdinand VII of Spain. He simply dismissed the scientific professors from the University of Salamanca, and until a recent period there has been general exclusion from Spanish universities of professors holding to the Newtonian physics. So, too, the contemporary Emperor of Austria attempted indirectly something of the same sort; and at a still later period Popes Gregory XVI and Pius IX discouraged, if they did not forbid, the meetings of scientific associations in Italy. In France, war between theology and science, which had long been smouldering, came in the years 1867 and 1868 to an outbreak. Toward the end of the last century, after the Church had held possession of advanced instruction for more than a thousand years, and had, so far as it was able, kept experimental science in servitude—after

---

* For a reappearance of the fundamental doctrines of black magic among theologians, see Rev. Dr. Jewett, Professor of Pastoral Theology in the Prot. Episc. Gen. Theolog. Seminary of New York, *Diabolology: The Person and Kingdom of Satan*, New York, 1889. For their reappearance among theosophists, see Éliphas Lévi, *Histoire de la Magie*, especially the final chapters. For opposition to Boyle and chemical studies at Oxford in the latter half of the seventeenth century, see the address of Prof. Dixon, F. R. S., before the British Association, 1894. For the recent progress of chemistry, and opposition to its earlier development at Oxford, see Lord Salisbury's address as President of the British Association, in 1894. For the Protestant survival of the mediæval assertion that the universe was created out of nothing, see the Westminster Catechism, question 15.

it had humiliated Buffon in natural science, thrown its weight against Newton in the physical sciences, and wrecked Turgot's noble plans for a system of public instruction—the French nation decreed the establishment of the most thorough and complete system of higher instruction in science ever known. It was kept under lay control and became one of the glories of France; but, emboldened by the restoration of the Bourbons in 1815, the Church began to undermine this hated system, and in 1868 had made such progress that all was ready for the final assault.

Foremost among the leaders of the besieging party was the Bishop of Orleans, Dupanloup, a man of many winning characteristics and of great oratorical power. In various ways, and especially in an open letter, he had fought the "materialism" of science at Paris, and especially were his attacks levelled at Profs. Vulpian and Sée and the Minister of Public Instruction, Duruy, a man of great merit, whose only crime was devotion to the improvement of education and to the promotion of the highest research in science.*

The main attack was made rather upon biological science than upon physics and chemistry, yet it was clear that all were involved together.

The first onslaught was made in the French Senate, and the storming party in that body was led by a venerable and conscientious prelate, Cardinal de Bonnechose, Archbishop of Rouen. It was charged by him and his party that the tendencies of the higher scientific teaching at Paris were fatal to religion and morality. Heavy missiles were hurled —such phrases as "sapping the foundations," "breaking down the bulwarks," and the like; and, withal, a new missile was used with much effect—the epithet "materialist."

The results can be easily guessed: crowds came to the lecture-rooms of the attacked professors, and the lecture-room of Prof. Sée, the chief offender, was crowded to suffocation.

A siege was begun in due form. A young physician was

---

* For the exertions of the restored Bourbons to crush the universities of Spain, see Hubbard, *Hist. Contemporaine de l'Espagne*, Paris, 1878, chaps. i and iii. For Dupanloup, *Lettre à un Cardinal*, see the *Revue de Thérapeutique* of 1868, p. 221.

sent by the cardinal's party into the heterodox camp as a spy. Having heard one lecture of Prof. Sée, he returned with information that seemed to promise easy victory to the besieging party: he brought a terrible statement—one that seemed enough to overwhelm Sée, Vulpian, Duruy, and the whole hated system of public instruction in France—the statement that Sée had denied the existence of the human soul.

Cardinal Bonnechose seized the tremendous weapon at once. Rising in his place in the Senate, he launched a most eloquent invective against the Minister of State who could protect such a fortress of impiety as the College of Medicine; and, as a climax, he asserted, on the evidence of his spy fresh from Prof. Sée's lecture-room, that the professor had declared, in his lecture of the day before, that so long as he had the honour to hold his professorship he would combat the false idea of the existence of the soul. The weapon seemed resistless and the wound fatal, but M. Duruy rose and asked to be heard.

His statement was simply that he held in his hand documentary proofs that Prof. Sée never made such a declaration. He held the notes used by Prof. Sée in his lecture. Prof. Sée, it appeared, belonged to a school in medical science which combated certain ideas regarding medicine as an *art*. The inflamed imagination of the cardinal's heresy-hunting emissary had, as the lecture-notes proved, led him to mistake the word "*art*" for "*âme*," and to exhibit Prof. Sée as treating a theological when he was discussing a purely scientific question. Of the existence of the soul the professor had said nothing.

The forces of the enemy were immediately turned; they retreated in confusion, amid the laughter of all France; and a quiet, dignified statement as to the rights of scientific instructors by Wurtz, dean of the faculty, completed their discomfiture. Thus a well-meant attempt to check science simply ended in bringing ridicule on religion, and in thrusting still deeper into the minds of thousands of men that most mistaken of all mistaken ideas: the conviction that religion and science are enemies.*

---

* For a general account of the Vulpian and Sée matter, see *Revue des Deux Mondes*, 31 mai, 1868, "Chronique de la Quinzaine," pp. 763–765. As to the result

But justice forbids raising an outcry against Roman Catholicism for this. In 1864 a number of excellent men in England drew up a declaration to be signed by students in the natural sciences, expressing "sincere regret that researches into scientific truth are perverted by some in our time into occasion for casting doubt upon the truth and authenticity of the Holy Scriptures." Nine tenths of the leading scientific men of England refused to sign it; nor was this all: Sir John Herschel, Sir John Bowring, and Sir W. R. Hamilton administered, through the press, castigations which roused general indignation against the proposers of the circular, and Prof. De Morgan, by a parody, covered memorial and memorialists with ridicule. It was the old mistake, and the old result followed in the minds of multitudes of thoughtful young men.*

And in yet another Protestant country this same mistake was made. In 1868 several excellent churchmen in Prussia thought it their duty to meet for the denunciation of "science falsely so called." Two results followed: upon the great majority of these really self-sacrificing men—whose first utterances showed complete ignorance of the theories they attacked—there came quiet and widespread contempt; upon Pastor Knak, who stood forth and proclaimed views of the universe which he thought scriptural, but which most schoolboys knew to be childish, came a burst of good-natured derision from every quarter of the German nation.†

But in all the greater modern nations warfare of this kind, after the first quarter of the nineteenth century, became more and more futile. While conscientious Roman bishops, and no less conscientious Protestant clergymen in Europe and America continued to insist that advanced education, not only in literature but in science, should be kept under careful control in their own sectarian universities and colleges, wretchedly one-sided in organization and inadequate

---

on popular thought, may be noted the following comment on the affair by the *Revue*, which is as free as possible from anything like rabid anti-ecclesiastical ideas: "Elle a été vraiment curieuse, instructive, assez triste et même un peu amusante." For Wurtz's statement, see *Revue de Thérapeutique* for 1868, p. 303.

* De Morgan, *Paradoxes*, pp. 421–428; also Daubeny's *Essays*.

† See the Berlin newspapers for the summer of 1868, especially *Kladderadatsch*.

in equipment; while Catholic clerical authorities in Spain were rejecting all professors holding the Newtonian theory, and in Austria and Italy all holding unsafe views regarding the Immaculate Conception, and while Protestant clerical authorities in Great Britain and America were keeping out of professorships men holding unsatisfactory views regarding the Incarnation, or Infant Baptism, or the Apostolic Succession, or Ordination by Elders, or the Perseverance of the Saints; and while both Catholic and Protestant ecclesiastics were openly or secretly weeding out of university faculties all who showed willingness to consider fairly the ideas of Darwin, a movement was quietly in progress destined to take instruction, and especially instruction in the physical and natural sciences, out of its old subordination to theology and ecclesiasticism.*

The most striking beginnings of this movement had been seen when, in the darkest period of the French Revolution, there was founded at Paris the great Conservatory of Arts and Trades, and when, in the early years of the nineteenth century, scientific and technical education spread quietly upon the Continent. By the middle of the century France and Germany were dotted with well-equipped technical and scientific schools, each having chemical and physical laboratories.

The English-speaking lands lagged behind. In England, Oxford and Cambridge showed few if any signs of this movement, and in the United States, down to 1850, evidences of it were few and feeble. Very significant is it that, at that period, while Yale College had in its faculty Silliman and Olmsted—the professor of chemistry and the professor of physics most widely known in the United States—it had no physical or chemical laboratory in the modern sense, and

* Whatever may be thought of the system of philosophy advocated by President McCosh at Princeton, every thinking man must honour him for the large way in which he, at least, broke away from the traditions of that centre of thought; prevented, so far as he was able, persecution of scholars for holding to the Darwinian view; and paved the way for the highest researches in physical science in that university. For a most eloquent statement of the opposition of modern physical science to mediæval theological views, as shown in the case of Sir Isaac Newton, see Dr. Thomas Chalmers, cited in Gore, *Art of Scientific Discovery*, London, 1878, p. 247.

confined its instruction in these subjects to examinations upon a text-book and the presentation of a few lectures. At the State University of Michigan, which had even then taken a foremost place in the higher education west of the Great Lakes, there was very meagre instruction in chemistry and virtually none in physics. This being the state of things in the middle of the century in institutions remarkably free from clerical control, it can be imagined what was the position of scientific instruction in smaller colleges and universities where theological considerations were entirely dominant.

But in 1851, with the International Exhibition at London, began in Great Britain and America a movement in favour of scientific education; men of wealth and public spirit began making contributions to them, and thus came the growth of a new system of instruction in which Chemistry and Physics took just rank.

By far the most marked feature in this movement was seen in America, when, in 1857, Justin S. Morrill, a young member of Congress from Vermont, presented the project of a law endowing from the public lands a broad national system of colleges in which scientific and technical studies should be placed on an equality with studies in classical literature, one such college to be established in every State of the Union. The bill, though opposed mainly by representatives from the Southern States, where doctrinaire politics and orthodox theology were in strong alliance with negro slavery, was passed by both Houses of Congress, but vetoed by President Buchanan, in whom the doctrinaire and orthodox spirit was incarnate. But Morrill persisted and again presented his bill, which was again carried in spite of the opposition of the Southern members, and again vetoed in 1859 by President Buchanan. Then came the civil war; but Morrill and his associates did not despair of the republic. In the midst of all the measures for putting vast armies into the field and for saving the Union from foreign interference as well as from domestic anarchy, they again passed the bill, and in 1862, in the darkest hour of the struggle for national existence, it became a law by the signature of President Lincoln.

And here it should not be unrecorded, that, while the vast majority of the supporters of the measure were laymen, most efficient service was rendered by a clergyman, the Rev. Dr. Amos Brown, born in New Hampshire, but at that time an instructor in a little village of New York. His ideas were embodied in the bill, and his efforts did much for its passage.

Thus was established, in every State of the American Union, at least one institution in which scientific and technical studies were given equal rank with classical, and promoted by laboratories for research in physical and natural science. Of these institutions there are now nearly fifty: all have proved valuable, and some of them, by the addition of splendid gifts from individuals and from the States in which they are situated, have been developed into great universities.

Nor was this all. Many of the older universities and colleges thus received a powerful stimulus in the new direction. The great physical and chemical laboratories founded by gifts from public-spirited individuals, as at Harvard, Yale, and Chicago, or by enlightened State legislators, as in Michigan, Wisconsin, Minnesota, California, Kansas, and Nebraska, have also become centres from which radiate influences favouring the unfettered search for truth as truth.

This system has been long enough in operation to enable us to note in some degree its effects on religion, and these are certainly such as to relieve those who have feared that religion was necessarily bound up with the older instruction controlled by theology. While in Europe, by a natural reaction, the colleges under strict ecclesiastical control have sent forth the most powerful foes the Christian Church has ever known, of whom Voltaire and Diderot and Volney and Sainte-Beuve and Renan are types, no such effects have been noted in these newer institutions. While the theological way of looking at the universe has steadily yielded, there has been no sign of any tendency toward irreligion. On the contrary, it is the testimony of those best acquainted with the American colleges and universities during the last forty-five years that there has been in them a great gain, not only as regards morals, but as regards religion in its highest and best sense. The reason is not far to seek. Under the old

American system the whole body of students at a university were confined to a single course, for which the majority cared little and very many cared nothing, and, as a result, widespread idleness and dissipation were inevitable. Under the new system, presenting various courses, and especially courses in various sciences, appealing to different tastes and aims, the great majority of students are interested, and consequently indolence and dissipation have steadily diminished. Moreover, in the majority of American institutions of learning down to the middle of the century, the main reliance for the religious culture of students was in the perfunctory presentation of sectarian theology, and the occasional stirring up of what were called "revivals," which, after a period of unhealthy stimulus, inevitably left the main body of students in a state of religious and moral reaction and collapse. This method is now discredited, and in the more important American universities it has become impossible. Religious truth, to secure the attention of the modern race of students in the better American institutions, is presented, not by "sensation preachers," but by thoughtful, sober minded scholars. Less and less avail sectarian arguments; more and more impressive becomes the presentation of fundamental religious truths. The result is, that while young men care less and less for the great mass of petty, cut-and-dried sectarian formulas, they approach the deeper questions of religion with increasing reverence.

While striking differences exist between the European universities and those of the United States, this at least may be said, that on both sides of the Atlantic the great majority of the leading institutions of learning are under the sway of enlightened public opinion as voiced mainly by laymen, and that, this being the case, the physical and natural sciences are henceforth likely to be developed normally, and without fear of being sterilized by theology or oppressed by ecclesiasticism.

END OF VOLUME ONE.